WHERE WILL LIGHTNING STRIKE NEXT?

"Uranus [is]…a difficult energy for anyone to take personal responsibility for, since Uranian outcomes seem so insane and ridiculous at times, often punctuated by jaw-dropping surprises."

—from Chapter One

"Uranus lightning strikes without warning—zap!…If we prefer a smoother, meandering stream of intuition, we'll have to wait instead for a major Neptune transit. Uranus is more a wild-eyed god whose intuitive power crackles with electrical vitality."

—from Uranus in the First House

"Did you know that mythological Uranus was Jupiter's grandpa? It's not too clear if they ever had any real family contact. Astrologically, however, Uranus and Jupiter share a lot in common. Uranus is a fiery kind of air planet, while Jupiter's an airy sort of fire planet. Both can get charged up with excitement, both think that small-mindedness is a big waste of time, and both need lots of elbow room to explore life's greater potential."

—from Uranus/Jupiter Transits

"Uranus can be a crazed planet, capable of erratic behavior in matters of the heart. We'll need to first look at our chart for natal clues regarding how much resilience and emotional fortitude we have before we start to boldly experiment in love's laboratory. Let's not forget the value of Uranian detachment and objectivity. Otherwise, some of us can fall hard for some person we just 'know' is perfect for us, only to find later that nobody gets to hang on to such perfection for long."

—from Uranus in the Seventh House

"Think of Neptune as a beautiful soap bubble, with its swirling colors and its ability to float effortlessly in the air. Then think of Uranus as a huge, sharp, gleaming straight pin suddenly coming out of nowhere to prick that iridescent bubble—pop! This is Uranus' basic role at this time: to burst our own self-created bubbles of illusion or those set up by our environment that lull us into a false sense of serenity, security, or complacency."

—from Uranus/Neptune Transits

ABOUT THE AUTHOR

Bil Tierney has been involved with astrology for over thirty-two years. As a full-time professional, he has lectured and given workshops at major astrological conferences throughout the United States and in Canada since the mid-1970s. He has a special interest in studying the birthchart from a practical, psychological level that also encourages spiritual growth. He is a longtime member of the Metropolitan Atlanta Astrological Society (MAAS), and has served as its newsletter/journal editor several times. Bil's work has also been published in astrological publications such as *Aspects* and *The Mercury Hour*. Other books he has written are *Dynamics of Aspect Analysis*, and the Llewellyn publications: *Twelve Faces of Saturn, Alive and Well with Neptune,* and *Alive and Well with Pluto.*

When Bil is not busy with client consultations, lecturing, tutoring, and writing articles and books, his other big passion is computers. Clients enjoy Bil's animated, warm, and easy-going style, and they are impressed with his skillful blend of the intuitive and the analytical. His humorous slant on life is appreciated as well. Readers enjoy his thorough, insightful approach to astrological topics.

TO CONTACT THE AUTHOR

If you would like to contact the author or would like more information about this book, please write to him in care of Llewellyn Worldwide. All mail addressed to the author is forwarded, but the publisher cannot, unless specifically instructed by the author, give out an address or phone number. Please write to:

Bil Tierney
c/o Llewellyn Publications
P.O. Box 64383, Dept. K713–7
St. Paul, MN 55164–0383, U.S.A.

Please enclose a self-addressed, stamped envelope for reply or $1.00 to cover costs. If ordering from outside the U.S.A., please enclose an international postal reply coupon.

Llewellyn Worldwide does not participate in, endorse, or have any authority or responsibility concerning private business transactions between our authors and the public.

Alive and Well with URANUS

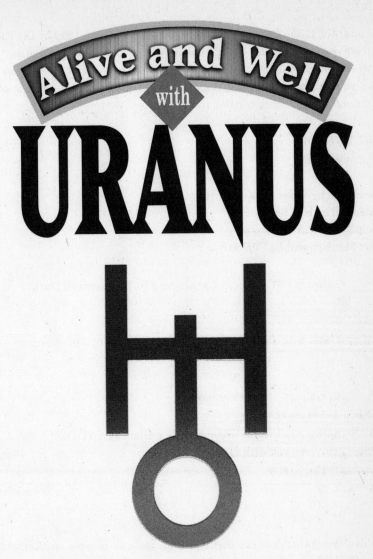

Transits of Self-Awakening

Bil Tierney

1999
Llewellyn Publications
P.O. Box 64383–0383
St. Paul, MN 55164–0833

FIRST EDITION
First Printing 1999

Cover design by William Merlin Cannon
Book design by Ken Schubert
Editing by Marguerite Krause
Project Management by Eila Savela

Library of Congress Cataloging-in-Publication Data

Tierney, Bil.
    Alive & well with Uranus : transits of self-awakening / Bil Tierney. — 1st ed.
        p.   cm.
    Includes bibliographical references.
    ISBN 1–56718–713–7
    1. Astrology. 2. Uranus (Planet) — Miscellanea. I. Title.
II. Title : Alive and well with Uranus.
BF1724.2.U7T54   1999
133.5 ' 38--dc21

                                                                99–34814
                                                                   CIP

Printed in the United States of America

Llewellyn Publications
A Division of Llewellyn Worldwide, Ltd.
P.O. Box 64383, Dept. K713–7
St. Paul, MN 55164–0383, U.S.A.
www.llewellyn.com

DEDICATION

I dedicate this book to Edith Custer, a true
Uranian ambassador of astrology, whose vision
and humor have brought people together in
love and hope throughout the years.

ACKNOWLEDGEMENTS

A thanks goes to Llewellyn's Acquisitions Manager Nancy Mostad for accepting my Uranus manuscript for publication. I'm also very grateful to Ken Schubert for his fine book design, William Merlin Cannon for the striking cover he created, and hard-working Marguerite Krause for her fine editing skills. Moreover, I really must applaud the efforts of Project Manager and Astrology Editor, Eila Savela (thanks, Eila, for keeping all of my book's offbeat humor intact, and for smoothly handling last-minute details). My thanks also goes to Lisa Braun (Publicity), and Tracy Whartman (Marketing), and everyone else who helped in the production of this book.

And last, I wish to express my gratitude to my colleagues Donna Van Toen and Gloria Star, who read pre-publication copies of *Alive and Well with Uranus* and submitted their comments for the back cover.

Oh, and thanks, Uranus, for zapping me with your enlivening energy. You enabled me to brag about the wonderfully exciting force of self-awakening you really are!

TABLE OF CONTENTS

Chapter Eight

Chapter Nine

Chapter Ten

Chapter Eleven

Chapter Twelve

Chapter Thirteen

Chapter Fourteen

Chapter Fifteen

INTRODUCTION

Old Sky God

Uranus, as Ouranos, was considered in Greek mythology to be the first of the sky gods—deities belonging to the heavenly realms—although he was never the superstar that more human-like Zeus became. The myth of this original sky god takes place in a distant time, long before the social drama of the gods and goddesses of Mount Olympus that we are most familiar with (and way before the creation of mortal humans). Uranus was symbolic of the cosmic mind-spirit energy needed to create life by interpenetrating the primal emotional-physical energies of Earth itself (personified by his mate, Gaia, who also symbolized the cosmic womb). More about his myth later. What needs to be emphasized here is that Uranus represents purer states of unfettered, abstract ideation—the pristine condition before any concept is challenged to materialize and face the limitations of tangible form. Uranus therefore depicts the divine mind and the unearthly world of archetypal ideas, echoing Plato's philosophy. (An archetypal idea of a microwave oven theoretically exists even before anyone gets to brainstorm and design one physically. In other

1

words, all that exists is first an unmanifest idea waiting for the right time to emerge—to literally be plucked from the ethers.)

When we keep creative ideas solely in our head, they can live indefinitely in an unconditional state. Yet once these ideas are applied and tested in the concrete world, where a sense of time and other earthly limitations are imposed, such ideas are no longer sheltered by the "heavenly realms," i.e., our inner world. Our thoughts start to flesh out in ways that may either expose how impractical they really are—which can be very disillusioning—or how brilliantly they can be executed in the "real" world if worked on a bit more (something Saturn takes charge of). Even when these ideas materialize successfully, Uranus still "knows" that much has been lost in the translation, because the world of matter is relatively dull, dumb, and slow compared to the astonishing light-filled universe that Uranus calls home. In addition, Uranus is afraid that exposure to the gross world of manifestation leaves its perfect ideas tainted by human emotion and desire, resulting in something inferior and full of flaws. This explains why true Uranians are never completely satisfied with their creations (although everyone one else is amazed by them).

My approach to Uranus' transits won't be an abstract, metaphysical one, because I try to bring this planet's energy down to the level of our own experience as much as possible without trivializing what Uranus means. This won't be an intellectually detached look at a planet that already seems too mentally removed from the routines of mundane living. Although Uranus, the archetype, may observe life from a safe distance rather than participate directly in worldly affairs, our challenge is to get our inner Uranus more fascinated with and thus more involved in what goes on in our personal lives. We do that by learning more about this Outer Planet's psychology and by becoming more willing to adopt certain Uranian values while still retaining many of our very human traits. It's not easy to embrace Uranus, but we need to be more open to this experience and willingly take risks in the name of self-awakening.

THE NAME GAME

To demonstrate how Uranus symbolizes individualistic and sometimes contrary thinking, note that its name can be pronounced in several ways. Two of those ways are somewhat embarrassing to people who find in them distasteful connotations, even though these are the pronunciations preferred by astronomers and most used by stand-up comedians and the general public as well. However, one workaround (which I don't use since I instead side in with the astronomers) is to call this planet Ur-RAH-nus, as if to give it an air of sophistication—although that sounds a bit affected to some ears (and remember, we're dealing with a planet known for enjoying words and concepts that have "shock value"). Still, you can't be accused of sounding too rude when you let "Ur-RAH-nus" trickle off you tongue! In fact, you sound like you know something the rest of us don't know. Therefore, be prepared to hear this planet pronounced this way at many an astrology conference. But there's even another option that instantly conveys to listeners that you probably have a scholastic background (or that maybe you've got a superiority complex), and that's when you call this planet "OOR-ron-noose"—supposedly just like the ancient Greeks did.

HOW TO USE THIS BOOK

This book is strictly about transiting Uranus contacting various factors in our birth chart. Let's say you just discovered that you are having transiting Uranus square your Sun (and to confirm it, you're probably already feeling possessed of an excitable energy and a strange sense of anticipation, telling you something different may indeed be unfolding for you at this time in your life, something that possibly may shake up your familiar life patterns). Well, you'll obviously need to read what I've suggested this transit could mean in Chapter Ten.

But in that same chapter you may also wish to read about transiting Uranus going through the Fifth House (since the Sun and the Fifth, along with the sign Leo, are manifestations of the same life principle—Principle Five in this instance; they thus all share similar themes). There may be a few comments I've made about Uranus passing through the Fifth that will make even more sense to you than some of the Uranus/Sun material. You also may want to think about your natal Uranus house position and read what I've said about this planet's transit of that same house; this could give you further insights into some of your natal issues. And please read various planet-to-planet transits to gain an extra understanding of your natal aspects involving those same planets.

Also realize that not all of what I've stated about any transit is supposed to fit your current life pattern to a tee. It would be very rare if this happened. Transits can have unlimited ways of manifesting for us, because our psyche is immensely fertile and boundlessly creative when it comes to finding suitable, meaning-ful circumstantial affairs that mirror our internal self at any par-ticular point in our lives. No astrologer can write a book tapping into each and every scenario possible. It's actually the general tone of each Uranus transit that I wish to convey, rather than try-ing to achieve a high score on accurately predicting trends and events. Keeping all the above in mind should help you enjoy this book very much and hopefully put what I've said to best use.

MERCURY AND VENUS TIMES TWO

In case you haven't read my book *Twelve Faces of Saturn*,[1] you may think that references in Chapter Eight to "air Mercury" mean Mercury in any air sign. Actually, as Mercury rules two signs—Gemini and Virgo—its airy side is linked to how Gemini processes experience, while its earthier face belongs to Virgo (see Chapter Eleven). I cover the transit of Uranus to both air Mercury and earth Mercury in separate chapters to show how dif-ferently these two facets of Mercury react to life, regardless of what element/sign our natal Mercury actually is in. A Pisces Mercury can therefore respond according to either its air or its earth potential, since Mercury is capable of expressing both modes. It would be interesting in this instance to figure out whether Pisces would relate more easily to Mercury's earthy or airy quality; a case could be made here for both sides of the Mercury profile.

The same goes for Venus—ruler of Taurus and Libra (again, an earthy Venus and airy Venus work together to fully define the Venus principle—see Chapters Seven and Twelve, respectively). Some transits readily evoke the air side of Venus more than others; some natal house positions favor Venus' earthy side more (such as Venus in the Second; yet if that Venus happens to be in Libra, it's an intepretational toss up). What's most important to keep in mind is that we probably need to be (and actually are) doing both sides of Mercury and Venus to get a fuller picture of how these planets operate and what they can do to enhance our growth. So please don't get too hung up wondering if you're in an air Venus or earth Mercury mode when an Outer Planet is making a major transit to these natal planets (especially with Neptune, a planet that defies categories and clear-cut definitions).

FUNNY BONE

I'm a firm believer in the teaching power of humor. I certainly don't think the world is one big cosmic joke, but it sure can seem to be quite an odd and funny place in which to evolve. I wouldn't be me if I didn't bring a gently irreverent slant on life into my writings. And I like to think the Cosmos has a healthy sense of humor as well. (After all, Divine Intelligence—probably with the help of Uranus—did create the platypus!)

Although I probably won't offend anyone, I still ask that readers take some of my comic remarks in the playful way in which they were intended. We sometimes get too long-faced and holy about astrology, especially those of us seeking perfection on the spiritual track. Remember that there's a planet revered for its philosophical wisdom and broad understanding that is also known for enjoying a good belly laugh—Jupiter (and it's certainly got the belly for it). While I won't have you roaring in the aisles, you may find yourself grinning from time to time. Humor that has something relevant to say about our life condition contributes to the joy of learning astrology—always a fascinating but nonetheless very human subject.

Some planets are easier to poke fun at than others. Uranus is a breeze because without this planet there'd be no living theater of the absurd! With Uranus it's easy to come up with colorful and unusual commentaries about life because this planet supports outrageous, brash viewpoints. What often makes humor work is how it exaggerates an evident truth about human nature and goes as far with it as it can to draw plenty of laughs from the crowd—which is probably why humor is *not* associated with Saturn. Although leave it to Saturn to convince some of its devotees that real humor can still be "dry." Well, yes, and I suppose it can be flat as a tire as well. Anyway, Jupiter's surely not buying that line of thinking.

I'm glad to see that other astrologers have been fine-tuning their astro-humor for years. Let me just mention one of the best: Kim Rogers-Gallagher, a gal who has done her darndest to take the starch out of astrology (keep it up, ole wise Uranus-inspired Sag). You can read her witty insights in her entertaining *Astrology for the Light Side of the Brain* and in her equally fun *Astrology for the Light Side of the Future.*[2]

THAT OLD MYTH REVISITED

Ouranos was a "ready-for-primal-time" god in Greek mythology who wasn't quite portrayed in easily identifiable human terms, unlike Mars or Jupiter (two archetypes we still run into at soccer games). Let's explore the Ouranos myth a little to get a better feeling for the psychology of this planet (and I'll do as the Romans did and just call him Uranus—and I ain't saying which embarrassing pronunciation I prefer). From the moment of his debut, Uranus was a god associated with the sky above, the heavenly expanse. He emerged from the womb of the great primal mother Gaia, who herself was born out of that hard-to-define cosmological state called Chaos—let's just say this was sometime long before the Universe took on differentiated physical form. Gaia personified matter and Earth itself (she thus symbolizes that which we affectionately call "Mother Nature"). Uranus rather quickly became her consort (just as the sky and the land—the celestial and the terrestrial—appear to merge together in partnership at some point).

It seems that the myth involving Uranus is not too rich in details describing his personality. Already it's a clue that this god represents something more impersonal about existence, something that's not very human or accessible in expression. He's not one to instinctively touch the ground we walk on. He's more an abstract force than he is a familiar identity we can relate to. If anything, he's a lonely, tragic figure compared to many of those more colorful soap-opera stars later associated with the comings and goings on Mount Olympus! However, he did mate with Gaia, and from their union was born an assortment of strange progeny.

Still, how nice—we innocently assume—for Uranus to hopefully have the humanizing opportunity to become a somewhat more earthly daddy. Yet, unpredictably, he reacted adversely to his kids right from the start. And that's partly because, much to his surprise and dismay, he sired offspring that were freakish—especially the three Hecatoncheires (giants with fifty heads and a hundred hands each) and the three Cyclopes (also gigantic and, at least to Uranus, oafish and hideous-looking. Interestingly, their TV wrestler-sounding names—Arges the "Thunderbolt," Brontes the "Thunderer," and Steropes the "Lightning-Hurler"— all confirm that they indeed were their father's turbulent sons). Yet considering how well astrologers would say their nicknames now distinctly apply to him, Uranus was obviously in some sort of major denial at the time, projecting all that wildness and weirdness onto his boys. Uranus also fathered the relatively normal-looking but very earthy Titans (a little rough around the edges), who also had disruptive, quarrelsome energies. It seems there was bad blood shared between this emotionally-distant parent and each child. But heck, Daddy Uranus surely didn't go out of his way to make himself lovable.

It seems that in sky god Uranus' aloof, self-absorbed world of creative abstractions, only the ideal version of whatever his divine mind designed was deemed perfect and therefore accept-able to him, not the disappointing concrete manifestations of that ideal. This is forever a problem with the Uranus archetype—its strong disgust and discontent with any earthly form that is flawed. Once an idea becomes real—is made tangible—it loses its special status and thus its appeal. Uranian detachment is born of this as is, in this father's case, the shunning of one's commitment to parenthood (with Uranus in disbelief that these children could be the byproducts of his mind). Part of the reason mythic Uranus was so turned off could be that Gaia herself, at least in these first attempts at birthing, was new to this and had not mastered the refinement of matter essential to producing harmony and beauty. Similarly, early forms of life on Earth may not have been all that pretty to look at, seeming to resemble instead evolution's failed experiments.

Now, if all this was the extent of the family dilemma—that Uranus created a wall of emotional distance from his "ugly" chil-dren and that they in turn suffered from getting his cold shoul-der—things would be bad enough. After all, maybe he himself was scarred because he never had a dad to role-model for him the positive aspects of fathering. Luckily, Gaia had no trouble emo-tionally bonding with her children and allowing herself to stay in touch with her maternal feelings. She'd never think of rejecting them. But Uranus, unable to cope, decided he couldn't face these towering monstrosities one more day, and so he coldly shoved them deeply into Gaia's womb which—because she is Earth per-sonified—actually turned out to be Tartarus, the lowest level of the darkest bowels in Hades, the Underworld; there they remained imprisoned and totally cut off from their nurturing con-nection to their mother. But all that Uranus cared about was that these creepy-looking kids were out of his sight for good! The Titans were spared this fate, but who could say for how long?

As perhaps only a powerful primal mothering force could, Gaia intensely grieved to the point of rage and revenge (obviously, her pleading for the return of her children meant nothing to "off-in-his-head-somewhere" Uranus—he probably dismissed her emotional state as needless hysteria, which likely further infuriated her). Yet one day she woke up feeling rotten, smelled the coffee, and thought, "Why in Hades should I continue to give birth to babies just to have this cold-hearted tyrant make them disappear as if they never even existed, and all because they turn out a little big for their age? No more of this!" The time for taking drastic action was near. Her youngest-born Titan, Cronus (it's "Mr. Saturn" on his business cards) sought to avenge his mother's mistreatment and the sad, unjust fate of his captive siblings. Leave it to Saturn to take full responsibility for the amending of any grievous wrong. This horrible situation simply could not go on.

NOT FOR THE SQUEAMISH

But how to go about doing it? His was a most serious mission, leaving no room for failure, but at least Cronus had a reputation for being crafty and ambitious. He wouldn't chicken out. His mom, probably out of desperation, came up with a daring solution: her plan was to hide her son and have him wait in ambush. Cronus knew just what weapon to have on him when the moment to strike was right. And as clueless Uranus began to lay upon seething Gaia that night, Cronus jumped out of hiding and, brandishing a sharpened sickle in his left hand, swiftly castrated his father! Although the myth tells us nothing about Uranus' immediate reaction, it does say he was badly wounded. We should assume he was stunned and in severe shock. However, once drops of Uranus' blood from his severed genitals fell on Gaia, they created the snake-haired Furies—finally, some raw, repressed, hostile emotion spilling out! These three Furies went on in Greek myth to develop a reputation as formidable goddesses of vendettas and "payback" time.

Cronus, probably in a frenzied state, tossed those genitals out into the sea. Uranus would never see them again nor feel the creative vitality he once had. His days of willful idealization having the power to suppress physical embodiment were over, since the source of his potency was now destroyed. Actually, transformed is more the case. Those genitals (still filled with highly potent seed) energized the waters and created a regenerative foam from which sprung the very beautiful and very well-formed Aphrodite (Venus). How ironic that, once Uranus' creative juices were lost to him, what resulted was a being of rare loveliness. Of course, Aphrodite was an earthy archetype of beauty, and a lusty gal, and so maybe Uranus would have found fault with her, too. But then again, he might well have learned to better appreciate the beauty of human love and passion.

So, what happened to this ole, neutered god? It's not too clear—did he die or was he exiled to Tartarus, perhaps? It seems once his generative power was severely weakened, his story was abruptly over—except that his last dying or parting words may have been shocking ones: it's rumored he prophetically warned Cronus that someday a rebellious child of his would also overthrow him. Upon hearing this, a shaken Cronus must have thought that his old man was a disruptive force even unto the bitter end! Whatever his fate, this once-mighty sky god was now powerless to block Gaia's needs ever again.

Cronus then became the supreme ruler of Heaven and Earth—was he having his Saturn Return? He ushered in a Golden Age and made "time" essential to the order of life. From here on, material manifestation were seen as a good thing, offering security when well-organized. Everything clicked, with Nature and humankind now in harmony. Earthly existence was here to stay—warts and all—and ideal perfection was something to slowly work at rather than to try to instantly achieve by sheer will alone. Cronus wanted a world where nothing would remind him of his hated father—thus, his was an era of no more unexpected crises.

Uranus quietly faded away, and life on the planet marched on without missing him one bit—that is, until he reemerged in 1781, now as a new planet of our solar system (disrupting the old astronomical model, of course, which had given "last known planet" status to Saturn). Revolutions in the name of individual freedom versus monarchical oppression broke out around the world during that period. Apparently, the Uranus archetype had time to evolve and transform for the better (in fact, now Ouranos would actually enjoy "high-fiving" his freaky kids). And ever since, Uranus has reclaimed its creative power and today is an exciting force that we're learning to better integrate into our nature, and consciously so.

THE MORAL OF THE STORY

In this mythic tale, amid its bizarre imagery, is a message for us about Uranus in our charts. This myth warns us that our inner Uranus, when unawakened to its true nature, represents an unattractive part of us that is cold, unfeeling, unsympathetic, and too mentally remote to be able to empathize with others. Although we can visualize our ideals on abstract mental levels, we run the risk of being disillusioned when our visions are brought down to earthier, human levels where imperfections abound. The myth also makes plain that we cannot afford to treat children in exclusively Uranian ways, devoid of emotional nurturing and physical touch. A hostile child can later grow up, turn around, and rebel against all we stand for—with great mutual alienation resulting (even tragic violence).

A painful psychological dilemma that today still has much power within the collective unconscious involves "the rejecting parent" archetype. Children with such parental experiences universally feel they can never live up to the impossible, unrealistic standards expected of them. The punishment for failure is that a mother or father creates as much distance from the child as possible. This tale also suggests how difficult it can be for a hard-core Uranian to cope with how things really are on the planet. Becoming remote and untouchable are not valid coping mechanisms—they're defensive retreats. We with powerful natal and transiting Uranus patterns will have to watch out for that. Well, there will be more to say about all this soon. Let's just begin this book and hope we all learn a lot about how to better handle our amazing Uranus transits!

NOTES

1. Bil Tierney, *Twelve Faces of Saturn,* Llewellyn Publications, St. Paul, MN, 1997.

2. Kim Rogers-Gallagher, *Astrology for the Light Side of the Brain,* ACS Publications, San Diego, CA, 1995, 242 pp.; *Astrology for the Light Side of the Future,* ACS Publications, San Diego, CA, 1998.

PART ONE

WAKE UP AND SMELL THE FREEDOM

DON'T BLAME THE MESSENGER

SOME GOOD ADVICE

When we begin to study astrology, some of us may unfortunately get the impression that the Outer Planets, especially when transiting, are single-handedly the cause of the thorniest problems that we face in life (transiting Saturn gets labeled a nasty trouble-maker as well). We might also suspect that actual planetary rays unfairly blast away at us, using mysterious forces to sometimes directly block even our best intentions—and that the basic purpose of astrology, then, is to learn how to always stay a couple of jumps ahead in outsmarting these tricky, security-threatening forces whenever possible. We simply stick to picking the "good" days to do certain desirable activities while we totally avoid, at all cost, taking action on the "bad" days. Such simplistic thinking reflects an older school of traditional astrology at play, one that puts a sense of fear and foreboding into our hearts—only now by exploiting Uranus, Neptune, and Pluto as the new "heavies" to worry about! Yet in olden times, Mars and Saturn were the perennial "bad boys," made to carry the shadows of "sin" and "evil" and their all-too-human consequences. Now, it seems, new celestial members with somewhat more frightening faces have joined the club.

Armed with this less enlightened attitude, some of us may not be as ready or willing to accept transits as symbolic reflections of conditions triggered within our psychological selves—probably because this is a less satisfying option, one that places more responsibility for our life's unfoldment on our shoulders in a manner that leaves us with few real outside enemies to conveniently blame. After all, it's so much easier to point the finger at a red and angry old Mars whipping up a fire storm in our chart while roughing up our Moon, which then "causes" an imbalance in our emotional chemistry—one that makes us the perpetrator or victim of aggression and violence. We could therefore justify our predicament by saying, "Hey, it wasn't my fault...it was that wicked square from Mars in my Twelfth that made everything spin out of control!"

Few astrologers still go along with this dated script: the idea that planets and zodiac signs solely represent good and bad external influences in the form of powerful energy fields that perhaps astronomers might one day discover and scientifically validate—hoping maybe then this could allow astrology to finally join the ranks of other true sciences. If so, according to this assumption, astrology would at last be more clearly understood and valued as "real," including the claim that "bad" transiting planets in "hard" angles in an already "difficult" chart are literally—not symbolically—responsible for the rotten days of our lives (and therefore, such transits are just as guilty of directly causing personal disaster as is lightning striking a house).

Even if such a mechanical approach to an apparently uncaring Cosmos holds some scientific truth, at what expense to our psyche? After all, we'd thus be admitting that we're trapped in an ongoing adversarial relationship with a cold-hearted solar system whose planets randomly besiege us—and all because we apparently were born on the "wrong" day or have dared to begin a project or a relationship at an "inauspicious" time! And so, it would appear that the cosmic forces at work don't care a whit if we're "just human" and can't help but to give in to our urges and impulses at times. We're still likely to be penalized by such impersonal agents for taking action when our timing is off, as indicated by certain planets in tensional alignment.

Yet to adopt such an attitude about astrology could make us more fearful than ever of our birth chart patterns and their assumed power to make or break us. We'd feel ourselves to be easily targeted victims of the whims of an unsympathetic and soulless Universe—in almost the same cold and heartless way we humans treat the lowly insect kingdom. From this disheartened view of the Universe, Mars and every other heavenly body indeed get to "attack" planet Earth again and again, leaving us vulnerable earthlings forever insecure about the twists and turns on our life path. In addition, it sounds like we'd be blindly groping our way through an existence devoid of ultimate meaning. Is this the chilling path we want to take astrology and humanity down? I think not.

Sure it's sweet news when our Jupiter and our Venus transits roll around and everything's coming up roses, because here the outer world caters to nearly our every wish (of course, we're not thinking about the damage that too much of a good thing can create). But there's no joy in our heart when Saturn and the Outer Planets come barging in on the scene to "cause" major grief by ruining any future plans for personal happiness. And so, by adopting the above materialistic outlook on astrology (with its flat-out belief that all basically originates from sources outside of us), we're made to feel like pawns of dumb fate or captives of some darker destiny we seem powerless to change. The best we can do is just learn to duck as much as possible and avoid being hit broadside by any antagonistic cosmic powers-that-be. Yet, realistically, how long can we expect our "duck" to last? This quickly becomes an anxiety-provoking way to work with astrology, one that tosses any sense of personal empowerment right out the penthouse window and smack dab onto the unforgiving pavement below! It simply is not the way I choose to experience the birth chart.

By the way, Uranus is one planet that's real easy to blame for any and all screwy external predicaments that seem to happen *to* us rather than *because* of us. It's a difficult energy for anyone to take personal responsibility for, since Uranian outcomes seem so insane and ridiculous at times, often punctuated by jaw-dropping surprises. How could we possibly have anything to do with co-producing such oddball Uranian scenarios, we wonder, especially when they often only get crazier by the minute and leave our life in a temporary tailspin? We reason that no intelligent person would intentionally opt for such a chaotic turning of events. No one really relishes being that caught off-guard in life.

I suggest it makes more sense to regard Uranus, Neptune, and Pluto as cosmic messengers providing us with spanking new information about our greater human potential; and that we should savor all personal insights gained as if we're being served nourishing soul food (exotic to the taste). How much of this food we're willing to consume and digest at any one time is up to us and our ability to overcome family programming from the past, as well as social conditioning in the here-and-now. Some of us are ready to eat a big, hearty bowl of such transcendental soup, while others will take only little spoonfuls, swallowing reluctantly; and then again, some of us are just not hungry enough to sample anything that tastes unfamiliar.

Each Outer Planet reveals something valuable to us regarding how our true Self best functions at deeply integrated levels (since each planet deals with helping us see things in terms of greater wholes rather than of fragmented pieces). If anything, personal events probably only happen once such planets have been given the go-ahead by our spirit—that eternal part of us forever in co-creative alliance with greater universal designs. Of course, this core spiritual Self of ours operates in ways that seem incomprehensible to our normal waking consciousness. Thus, we seldom readily comprehend life's game plan for us when an Outer Planet transit is active in our chart.

Obviously, we certainly don't feel we ever deliberately give Uranus permission to shake up our foundation overnight and leave us in a state of disorientation and uncertainty. However, to exclusively condemn a Uranus transit for throwing our life into a temporary state of uproar and disarray is to miss the point. It is our central Self who internally signals the Universe to willingly provide us with "whatever" we need for a major life overhaul. The transits just show us in measurable, graphic terms when the timing is ripe and when a fresh cycle of rapid growth is at hand.

STAYING CONSCIOUS

It's always a good idea to work consciously with our astrological transits—defined as the current positions of the planets in the sky that signal natal planet response in our birthchart. Being conscious and alert allows us to sample more from life's menu of situational choices. Recognizing our options for action during any transit empowers us; we become less locked in and limited by circumstances.

When certain life events feel fated and thus unalterable, it suggests we're probably unconscious about many underlying elements of the matters in question. Our sense of choice seems nonexistent. Remaining unconscious about issues also means we're less able to evaluate all sides objectively, least of all the side that probably bothers us the most. We don't sense we have choices because we have little psychological insight into what our conditions are all about. Why are they occurring this way, and why now? And most of all, why are they happening to us? That's something we typically ask when things are turning out badly or seemingly beyond our control.

Thus, the less aware of our psychological make-up we are, the more bewildered we can be when fate decides to take a sudden or tragic turn (Uranus is especially notorious for its stunning turn-abouts). Of course, few of us seem to care how unconscious we are when things are going fantastically well—that's usually not the time when we wring our hands and cry out, "Why me?" Still, it's probably good to get a psychological handle on those situations that directly involve us as soon as possible. Astrology can assist us well in this regard.

Our transits can quickly get us in touch with important issues of the moment that require our attention and skillful resolution. That often means that we'll need to flexibly negotiate with the Cosmos (and with our loved ones) without getting too hung up on the fact we're actually compromising here and there. Actually, compromise is a Libran way of dealing with the world. It works well in many circumstances and keeps our relationships running smoothly, but most so when we are conscious of this process and are making our adjustments to others willingly. To do that requires our understanding that what's going on inside us is also being reflected outside of us, projected onto an environment that will offer us a better look at the inner themes we are working with. That's why we shouldn't be so quick to wish for certain transits to be over and done with (good riddance!) before we've even figured out *why* they are visiting us in the first place.

Transits point out potential life problems, but—thanks to astrology's rich and multileveled symbolism—our transits also provide creative solutions for such problems (something more than just that old "duck-and-avoid" routine). The answers we are looking for are to be found in the transit pattern itself, played out within the context of current circumstantial probabilities. We can either ignore solutions implied, maybe because they require more inner and outer change than we care to undergo, or we can take the ball and run with it as far as possible. The key is knowing how to effectively put astrology's symbolic building blocks together (shown by the natal/transiting planets, the signs, and the houses involved), and then creatively construct something satisfying out of them.

Let's say that transiting Uranus is about to square our natal Venus in the Second House. If we've studied astrology for a while, we already know what Uranus is basically telling us here: something needs a quick and maybe surprising change; something's becoming stale and freedom-inhibiting; we've outgrown a condition that's now holding us back. We also can figure out what the earthy side of Venus in the Second can mean (all those things we're very attached to for our material and emotional security— things that ground us, pleasure us, and make us feel content). We then need to consider the distinctive traits of the aspect involved, the psychology it uses to take action. (See *Dynamics of Aspect Analysis*[1] for a detailed review of each aspect's "personality".)

We can also try to link this Uranus transit with other contrasting or reinforcing factors operating in our chart as we look for even broader patterns of meaning. Still, at some point, we'll need to question what we can do to better direct this and any other transit toward areas appropriate for our development, and according to the growth requirement of that transit (conjunctions have different agendas than do oppositions, for instance). Otherwise, the Universe has no choice but to come up with a few of its own Uranian scenarios that we're bound not to like (and then everything seems to indeed come at us from the outside). Why not instead attempt to co-create a better outcome?

It's wiser to first come up with a few common-sense solutions that are well within our power to attain, even though Uranus has a taste for unanticipated resolutions that seem more a matter of using "uncommon" sense. Life is telling us to at least dump those Second House belongings we've underused for years (especially those fad purchases we made during some other zany, carefree Uranus transit). We then might instead relish the experience of freedom from the clutter of possession overload, at least for a while, as we reform our material structure. Rather than impulsively toss items into the trash bin (a quick but mindless Uranian solution), why not give some away to charitable organizations that recycle them back into the community (particularly so if our Venus is in Scorpio)? Every transit is trying to tell us a personal story about our subjective reality and what we can do to change the plot if we don't like how the drama is going. We get to take a shot at being our own director and screenplay writer as well.

Note

1. Bil Tierney, *Dynamics of Aspect Analysis,* CRCS Publications, Sebastopol, CA, Rev. 2nd Edition, 1993.

CHAPTER TWO

OUT OF THIS WORLD

WHEN EGOS ROAMED THE EARTH...

Don't let my book's upbeat, encouraging title mislead you. "Alive and well" may be more what we *hope* we'll feel like after Uranus has made sweeping changes in our lives and has rewarded us with personal enlightenment, liberation from social indoctrination, and a greater awareness of our mind's vast potential. After all, Uranus and the other Outer Planets are here to show us that we are and have always been more than just human (of course, we're still fixated on proving to ourselves we're more than just mammals). Metaphysicians like to remind us that we hold the keys that unlock the "god-within" spark of our being, enabling us to safely experience increased degrees of transpersonal (cosmic) consciousness—but only when we have psychologically prepared for this and are truly ready. Besides, it's tricky to operate exclusively on such otherworldly levels and still hold down a job, pay our bills, and do all those other mundane things we must do in order to carry on and be regarded as reasonably sane.

Too much Outer Planet energy, if prematurely activated, could create great unrest and discontent for us should suddenly we balk at being tied down by our all-too-earthy physical body. Even worse, we may loathe being limited by the bio-mechanics of our brain, which some days seems to work as slowly as a floppy drive. In this state of affairs, if given a choice, some of us would rather be pure consciousness. Some of us would even rather be "all mind" anywhere else but on this "backward" planet (any "advanced" solar system will do)! That's what it's like—a touch of divine madness—when we become spiritually accelerated too soon (or when we think we are). Yet Saturn won't allow us true transcendence until we've learned to better ground ourselves in matter with greater self-awareness (even a superbly healthy body is needed for deep spiritual awakening, according to yogic masters of the East).

Our strong urge to fly high and free ourselves from our earthly confines is, however, met by the restrictive heaviness of physical gravity and any hidden resistances coming from the shadow side of our psyche. We're angels wearing lead wings and we're stuck on the runway! Blind escapism, driven by unconscious compulsions, can be an Outer Planet's defense against our mortal limitations. It's a response that is often destructive to our ego's health and to our ability to function in Saturn's real and very concrete world.

Therefore, when our otherworldly fantasies crash and burn, and when "gurus" we've devoted ourselves to fail us, a few of us could fall into deep depressions or even undergo psychological breakdowns. Without mental poise and emotional balance, an overload of poorly integrated Uranian energies could turn us into agitated, unstable nervous wrecks. This is why a healthy sense of ego is very important to have, along with a capacity to objectively analyze one's inner motives and also to learn to use all incoming energies in moderation.

Ego is not a dirty word, especially in Saturn's dictionary. In fact, Uranus has great plans for our ego once we first are willing to undergo a few needed make-overs. Often quite street smart, our Saturn-protected ego teaches us how to thrive on the material plane by knowing the ropes of survival. This is something which provides Uranus with valuable tips and experiences it cannot otherwise appreciate in its own loftier realm. After all, the world of the human ego is rather exotic to the Outer Planets, primarily because it's so darn alien to their being. Therefore, learning how to make it in the material world can teach Uranus a few new tricks, which in turn helps rouse this planet's inventive streak in ways that make our physical experience more exciting.

Our solar ego—the commander of its own, limited realm—helps Uranus to creatively manifest its timeless, abstract essence in more material ways. Unfortunately, our ego doesn't really comprehend "timeless" states of being, since it's so used to operating in terms of a finite reality where promising beginnings can have successful endings, depending on how ambitious our strivings are from start to finish. Planets that point to our ego and its development are the Sun, our internal "big shot" who comes up with grand and self-glorifying plans; Mars, the prime instigator who provides the fuel to burn, which then speeds up goal attainment; and Saturn, the qualified headmaster of finite reality who must, after much careful inspection, authorize final approval of all ego-driven projects, since "Them's the rules," he says.

Now give each of these planets a strong Uranus transit and watch how we start to feel restless, speed up our tempo, and attempt to break out of our habitual expression in favor of doing something more experimental. Yet the process is all a bit nerve-wracking for us as soon we realize that, with this sky god, there's no way to know for certain whether satisfying outcomes will result from all this acceleration of energy and all this bold risk-taking.

Nonetheless, our ego's willing to play the great host to this unusual out-of-towner from afar, if it can receive something outstanding in return, something that will put its shining solar face on the map, so to speak, and make it a mighty force to reckon with in this world. Well, guess what? Our ego eventually finds out the hard way that Uranus won't make such quickie deals with the Devil (although Pluto might be tempted); in fact, Uranus will guarantee us nothing but an "interesting" time that might just also leave us with more questions about life than with solid, workable answers. Our ego had better learn fast how far it can and cannot go with Uranus, Neptune, and Pluto—each armed with awesome powers we only dream about; we might be treated to a few of their divine gifts once we stop wanting them so badly for our own self-centered purposes. But we'll first need to understand and appreciate the wisdom of true Outer Planet detachment—and Uranus certainly can give us a real taste of this.

SURRENDER OR ELSE!

Our solar ego, feeling self-deserving, is determined to marshal the energies and assets symbolized by Mercury, the Moon, Venus, Mars, Jupiter, and Saturn. These can become valuable resources to be harnessed by the execution of our will's desire. This is perhaps why the ego feels so capably in charge of our total waking reality—it has back-up support at its command. It thereby considers itself well-fortified and able to do battle with all Outer Planet energy during those times when ego-preservation is at stake (our ego will do desperate things to keep its central seat of power). We humans will therefore have to contend with this defensive element within our psyche that won't let our ego be taken over by a planet representing the very things unstabilizing Uranus does. Yet if we refuse to surrender, we cannot grow beyond a certain established point. In time, our defiant ego often ends up being invaded anyway by the overwhelming forces of the Outer Planets, but all parties are angry and disgusted at this point. By then it's a case of a hostile takeover!

We also have internal complexes dwelling in our personal unconscious that often fight against our transcendental development (complexes that have a potent life the ego knows nothing about). Should they get a foothold into our emotional territory, as they often do, they'll manipulate us using fear and distortion to arm-twist our instincts into blocking Outer Planet integration. Such complexes will feed us a warped interpretation of trans-Saturnian principles. We've seen throughout history how the collective unconscious' shadow-parts (made up of a greater and more potent mass of these internal complexes) have misused Uranus, Neptune, and Pluto energies in twisted ways, causing many people to suffer as a result.

However, considering all such combined opposition they run up against, Uranus, Neptune, and Pluto still manage miraculously to come out as winners. They know what the Master Plans of Spirit are about and will not give up on their ultimate goal of achieving our evolutionary flowering (I guess they love mind-boggling challenges)! If too much ego gets in the way and tries to obstruct such planetary flow, this only postpones but never fully obliterates our cosmic unfoldment. Uranus, Neptune, and Pluto in the end prevail, and for that we will someday be "eternally" grateful.

MAKING HOUSE CALLS

Outer Planet transits can last quite a while, which already confuses some of us, because we're taught that such "Higher Octaves" are all about *quickening* our energies and speeding up our evolutionary journey. So it must be Earth's gravity that weighs these planets down and makes them move about so slowly—something only Neptune actually seems to enjoy. Actually, "slow" is good when it comes to in-depth development; slow means "deliberate" and "meaningful." It's not all that uncommon for Uranus to spend almost a decade in some of our natal houses (and typically even longer than ten years for Neptune and Pluto—the two who'd most want us to ponder our life's ultimate purpose with greater depth). These planets, remember, think nothing of time going by. Time is seen as Saturn's big, boring fixation—not theirs.

Yet things can get tedious and testy when, like guests who've overstayed their visit, these planets refuse to leave us alone—they hang around way too long. "How much Pluto transiting Venus must a decent person endure?" could be our lament by the second year of such interpersonal intensity (actually, it's the *in*decent folks that Pluto's more interested in). It's typical that when we don't see quick results solving our problems, and when new complications set in instead, that we're ready to do a little Outer Planet bashing. (Yes, even professional astrologers sometimes do it in the privacy of their own homes, probably while taking a shower when no one else is around to hear the blood-curdling screams.) We collectively are not sure if we even *like*—much less trust— Uranus, Neptune, and Pluto, each of whom can come off at times like a certifiable weirdo from the other side of the cosmic tracks.

And so I say it's perfectly all right to get ticked off by a nerve-racking transiting Uranus opposition or a hair-pulling quincunx; we're human and we must blow off a little steam to keep from really cracking up. But even when we've had enough—and naively say, "Poof, be gone!"—Uranus won't simply disappear; in fact, stranger-looking rabbits start coming out of more hats! Therefore, it's best that we deal with the matters of our Uranus-transited house squarely and not waste energy moaning or getting all paranoid about our future. This planet will be there for the long haul, and so our adaptability is vital to the enjoyment of productive outcomes.

Over the course of time (which is when Outer Planet efforts really shine), we'll realize we have a new and improved natal house to play in, a place where we can entertain our friends more creatively and even invite the world-at-large to see the stylish renovations made. We probably also have more room here to breathe and stretch. We thus shouldn't be so fast to kick out these three often disruptive "interior redesigners," because they all have a talent for being far-sighted and visionary (well, Neptune does need special reading glasses now and then). Uranus, especially, knows what we'll need to have around us in our environment in order to make quantum leaps in psycho-spiritual growth. And it gets its orders directly from the Great Beyond, which tells us we should let those people and things suddenly ushered into our lives fall into place, even if it takes a while for the dust to settle. A greater Intelligence than ours or society's is rearranging the scenery and revising the script every so often; all we have to do is bravely make our entrance and let Uranus' exciting action begin.

A TRANSITING PLANET'S ROLE

This is a good time to clarify terms. A transiting planet functions differently than does a transited natal planet (the one receiving

the transit). Our natal planets are part of our character unfoldment process. We'll have plenty of time to develop the strengths
they symbolize while we work on recognizing their weaknesses
when mishandled. Time allows us to refine these energies and
custom-tailor them to our growing needs; they feel very close and
natural to us at some point, for better or worse. Natal planets
don't seem as if they are coming from external sources, although
they sure do when we habitually project them onto others (the
natal Outer Planets can be readily projected in our early years, or
even for a lifetime for some of us hard-core Saturnians).

During a transit, our natal planet takes on a passive-receptive
role as it initially absorbs the energies of the transiting planet. It
may absorb and then try to quickly expel such incoming energies
(not everything tastes so good at first), yet it still has to take
something of that other planet inside itself before it can accept or
reject anything. We therefore have an opportunity to see how well
that planet's new pieces fit into the familiar world of the natal
planet we already know and love or loathe. However, many times
what we "already know" is limited or obsolete; we're also not seeing our perennial blind spots in this area. A Uranus transit, especially, becomes a time for us to open up to new information about
becoming more self-aware.

But in the long run, it's the *natal* planet's principles that must
be satisfied by this transit if this passage is to be deemed successful. The natal planet in question is not a puppet in the hands of a
dictatorial transiting planet, but hopefully it will act as a co-cre-
ator of our unfolding consciousness.

For example: transiting Uranus may be opposing our natal
Mars. This tells us that a lot of Uranian people and circumstances will work together to make an impact on those personal
facets of our being represented by our natal Mars, with emphasis
placed on our customary ways of responding to outside stimuli
whenever our Mars is opposed by any planet—are we usually
aroused or enraged? We'll have our typical ups and downs and
surprises coming from "outta da blue" during this period. Still,
how creatively we resolve this transit will have much to do with
how well we've fulfilled our Martian urges. Even if our Mars

takes in a bit more of Uranus' vision, it's still our Mars that has to go forth in life doing what it does best (only now with a different perspective than before regarding venturesome action). The bottom line is that any transit's end results must highlight the nature and needs of the transited planet; thus, if we start with a scary-looking Pluto aspecting our oh-so-sweet Venus, Venus will have the last word in the matter—hopefully a deeply transformed Venus, that is!

A FEW TIPS

If you don't have and never will have transiting Uranus going through your Third, let's say, it still will transit your natal Mercury a few times during your life span. In this case, determine when those Uranus/air Mercury transits will occur, read my material about them, but then also read what I've said about Uranus transiting the Third. Air Mercury and the Third House are two examples of Principle Three, one of the twelve principles that hold astrology and life itself together. The material I've offered is thus fairly interchangeable; the house transit reports at least supplement the transited natal planet information. This is why it's best to read and compare both sections. I also feel that my material could be applied to just natal interpretations, keeping in mind what I said before about how character-developing themes of our natal planets take longer to unfold and are less urgently resolved.

ROCKET AWAY

I think you've been sufficiently briefed. Your travel papers appear to be in order—but not for long, I predict! You've taken all your shots to protect you from coming down with rare states of mind, although little good that will do you on this long and very strange trip you're about to take. Your earthly insurance policy is covered, but no guarantees are provided once you leave Saturn's orbit. "Them's the rules," he says! All geared up in your shiny new astro-sensor body suit, there's really no time to bail out now.

The launching pad is has been cleared, there's a flash of lightning in the distance, the space ship is about to take off, and *you* need to be strapped into your mind decompression module before we take off in ten seconds, nine seconds, eight seconds...one second...blast off!

CHAPTER THREE

A URANIAN TOUR OF OUR NATAL PLANETS

THE USUAL SUSPECTS

Before exploring the details of the gravity-defying world of our Uranus transits, it might be smart to first preview each planet and house that Uranus technically gets to contact and maybe revolutionize—including itself—during the course of its whirlwind eighty-four-year spin around our natal (birth) chart. Astrologers call one such complete Uranus cycle a "Uranus Return." It seems the old wild-eyed freedom-fighter, after defiantly leaving home at a tender age to seek enlightened independence, eventually gets to head on back again to earlier natal stomping grounds—that is, *if* we've eaten enough fruits and veggies and have managed to live long enough to experience this transit. Hopefully, we arrive late in life at this special passage less weighed down by cumbersome, conflictive, psychological baggage.

By the time of its Return, Uranus has helped to shake up many of our life's duller, socially conditioned structures throughout decades of sometimes dramatic and unexpected change—usually all for the better. We typically want to feel free from the constraints of society's conventional expectations of us by the time our Uranus Return is official—at this point, no one can tell us how we *should* act and behave at our age, because we simply don't care to submit to set social standards any longer (besides, we figure we've paid enough dues to society already). Many of us by now are impervious to the narrowly defined rules of the Establishment—and not just at this seasoned point in our lives, but perhaps ever since our mid-life crisis years when transiting Uranus opposed our natal Uranus (a lively time of bolder inner questioning, when we began to throw off a few oppressive shackles of conformity in favor of exciting self-discovery).

At the Uranus Return, as well as at other pivotal points during this planet's life cycle, we are inclined to ignore traditional social guidelines and external authorities and, instead, willingly detach from society so that we may follow the lead of our own "inner awakener." Not all of us will turn out to be so headstrong and defiant by our mid-eighties—but try getting in our way and pushing us around during our well-deserved Uranus Return. You're just asking for trouble!

Let's start with a planetary review of our birth chart's cast of colorful characters—especially when seen from Uranus' uncommon perspective. Doing so may help us get a better feel for just what's at stake during our major and more memorable Uranus transits to come:

THE SUN

Here's where the heartbeat of our chart can be heard, letting us know that our our ego—backed by our will—is ready and able to pump vital energy into our lives so that we can be all that we can be (at least from our ego's limited perspective). Our Sun helps us to realize—in a positive, confident way—that we have a right to be naturally self-focused when it comes to our personal development. Nobody else can do our lives as convincingly as we can, once we follow our heart and allow ourselves to shine brightly in the world. We're not to hold back from displaying our creative potential, and certainly not because we doubt our innate worth.

Our natal Sun symbolizes the part of ourselves that honors our individuality, proudly so at times. The Sun encourages a healthy and creative expression of "me" consciousness. Our Sun's sign and house positions thus become vital to our unfolding growth as we mature. Here's where we radiate our essential being and where we are challenged to display our admirable strengths to the world in good faith and with an sense of optimism.

Right from the start, Uranus is quickly attracted to the display of high self-regard and courage typically shown by the Sun. Uranus admires this fiery planet's willingness to individualize itself, because this prompts this planet to stand out in a crowd by boldly showing off what it's got that makes it so special. After all, that's what Uranus itself does all the time—stand out as being different, although not always in the popular, crowd-pleasing manner of an impressively well-managed Sun!

The Sun is no coward or wimp. It has a strong enough psychological backbone to gain Uranus' attention and respect. This is good, because Uranus is more apt to suddenly shatter any self-limiting sense of weakness or feelings of victimization a planet may have. Therefore, it helps when the planet in question already has a positive sense of its own strengths and merits. In fact, the self-congratulatory Sun always wants to come up a gold-medal winner in life, a figure to be held in high esteem. As a rule, with the Sun, Uranus doesn't have to address destructively self-defeating issues that often torment us regarding some of the other planets in our chart. It's hard to imagine that our Sun would ever see itself as having inner demons to grapple with (the kind Neptune and Pluto have us struggling against).

If anything, we may feel at times like a frustrated actor waiting to play a major starring role in life, and yet our Sun isn't going to be too traumatized by anything—basically because it's not a planet that deeply internalizes its experiences. It's an extrovert at heart, with its sights set on commanding what's "out there" in life, not plumbing our inner depths.

In many ways, Uranus already acts very much like the Sun. It is highly self-expressive and desirous of all sorts of colorful adventure. It is brave and up front when it comes to facing up to life's obstacles, and it always does what it takes to live autonomously, free from the dictates of others. All these are traits we apply to the

Sun. However, the gift of detachment belongs to Uranus, not to the Sun, who is instead prone to take offense and get all worked up when not given the royal treatment. Being worshiped and adored is something Uranus couldn't care less about; in contrast, the attention-hungry Sun is hooked on receiving glowing reviews from the world. Still, Uranus knows our Sun is willing to support the freedom to individualize at will. This appeals to Uranus.

Later on, when we begin exploring our Uranus/Sun transits at length, we will find that they represent stimulating periods in life when we are driven to pull away from mindless identification with others on a collective basis. We feel impelled instead to "do our thing" and separate from the masses. Sometimes we may do too good of a job of individualizing and thus may alienate our environment by showing an uncooperative streak of rebellion. However, for the most part, Uranus/Sun transits help us to open our eyes wider in order to better see our true self and our brilliant but untapped creative potential.

If there's going to be a problem here, it's due to the Sun assuming that it can and will run the show single-handedly; that it has unlimited license to dictate over and control everything and everyone within its grasp. Our ego tends to enjoy running amok with unchecked power whenever Saturn's out of town on business and therefore unable to regulate our Sun's energy. Uranus won't go for that, and will conspire to break our (self-centered) will by introducing swift but fateful changes in our life that demand we let go of the reins we hold so tightly. Our Sun's perspective is seldom broad or experimental enough to satisfy Uranus, and so we are forced to redirect our will power and instead learn to be more adaptable and curious enough to accept the sudden alterations required of us. The natal house of our Sun should give clues as to where this Uranus transit can be most effective, even if jolting at times. Here we can go through a renaissance period of sorts, one that can prove to be very exciting regarding the Sun's creative outpourings. See Chapter Ten for further discussion on this topic.

THE MOON

The Moon is not given enough credit in astrology for symbolizing a wise and knowing part of our basic nature. Calling it merely a planet of "instinct" tends to imply that our lunar self is something primal, primitive, less accessible, and less spiritually evolved (after all, even the lowly amoebae has blind instinctual reactions to the simplest of outer stimuli). Meanwhile, Mercury has traditionally been touted as being the smart and clever one with a "real" brain in its head. We figure we can always count on Mercury—armed with relevant facts and figures—to know what's going on in life, to have ready and reliable answers if need be.

However, we forget that it's our natal Moon that securely knows its way around the block, more so than does any other planet in our chart. The Moon senses that it has already "been there, done that," and now has an inner feel for the territory it had once covered so well in the past (i.e., past lives). It doesn't even have to first analyze any situation in order to appropriately respond to it. Its emotional take on reality can be quite astute and on target (not just "overly subjective," as its critics would claim). The Moon's special intelligence about human behavior is a matter of a deep gut-knowledge that goes beyond what books and classes can teach us about life. Our lunar-sensitive solar plexus probably functioned more as our "animal brain" when we were first developing our human traits as a race in those earliest of times on the planet—in days before spoken language and our ability to rationalize our feelings were born.

It would seem that Uranus would enjoy working with any planet that claims to have a good sense of "knowing" and that comes with its own built-in radar system (no batteries needed). That's because Uranus is also pretty intuitive about things—it can size up people and situations fast. Therefore, Uranus is very curious about our Moon's potential to further develop its sensitive ability to pick up subtle data quickly from the environment. Uranus is also a lightning quick scanner of what's happening around it (although not in the same observational way that

Mercury scans). However, the Moon can sponge in its environment rather indiscriminately. It's so good at absorbing life that it doesn't filter out the undesirable elements of its surroundings as well as it needs to. Thus, it gets waterlogged periodically.

Remember, one of the most important assets a planet can have—according to Uranus—is adaptability (and probably this is so valued because Uranus is itself such a quick-change artist, never one to be mired in routine and habit for long). Any planet too easily stuck in ruts because of a stubbornly inflexible mindset gets Uranus all riled up and ready to revolt! Uranus hates static conditions. Thus, it hopes the Moon is hip and "with it" enough to quickly go with the flow by being able to emotionally embrace radical change, overnight if need be. Even Uranus never knows ahead of time when such abrupt change must take place!

However, the Moon has a big problem with this—sudden change can be scary and security-threatening. Any imposed outer change can also unexpectedly remove from us those people and things we've grown accustomed to (in addition, Uranus seldom offers the warning signs of what's to come that the insecure Moon typically needs). Wherever our natal Moon is can describe how and where we instinctively do whatever it takes to keep things familiar, repetitive, stable, and feeling safe. It's the lunar side of our nature that likes to cling to whatever we have valued as a long-term source of emotional comfort and physical gratification (this gives our Moon strong earthbound appetites that seek predictable fulfillment). Uranus doesn't understand this "dumb" urge to cling like a barnacle to anybody or anything for dear life, especially for long periods of time. This is a sharp point of contrast and a major challenge for both planets to resolve.

During our Uranus/Moon transits, the house in which our natal Moon falls may be ripe and ready for an overhaul. Of course, our Moon doesn't think so and would rather things stay just as they are. The nature of the transiting aspect involved determines whether this time in our life feels like a welcome renovation or an explosive attack on our cherished patterns of behavior. To Uranus, it's plainly obvious that certain life conditions have grown stale and are now failing to offer us true emotional support. It's time to give it up, let go, release our grip, and walk away from on old but now obsolete source of security if we wish to awaken to our greater potential.

Uranus knows the Moon can be a deep assimilator of fresh patterns once it feels at home with its new situation. After our Moon has emotionally accepted and digested these experiences, Uranus feels more certain that its own sweeping visions of reform will take hold for a while on very personal, tangible levels (something Uranus can't seem to accomplish as well on its own, but that the Moon has little trouble achieving). Read more on this in Chapter Nine.

MERCURY

Natal Mercury is one planet that won't waste much time wanting to get to know this oddball stranger—the one calling himself Uranus—who's just entered the Busy Brain bookstore (Uranus' name alone makes Mercury want to come up with a few wisecracks). Mercury immediately has tons of questions to ask of Uranus (such as "What's makes your eyes strobe rapidly like that? Does it involve the use of batteries?" or "Why do you walk in such a fast zigzag motion? Is that some kind of rare hereditary condition?").

Uranus instantly likes this inquisitive and apparently forthright Mercury, with its nosy but innocent curiosity. Mercury is here to learn as much about life as possible, and Uranus has some pretty wild theories that it's ready to share. It seems that Mercury will listen to anything if it's stimulating enough—therefore, Uranus makes sure that what it tells Mercury is nothing but stimulating, even mind-blowing at times. Uranus gets a big kick out of watching Mercury's jaw drop in amazement!

Our natal Mercury house is a place in our life where we are always knowledge-gathering and looking for smart solutions to living. Here, our mind's antennae are always up and ready to intercept the signals that travel through the busy air waves of our immediate environment. We are sensitive to life's little fluctuations and changes—which is perfect, because Uranus will bring only change to our lives, not dull and uninspiring stability. Uranus' transits are nutty at times and too nerve-jangling for even Mercury to handle, but it's a sure bet that this little info-junkie planet will not be bored by what it hears coming out of Uranus (boredom is one thing Mercury can hardly tolerate).

As we are bombarded by new Uranian-inspired information, we learn to rapidly adjust our thinking while we put our analytical abilities on the shelf for a while, because this is not the best time to use them to understand our current life matters. Instead, it is an excellent time to expose ourselves to startlingly new ways to look at the world we live in, while we learn to quickly adjust our opinions accordingly, especially when viewing things from odd angles for the very first time. Uranus is eager for us to adopt a few peculiar perspectives. Logical Mercury will simply need to suspend disbelief and critical analysis until Uranus has had at least a fair chance to reveal its truths about reality's unusual and unpredictable nature. A little friendly mind-bending is just what the Cosmos has ordered for us.

During Uranus/Mercury transits, we often feel as if the old explanations of life we've stuck to without question are now up for a truth-seeking review. This will have a disruptive quality to it that unnerves us in the beginning—we start noticing sharp contradictions between what we were taught versus what we now intuit to be "the real deal." Knowing that Mercury is one fast learner (unlike old "slow-boat-to-China" Saturn), Uranus never worries about overloading this most open and flexible of planets. Actually, our own chart will hint at just how flexible or not we really are: Fixed Mercury signs—except for Aquarius—are likely to struggle at first to accept Uranian concepts, because such ideas seem so contradictory to what Mercury stubbornly upholds as factual and therefore valid. Cardinal Mercury signs, and especially Mutable Mercury signs—perhaps the least resistant of all, except for Virgo—seem more willing to go on the wild mental ride Uranus has to offer. They need the excitement of newness and novelty to keep their intellectual energies running on high.

In general, when Uranus meets Mercury, intellectual sparks fly. Such transits keep us on our toes mentally and won't let us get too down-and-out about anything discouraging that may be happening in our lives (just detach and create some emotional distance, advises Uranus). However, sometimes Mercury proves to be a little too nervous, flighty, and flaky for Uranus (not so calm and grounded itself). This can give Uranus' brilliant flow of meteoric energy a "here today, gone tomorrow" quality that stirs up much noise and excitement for the moment, but then is gone in a flash! That's not good for materializing progressive Uranian ideals that need to last at least long enough to produce real benefits to society (which is why I say the Moon has a better chance at making something more permanent out of Uranus' insights and illuminating revelations—that is, once the Moon can be convinced it has nothing to fear). Unfortunately, Mercury can talk up a good revolution for a short while without ever really intending to take the radical action needed to make Uranian dreams come true. It seems that Saturn will have to come to the rescue to makes such dream real. Chapters Eight and Eleven will cover this in detail.

VENUS

Calm and collected Venus loves its pampered life of graceful moves, smooth rhythms, affectionate displays, idealized romance, and all of those creature comforts that make the living easy. Venus also embraces the steady security of Earth's gravity, which keeps everything we value in place—that is, just as long as dear Venus doesn't end up feeling too weighed down in the process. Stability is very important to this planet, and so wherever Venus is found in our natal chart, it depicts where any sort of sudden disruption would certainly not be well-taken. ("Uh-oh," says Uranus, "this could be tricky!") How is the famous "Inner Awakener" to successfully arouse a sometimes lazy planet that likes to sleep in late, all comfy under warm and protective covers?

Actually, Uranus knows that sociable Venus responds well to meeting fascinating new people who are as sparkling mentally as they are appealing physically. Venus doesn't want to sit around and debate theoretical cosmology or some other such abstract thought (the stuff that excites Uranus and Neptune). Venus would like to know more about that scintillating guest with the fast-strobing eyes whose unexpected wit keeps everyone in stitches at the party. *Very interesting,* thinks Venus. Uranus is a little odd-looking to Venus, but in an unexpectedly appealing way. It must be that sexy brainpower presented in such a bold and innovative manner (usually not the first thing we'd expect to turn on Venus).

With Uranus/Venus transits, fresh new faces coming on the scene are part of the social education we are to receive. This sounds nice, except that Uranus is really not interested in indulging us in a world of earthly attachments and sugary love connections. The Venus part of us has no idea of what's in store, because it's anticipating possessing something or someone exciting and novel that it can then have around to enjoy for a long time. But that's not the future scenario that Uranus is brainstorming.

Our Venus symbolizes how we can achieve a state of balance in our lives, which is never a static condition, but is instead one of a dynamic interplay of contrasting but complementary factors. Yet even that approach is too careful and conservative for transiting Uranus, who'd love to upset the apple cart to see just how much fruit will fall. A steady and even rhythm is not as appetizing to Uranus as it is to Venus. Uranus can appreciate things that clash in an interesting way (like wearing stripes with plaids), but Venus always looks for similarities that help combined energies to blend (like wearing socks that match).

Therefore, these Uranus/Venus transits typically require emotional adjustments that we really don't enjoy making—a big one being the ability to love somebody in a more detached, non-possessive manner. During this time in our life, we may not even get to hold on to that which we like, much less love, for long before unanticipated factors come in to completely alter the arrangement. If we have mastered the Venusian art of clinging to our loved ones, while turning up the volume of emotional dependency, Uranus is ready to yank away a few of those people from our clutches. Uranus is trying to drum into Venus' head that the dreaded phrase "Let's just be friends for now" isn't necessarily the kiss of death. It usually means back off and allow room for another to breathe in this relationship.

Still, Venus is a very cooperative planet as long as it is treated fairly, not underhandedly. Uranus is not a devious planet, and will always be up front about its feelings. Our Venus may not want to hear the truth Uranus freely wants to say to us, but a part of us does appreciate the open honesty. It's just that Uranus is so darn inconsistent in Venus' eyes, something the love goddess has a real hard time tolerating.

With Venus, we seek self-expressive outlets—usually in the arts—that are reflective of the beauty and harmony we feel within. While Uranus is not particularly gifted in an artsy-craftsy way, it does have a superbly inventive mind which can help bring the spirit of innovation to any artistic project. Uranus helps us to grant ourselves permission to create unusual and at times bizarre beauty packaged in strangely appealing forms. Uranus and Venus can come together and give birth to styles that are very original and electrifying. Therefore, this transit works when we find suitable creative channels. But remember, it doesn't work as well when we are determined to grab onto an "ideal" lover whom we then try to exclusively own and not share with the world. For more on this subject see Chapters Seven and Twelve.

MARS

The thought of Uranus getting our Mars all roused up may initially sound like trouble. These are not mild-mannered planets blessed with any built-in sense of restraint and caution. And yet—perhaps because here's another dynamic fire planet—Mars' basic energy can be very appealing to Uranus (a bit fiery in its own reactions to life). Uranus is quite an activist when it comes to initiating change and ushering in new patterns for living a life of freedom. Mars is also bold and determined enough to act in pioneering ways that can help us plow new, unexplored territory for ourselves.

Uranus hates wimps and fraidy-cats, but that surely doesn't describe Mars. In fact, for any planet to insinuate that Mars is a yellow-bellied coward is to invite a black eye from a flying fist (and Mars always gets in the first punch in any dispute). This assertive, feisty planet is not scared to do something it's never done before. It's actually more than willing to challenge itself in this way. Therefore, transiting Uranus is eager to urge us to partake of a little healthy experimentation in Mars-related action, except we're not to get too hung up on the end results. However, Mars may need more than a little convincing that having everything turn out its way isn't the most important objective at this time. Mars thus needs to be less single-mindedly competitive.

Our natal Mars is a continuous source of vitality for us, most so when we keep busy and remain in motion. Mars finds taking too many pauses or rests to be energy-depleting (something Venus can hardly believe). This is a spunky planet that prefers to tackle the external world head on and in a high state of vigor, and yet at times it's also willing to put a lot of energy into internal matters—both mental and emotional—that require our courage and stamina. Mars hates to give in to weakness and would rather push itself (and us) hard.

Uranus knows that this gutsy planet is ready for practically anything at a moment's notice. That's just as well, because Uranus gives us only a moment's notice for the most part before it sets up unpredictable life conditions requiring our quickest reactions. Venus may not easily decide if it wants to play Uranus' quirky game of uncertainty but, "come hell or high water," Mars is ready to give it a go. It's fearless in dealing with whatever life volleys its way, even that which comes unexpectedly from far left field (where Uranus is mostly known to hang out). Uranus wonders what—so far—there is not to like about this speedy red fireball of energy!

When dare-devil Mars sides up with venturesome Uranus, we feel a strong urge to move about impulsively and to take a few wild chances (both planets are risk-seekers). We certainly can't stand being cooped up indoors with nothing to do, except to maybe suddenly trip and sprain an ankle! Taking some kind of dynamic action is better than taking no action at all for these two. And both planets would love it if what we engage ourselves in allows for moments of true spontaneity.

One big difference between these planets is that Mars won't relinquish its sense of personal power long enough to learn from the wisdom of airy detachment that Uranus offers (that's when Uranus wonders if Mars is actually a bit of a jerk after all). Impatient Mars forces issues even when it's highly inadvisable to do so. It tends to jump into things prematurely, with a headstrong determination that shows no real sense of strategic timing.

Actually, Uranus can be accused of all the above as well, but Uranus also has an additional secret weapon at its disposal: intuition that works at lightning speed. It's this highly intuitive side of Uranus that allows us to stop in our tracks and discontinue all activity—that old "screeching halt" technique—and it's based on some sudden hunch that tells Uranus to interrupt or even totally abandon a project. No further analysis of the situation is necessary for Uranus. It just quits! Mars alone would instead go at it feverishly until the bitter end, or until it runs out of gas.

During our Uranus/Mars transits, we can feel highly driven to take some sort of decisive action rather than endure delays or stagnation. But our efforts can seem short-lived or choppy as Uranus tries to periodically detach, leaving our Mars feeling suddenly less vitally engaged in activity (our level of interest plummets because our mind is elsewhere). There is a strange stop-and-go quality. Maybe it means we are to work quickly in short spurts of time, then take a break, and start up when we again feel the itch to give our project our very best get-up-and-go. However, Mars doesn't really like to stop once it's revved up. Nonetheless, Uranus demands periodic interruptions. Mars also has a temper that's a tad too emotional for Uranus, even though this is one cerebral planet that can get fired up due to its high-strung temperament. Therefore, expect hot Uranian lightning bolts to strike from time to time! Read more about this in Chapter Six.

JUPITER

Up until this point, all of the planets mentioned fall under the "personal" category. This means they have self-related special interests that revolve around obtaining varying degrees of ego-fulfillment. The answer to "What's in it for me?" is very important to all of them. None of these planets can be regarded as truly self-less, although the Moon does have its more saintly moments. However, by the time we move beyond the orbit of Mars we enter a different level of understanding that involves our growing awareness of society's more complex needs. This realm of experi-ence is symbolized by Jupiter and Saturn, each categorized as a "social" planet (societal is probably a better word). Communities of people who depend on one another in a broad sense are tradi-tionally built on the values of Jupiter and Saturn.

With Jupiter, we cannot afford to be too self-seeking, because this planet seems to reward those who most generously engage in social participation geared toward cultural betterment and moral uplift. Therefore, our scope of interest widens quite a bit, and we no longer revolve around ourselves alone—that small world of "me" and "mine." Here we learn to care for how all members of our culture are growing and improving themselves, even spiritu-ally. This is one reason why extended education and mind expan-sion are big on Jupiter's list of "must things to do" in a lifetime.

Jupiter is associated with our urge to reach out to our commu-nity and inspire others to build an ethical and humanely tolerant society. Uranus takes note of this benevolent Jupiterian urge, because the Great Awakener will attempt to spark the flame of social activism in every planet it contacts. Yet with Jupiter, such activism—although evident—is often in an early developmental stage. Actually, Jupiter is caught somewhere between its desire to be a philanthropic humanitarian and its wish to enjoy unlimited materialistic privilege. It wants to spiritually uplift the world by promoting global harmony, but it also wants a chance to cruise the freeway speeding along in the new silver Mercedes it bought after getting lucky in Las Vegas. Earthly detachment is not as strong in Jupiter as it is in Uranus.

Both planets tend to be dissatisfied or restless with the duller routines and responsibilities of mundane living. Each looks for inspiration and hope in a vast world of future possibility.

Exploring the past and rehashing old memories are not what interests these forward-thinking planets. Getting stuck in suffocating relationships with control-freaks are also avoided like the plague. Give them lots of wide-open spaces devoid of clutter and tedious detail.

Uranus/Jupiter transits can be joyous times when we feel the Universe is giving us permission to be super-adventurous. The cosmic toy store is now open and we are allowed to choose a few nifty gifts of the mind to play with—and no one's really paying attention to time. Being playful with such cosmic opportunity is part of the general tone of things to come. Because of that sense of play, we can innocently stumble onto something by chance that leads to major self-revelation and fresh new paths of meaning. Jupiter is open to whatever Uranus has in mind, within limits. (Yes, Jupiter has its limits, yet it still manages to cover a lot of expansive, fertile territory.) However, Jupiter is a much warmer planet in temperament, with buoyant feelings that seem more emotive than emotional (considering their characteristically exaggerated response). Jupiter has a talent for getting an enthused crowd to feel good. It can turn people into true believers who are ready to take up a worthy cause.

Uranus gets the crowd going too, but with mixed results. That's because what's often lacking here is warmth, sympathy, good cheer, and that winning Jupiterian quality of not harshly judging human flaws. Being less socially diplomatic, Uranus can get the crowd agitated enough to start a riot! But in general, Uranus sees Jupiter, with its beaming smile and its friendly handshake, as a willing agent of progress, a positive force eager to campaign for social reform. Both planets have to be cautious about fanaticism, however, and about trying to apply their vision to everyone indiscriminately, without regard for individual differences. Still, Jupiter is not afraid to explore a little of Uranus' exciting world, because it loves encouraging anything that enlarges our perspective on life. See Chapter Fourteen for more.

SATURN

You'll recall that mythic Uranus and Saturn had more to deal with than the usual father/son conflict. Uranus was a cold, remote, unfeeling parent who refused to relate to any of his offspring in a real one-on-one manner, simply because they each failed to meet his ideal standards of unearthly perfection (in fact, he found ways to remove them altogether from the scene). To recap the mythological highlights: In order to avenge his sorrow-filled mother who was deprived of nurturing her children, Uranus' youngest son, Saturn, caught his daddy off guard and castrated him (and psychologically the old man was never the same thereafter). The power to rule was thus transferred to Saturn. Saturn later became paranoid about a prophesy stating that one of his kids would oust him from power, so paranoid that he swallowed each child right after he or she was born—all, that is, except Jupiter. Jupiter eventually put Saturn in his place and then happily took over the seat of power, and had a much better time doing so than did his old man.

This is not just a tragic tale of enmity between father and son, but one that hints at a fundamental tension between astrological Uranus and Saturn. To Uranus, Saturn represents the dark face of earthbound existence, including the constant pull of gravity—keeping our bodies feeling heavy—and the entrapping limitations imposed by time and space. Uranus doesn't wish to get involved with Saturn's restrictive side (our inventive energy can be devitalized by too much Saturn blocking Uranus' spontaneous flow). Thus, there are things about Saturn that even the normally fearless Uranus finds a bit scary.

But Saturn also finds Uranus frightening, due to this airy planet's ability to create chaos and disorder in ways that mess up even life's most well-organized plans. Peace and quiet are not what Uranus brings us when it's suddenly awakening parts of our chart and filling us with a restless spirit of activism. Here is a planet that thinks nothing of shattering the solid forms Saturn so carefully constructs to ensure our self-preservation (forms that

Saturn expects to endure forever). Uranus is therefore seen as some sort of wacko by Saturn, and a potentially dangerous one at that. And yet, both of these planets need each other. What Uranus quickly realizes is that Saturn can readily harness energy quite well in the physical world, and it can patiently assemble reliable structures that won't fall apart at the seams. Gee, Uranus can't do that!

Organizing form with attention paid to every detail is not something that comes easy to this planet. Although Uranus has the ability to brilliantly invent concepts that seemingly spring forth from thin air at amazing speed, Uranus is not very good at manifesting the intangible products of its divine mind in practical terms. It needs Saturn to help bring its electrical spark of genius down to Earth. Saturn's even willing and able to attend to the "boring" details involved in helping Uranian innovation take shape.

Uranus knows it cannot afford to become too unruly or bizarre and thus risk completely freaking out and alienating Saturn. It can't afford to lose this valuable ally. Yet tension between these two can quickly mount, especially when Saturn starts to take over operations in too inflexible a manner. Uranus thrives on unexpected interruptions and unplanned detours, but Saturn can't stand too much of that and will quickly seek to enforce its rules and regulations—the big rule being that you always must finish what you start! (Saturn's into heavy commitment.) Uranus wants to bolt out the door and not look back when anxious Saturn starts to clamp down like that. Somehow Saturn knows that, when working with Uranus, it might be left all alone at an awkward time holding the bag! Still, once Uranus comes to its senses and stops the angry-rebel act, it realizes that Saturn is the planet it can count on most to make worthy futuristic ideals become workable realities.

Uranus/Saturn transits are times when we need to put sustained energy and focus into our visions of a better tomorrow. Natal Saturn shows us where we grow ever so slowly and sometimes with a great suspicion that unknown outer forces may someday conspire to overwhelm us and jeopardize our guarded security. Uranus is thus likely to be seen as a potential trouble-maker. However, the need to undergo a major attitude adjustment often coincides with a Uranus transit. Saturn must be shaken loose in order to quicken our development at this time. We'll have to trust that taking an offbeat path will offer us more security in the long run. We need to do whatever it takes to turn these two planets into co-producers of a personal game plan for living that fulfills our need for practical stability as well as our need for freedom of thought and action. More about this transit can be found in Chapter Fifteen.

URANUS

What is there about electrifying Uranus that's not to get excited about, wonders Uranus? Not much. Uranus is the big surprise factor in our chart, just waiting for the right moment to explode into action and come alive for us. If we insist on living too mundane or routine of a life, and if we let Saturnian authorities in society do the thinking for us (via the social indoctrination of both Church and State), then the Uranus energy within us remains uncomfortably stifled and unable to reveal the thrill of being a real individualist. Of course, with such individuality can come potential loneliness, outcast status, the likelihood of criticism for being too different, and an overall feeling of being out-of-sync with the times we nevertheless must live in. But it's sometimes the gamble we take to find our inner truth.

There are certain social comforts and protections to be had when gently rocked in the bosom of conformity. Those who seldom stir up waves in life also rarely capsize their boat! Playing things the safe way affords us many social advantages. However, Uranus is quite a dedicated wave-maker who'll rebel against doing things the "normal," expected way. Uranus doesn't go along with that old "safety in numbers" philosophy. It would rather operate alone and take its chances than risk being controlled by any repressive group mind-set. Transiting Uranus doesn't plan to lose any of its identity when relating to the other planets. Instead, it's those remaining planetary parts of our psyche that typically undergo sudden Uranian make-overs in the name of progress.

When Uranus is in such high focus, the tempo of our life speeds up. We don't have much time to mull over the opportunities at hand (assuming we view the sudden changes we're going through as opportune rather than exasperating). Uranus now must be confronted in our life, not ignored. How willing are we to let go of a few stale habits? Uranus now gives us the itch to do a few things differently. What about "going-nowhere" relationships? Uranus is ready to give us a needed push in the right direction—such as the nearest exit!

Uranus/Uranus transits are periods where our personal understanding of what true freedom means is tested. Because this is not a very conscious planet to begin with—it falls in the category of "transpersonal" planets, along with Neptune and Pluto—Uranus makes sure that our personal environment sets up fresh and hopefully exciting conditions that motivate us to switch gears and start looking forward to experimenting with new avenues of expression. Sometimes the environment offers chaos and instability if that's what it takes to force us to wake up and turn our life patterns around for the better. These are periods of personal reform and timely risk-taking. Later in this book—in Chapter Sixteen—I'll explore the major Uranus/Uranus life cycles we all technically can go through. For now, let's just say that Uranus digs being Uranus, and so these transits are examples of unfettered Uranian energy at work on altering our inner and outer realities—so hold on to your hats!

NEPTUNE

Uranus instantly recognizes that Neptune is also transpersonal in orientation, and is thus able to exercise an extraordinary influence on human consciousness. To Uranus, this happily means that Neptune doesn't abide by the confining rules of Saturn any more than Uranus does. Uranus loves to defy Saturn at every turn, that is unless the Ringed One can manage every so often to come up with well-designed concepts that brilliantly incorporate

Uranian energy. Still, Uranus is wary of being boxed in by too tightly packaged Saturnian structures.

The typical sense of time/space limitations that most other planets feel on Earth isn't a problem for ethereal Neptune, because it's a planet that doesn't even notice the workings of time and space. Neptune is symbolic of an emotionally expansive energy that helps us feel connected to everything around us on a soul level. This leads to our enlightened awareness of the unity of all life. Uranus likes the fact that Neptune's energy is so sweeping and far-reaching, because this suggests Neptune is willing to optimistically enter into realms of vaster possibilities and of greater unknowns. (Uranus appreciates any planet willing to go out on a cosmological limb like this.) It was probably dreams and deep faith induced by Neptune that, centuries ago, allowed explorers to cross the Neptunian seas to discover various lands of the New World. Uranus would love to take advantage of such fervent Neptunian faith, knowing it can help remove our sense of any and all barriers in life.

However, Uranus is more a product of the Universal Mind of the Cosmos, not its Divine Heart (more aptly Neptune's title). As it transits our natal Neptune, Uranus would rather de-emphasize emotionality and instead capitalize on the power of Neptunian imagination and vision. Uranus also has plenty of progressive vision, but Uranus has its sights always on the future, while Neptune can get sucked back into the unresolved issues of our psychological past.

Uranus rockets away into the far-reaches of outer space, while Neptune slowly sinks to the unfathomable depths of the ocean. Both gravitate toward worlds that are alien to our everyday waking consciousness. Yet it's hard to shake Neptune out of its enchanted dream state long enough to satisfy adventuresome Uranus and its quest for clarity and truth. However, if it's adaptability that Uranus always seeks in a planet, then Neptune is to be highly valued for its malleable disposition. Uranus just wishes Neptune had more focus and substance.

The secret is to first capture Neptune's often short-lived attention by appealing to its humanitarian idealism and even its rarefied quest for ideal beauty. Brilliant, cerebral theories and futuristic social ideologies may get Uranus fired up—but anything too intellectualized fails to move Neptune (we must always touch Neptune's heart before it will offer us its wondrous treasures). Uranus will need to realize that its awesome brain-power and fiery rhetoric are not enough to satisfy the Dream Maker, who seeks instead to be guided along the quiet pathways of the soul.

In many ways, Neptune is nothing but a source of aggravation and frustration for Uranus: it's too passive, too receptive, and too easily able to seduce or surrender to another planet's will, even if only temporarily. Uranus would hate to become like that, even for a second! It seems that Uranus, a planet championing our right to be a free-spirited, will probably never understand the non-dynamic ways of Neptune—a planet that urges us to lose ourselves in waves of non-resistance, so that we may ultimately discover who we really are at the level of pure Spirit (as vague and baffling as that sounds to Uranus and the rest of the planetary gang).

Uranus/Neptune transits are times when we had better not hold on too tightly to a rigid interpretation of reality, especially if it's someone else's short-sighted interpretation. Neither planet is interested in wanting things to remain the same—Uranus shatters and Neptune dissolves, while Saturn goes nuts! Where Uranus and Neptune can really have a field day is in the area of turning on our spiritual and psychic awareness. Even creative projects demanding high levels of innovation and imagination are excellent ways to channel these combined planetary forces. In truth, a lot of people are simply not interested in seriously delving into such outlets. That's too bad, because when we bring Uranus and Neptune energies down to our life's more mediocre levels of activity, we typically feel that our world is getting messy and hard to control. Read more on this in Chapter Seventeen.

PLUTO

Pluto is the only planet that behaves in even more extremist ways than Uranus does when driven by a great social cause, except that it certainly cannot be labeled erratic, as well. Uranus admires this powerful planet from a careful distance. (No planet seems to be easy and relaxed around intense Pluto.) Pluto is very much like a sleeping volcano that first appears dormant but that also is capable of unexpectedly erupting at any moment, given the right environmental and psychological triggers (transits time such triggers). Uranus is not a planet that is temperamentally suited to deal with powerful emotional eruptions, even when it nonetheless provokes such intense release, as in this case. Still, the experimental, thrill-seeking side of Uranus feels driven to keep pushing the buttons of the other planets just to see what will happen next—even Pluto's "Warning: Only Activate When Absolutely Necessary!" secret emergency button, the one that can set things into motion that irrevocably alter our life-course.

Like Neptune, Pluto can be a source of deep internal mystery—and mysteries are something with which Uranus is not at all at ease, especially unknowable or unsolvable mysteries. Pluto rules the darkest underground elements of our psyche—not the sky god's favorite place to hang out. Still, because Uranus loves to drastically overthrow stifling conditions in the blink of an eye, it appreciates that Pluto is also adept at wiping the slate clean of those life elements that have become too soul-suffocating for us. Therefore, Pluto's urge to eliminate what no longer works is something Uranus finds valuable and even intriguing—except that Pluto tackles all of this in an unrelenting manner that Uranus finds a little too grueling at times. (Well then, Uranus, stop messing with red flashing buttons that are accompanied by dire warning signs!)

However, one concern is that Pluto may eventually turn around and try to wipe out Uranus, who therefore has to stay alert at all times to the threat of a total Plutonian annihilation (especially during Pluto's transits to natal Uranus). Actually, Pluto will always try to let whatever is of true value survive and simply become reassembled or reused more potently. Uranus loves planets that are willing to change and attempt new ways of expression. But Pluto, like Uranus, does not easily bend to the will of another (it reads everything in terms of power plays). Uranian intentions can therefore be misunderstood and distrusted.

Here again, we have two transpersonal planets coming together, suggesting that what is to be accomplished during these transits bypasses the intentions of our ego altogether. With our ego out of the picture, so is any real sense of personal control over situational outcomes. Our emotional satisfaction in the matter is seldom considered by the Cosmos at this time (probably because our emotions would vote to keep things as they are for a lot longer). With Uranus/Pluto, our evolutionary acceleration is more important. We are to further empower ourselves spiritually and better direct our consciousness toward transcendental levels. However, this is only to be encouraged after a few unhealthy patterns in our lives have been destroyed.

Pluto is the master of depth awareness, something that Uranus doesn't have a degree in. But Uranus, with its laser-beam ability to pierce through illusion and confusion, figures it doesn't need to slowly plumb our murky depths in order to intuit where clogged internal pipes have thwarted our ability to liberate our consciousness. Uranus thinks that Pluto is too obsessed with controlling all facets of our inner development, and in much too serious and heavy-handed a manner. Pluto hardly smiles enough to suit offbeat, absurdity-promoting Uranus—a planet that would love to smash the thick walls of Pluto's underground inner sanctum and let in some bright strobe lights to replace Pluto's heavy sea of darkness. During our Uranus/Pluto transits, this is actually what we are trying to do, albeit unconsciously. Pluto gets to lighten up, becoming less suspicious about everything, and is more willing to release its hidden resources for us to inventively use. Read more about this in Chapter Thirteen.

Now that the planets of our life's drama have been introduced, we need a quick review of the next part of our birth chart: the twelve houses. Let's move on to the next chapter for a detailed look at each house and how Uranus might feel about transiting it.

CHAPTER FOUR

A URANIAN TOUR OF OUR NATAL HOUSES

THE USUAL PLACES

The following is a preview of transiting house themes and issues that will be covered at length starting in Part Two, Chapter Six of this book. Here's where most of the Uranian action will take place for us during this transit.

OUR FIRST HOUSE

Regarding transits, this is a sector of our birth chart where we are given the go-ahead to initiate new beginnings for ourselves based on any inner light and wisdom gained from our previous transit. The transiting planet in question has just spent meditative time moving through the more quiet, solitary depths of our Twelfth House in an attempt to sort out unwanted psychological debris left over from past failures and traumas. It's also possible that spiritually transmuted energy is now available to us, the result of a planet in our Twelfth having successfully undergone a process of soul refinement and redemption. It's as if that planet has taken a deeply rejuvenating bath in the healing waters of the spirit and is cleansed of toxic impurities accumulated from all previous house transits. Now it's ready to inaugurate a new cycle of experience for us as it crosses our Ascendant and enters our First House. We feel renewed on some level.

If that transiting planet is Uranus, we now have a powerful opportunity to use laser-sharp mind power to re-image and thereby reform our visible, surface identity—the person others think we are all about based on our surface presentation, due to our behavior and overall appearance. However, transiting Uranus didn't spend years and years soul-searching in our Twelfth House just to decide to have us emerge in our First wearing a nose-ring plus most of our hair chopped off, although that's certainly an option! Uranus now arrives in a state of wisdom (and grace) that we'll need to take advantage of, because we've been inwardly preparing for this special moment for quite a while, even if only on a subconscious level. We are more attuned to facets of Uranus at this point than we may realize at first.

We are capable and ready to make this effervescent planet come alive for us. Uranus spurs us to rapidly grow by leaps and bounds. Yet this will require that we show enough courage to venture forth into new, uncharted territory where we can make major breakthroughs concerning our self-image and the general way we display ourselves to others. It's a time to be very upfront with life and especially with letting the world know that we cannot be pushed around any longer or made to submit to any conditions that will hold back our potential.

The First House is an area symbolizing ways in which our life can be more fully in our own hands so that we may learn how to best steer ourselves toward new directions—without much interference from others. Uranus likes to hear that, because it's a planet that will never allow itself to be sidetracked by people for very long, unless they have something quick and brilliant to offer. Even when they do, Uranus still wants us to move on to new sources of stimulation.

Our First House is a life zone where pioneering personal beginnings can be set in motion by our self-starting impulses. We're the ones here who get the ball rolling. Uranus is all for making independent fresh starts, especially if they challenge the limitations of our old and familiar (but boring) personal identity. Yet the Uranian things we do or act on in our First often shake up the long-term patterns we'd laid down over the years with partners in our Seventh House. This could be good or it could signal the end of unions that we've outgrown.

In general, this is a house where Uranus can do a lot of what it does best: sweep out the old and bring in the new—all in exciting ways that we and others can plainly see happening. The First House is a high-visibility zone; people get to witness here how we make our personal changes. There's little about our First House that's hidden from view. Read Chapter Six for more.

OUR SECOND HOUSE

Once we have repackaged ourselves in ways that display more of our true, individualized self, our next challenge is to ground this new sense of selfhood in the material world so that our awakened identity can continue to survive and thrive. This helps us to better value such a reformulated self-image and to find innovative but practical ways to demonstrate its worth to others. In other words, let's make these new changes pay off for us! Let's further capitalize on all the exciting things we've learned about ourselves during our First House transit.

However, Uranus is less comfortable in this house, basically because it's ill at ease with fully materializing itself in dense physical expression—it fears being trapped, slowed down, and limited by too much earthy structure and predictability. It's true that the Second House is very form-conscious and quite dependent on sturdy, tangible expression. It wants something concrete to hold on to, not just something thrilling to dream about. Our Second House asks, "What good is Uranian brilliance if we can't make it part of our concrete reality, especially in a workable and even profitable manner on a day-to-day basis?" Actually, that question could be asked of any planet moving through our Second House. Just how well will this sky god function on solid ground amidst a calm and stable setting? Actually, the more Uranus feels trapped in routine, the more inventive it becomes in trying to free itself.

Typical matters of interest in our Second House—such as money and possessions—may seem at first too mundane for any Outer Planet. However, since so much of our consciousness can be wrapped up in our attempts to satisfy ongoing worldly desires, Uranus realizes that our adoption of false material values can stunt our development in the long run. Should we remain firmly attached only to physical securities, we'd fail to nourish our soul or broaden our mental awareness.

Historically, much spiritual seeking throughout the centuries has been tied to the renunciation of earthly desires in a devout attempt to overcome the appetites of the flesh. Well, in *this* house the flesh rules! Here, the body demands its creature comforts, and the sensory world of both our psyche and Mother Nature herself wield optimum power. Our Second House is not meant to suddenly transform into our otherworldly Twelfth House, no matter how spiritual our intentions are (each house has its evolutionary role to play out). Uranus, Neptune, and Pluto will therefore need to abide by certain "house rules" and show more consideration and respect for the ways that this Earth-appreciative Second House naturally pleasures itself and learns to grow.

Uranus usually tries to ignore such rules in favor of ushering in dramatic situational changes that destabilize our security base, and often in ways that seldom take our emotions or our sentimental feelings into account. It's true that we'll probably have to let go of a few "precious" things along the way that have outlived their usefulness, but that doesn't mean all that we have carefully built and valued throughout the years will be destroyed or taken away for no good reason. After all, Uranus doesn't really have full permission from the Cosmos to create whatever havoc it pleases!

The Uranian freedom to evoke "meaningful" chaos in the Second is still bounded by our life's practical realities. Therefore, it's no wonder that a tour of this gravity-loving house makes Uranus a little uneasy—there's too much darn physicality here to deal with, reinforced by the unyielding laws of matter. However, making it in the material world doesn't have to be so dull. Uranus will teach us to experiment and see what unusual Second House situations we can concoct for ourselves.

At least Uranus sees some of its original ideas translated into tangible form (but remember, it's not a planet that's easily satisfied with anything born of the world of matter). We'll probably have to remind ourselves that absolute physical perfection is not what we should be anticipating at this time. Material form has its limits. Still, this is a house that allows us to use vital Uranian energy to build our visions in realistic terms. Efforts we make can have enjoyable and surprisingly long-lasting results. More on this is found in Chapter Seven.

OUR THIRD HOUSE

Here's a house where Uranus can give us plenty of reasons to stretch our mental potential in order to see how far our childlike curiosity can take us, assuming that extended years of adulthood haven't all but killed off such inquisitiveness within us. Uranus wants us to take advantage of the nosy side of the Third House that Gemini and Mercury have always enjoyed—the side that gets us asking lots of direct questions about life: Why this, why that, and how come…? After all, we'll never know anything if we don't at least ask. Of course, the blunt questions Uranus wants answers for are often deemed outrageous and impertinent to those with chronically conservative Saturnian minds. (Uranus forgets that few of the other planets are as unshockable as it is!)

Transiting Uranus does very well with any house (or planet) that enjoys learning something new—information that's even "light-years ahead of its time." Now hopefully, after undergoing a few needed material shake-ups and financial turn-arounds, we have established smarter ways of working with our Second House resources (we feel less possessed by what we own and are now a bit freer to enjoy those things we have without insecurity or greediness spoiling everything). It's time to take a break from such worldly concerns and instead turn our attention to feeding our mind fresh, exciting concepts. We're ready to step up our mental tempo and accelerate our ability to learn as much as we can, most of it being unusual or offbeat subject matter.

Our Third House offers any planet passing through it lots to think about, plenty of options to choose from, and many refreshing ways to express ourselves. Remember, this is the place the airy side of Mercury loves to call home (and air Mercury doesn't care for anything that's dull and lifeless). Variety is important to the Third House. Uranus already intuits that it's going to have more fun here than it did in the Second. It's also obvious that we'll have more opportunities to think about all kinds of speculative topics in our Third than we had in our instinct-driven Second House (where, if you'd ask us why we really want to own something, we may not have thought about it too much—it just somehow "feels" right).

Constant change is also a major theme of our Third House experience—here's where we witness many comings and goings, all kinds of simultaneous movement in our immediate environment (like busy traffic). All this kinetic motion is something that excites Uranus, since it knows that here we'll be less likely to find ourselves stuck in rut-bound situations. It's not the nature of this house to want issues to drag on or remain unmodified. This is a life zone that encourages us to make quick adjustments in our day-to-day world and remain flexible to changes around us.

Uranus is an easily-triggered planet when it comes to mental exploration. The Third House loves to learn about everything going on in its surroundings or even about anything that's knowable in general. We get to develop our brain power here and find stimulating ways to educate ourselves. Uranus is very supportive of such efforts, except it often feels the Third House's standard approach to learning is too slow, especially when linear thinking and the old models of logic and reason are used exclusively. Uranus will bypass what it senses would be a time-wasting experience, including the linear process, which seems to be tedious to this planet!

In this house, we'll probably get to learn how to rely more on our intuition and to appreciate the sense of mental quickening it provides. Our Third House tries to point us in many directions that hopefully will vitalize our mind and keep us ever on the alert. This helps us to rely less exclusively on using those subjective instincts we used when approaching life in our First and Second Houses. In addition, Uranus thinks its intuition is superior to pre-programmed instinct, even if that's not necessarily true. Uranus loves alternatives in life and it already knows how to emotionally detach. Therefore, it's ready to fly high in our Third and give us a exhilarating, much-needed mental tune-up. Chapter Eight gives more details on this subject.

OUR FOURTH HOUSE

If all has gone well, by the time that Uranus enters this house we have learned to think independently. Thanks to our Third House experience, we've allowed ourselves to express Uranian opinions that perhaps have sounded at times either odd to others or else amazingly progressive. (Some of us have also learned how easy it is to shock the unsuspecting!) We probably have become less quick to rush to judgment regarding complex social matters, because we now view them from all sides and with greater tolerance and open-mindedness. We are more accepting of life's paradoxes and contradictions than ever before. This comes in handy when Uranus transits our Fourth House.

This is a critical life zone where the theme of emotional attachment is powerful, where childhood conditioning molds our root behavior in ways we don't even realize, and where we learn to eventually cut the psychological umbilical cord and separate from our parents. We must break away from their powerful influence over our less conscious inner nature—that is, if we wish to mature properly and develop an effective self-support system all our own. Severing ties that have an emotional stranglehold on us is very appealing to Uranus (not a very sentimental planet to start with, and one almost phobic about developing long-term close attachments).

No matter what our age and how well we appear to function outwardly as adults, we still may be trapped in various states of immaturity—as symbolized by our natal Fourth House condition—whereby some of us hunger unrealistically for a womblike sense of a security and protection. We may thus be afraid to face up to the harsh demands of the world. Therefore, whatever our true potential may be, a few of us turn out to be "late bloomers," if we ever bloom at all. Uranus, a planet that is not emotionally sensitive, transits this house with the intention of removing any thick tangle of roots that has blocked our path to true autonomy. For us to no longer want to be part of certain family dynamics, because they stifle our development, can trigger mixed reactions. Hopefully, by learning to analyze all sorts of contrasting life issues in our previous house, we can now better handle this delicate situation with greater self-honesty and courage.

Our Fourth House experience conditions us to need someone or be needed by them. So far, our first three houses haven't depended as much on intimate human connection for their overall success. But in our Fourth, we cannot live only for ourselves—we may even have trouble living *by* ourselves, because this is a house where we can first feel a sense of loneliness when missing the company of others with whom we've developed a dependent relationship. (We may at least need to have a few pets around to share our space.)

Uranus, however, is wary of human neediness and emotional dependency. It also thinks that living alone without the interference of others is one way for us to guarantee personal freedom. This is a planet that so easily can feel threatened by suffocation when caught up in intimacy-demanding relationships—its quick solution is to just hit the road and not look back! But that's not an easy option in this house. Pulling away from others is not what the Fourth is about, initially. Pulling together to create a warm and secure family experience is the objective. We only pull back and retreat from loved ones (especially parents) when they try to swallow us up.

The Fourth implies complications due to the fact that people with whom we've developed powerful security patterns are involved here. Uranus wants us to make a clean break from our past, yet the Fourth knows that we carry the past inside us wherever we go, especially when matters are unresolved. With Uranus, we may squirm a lot in this house while trying to establish breathing space, and yet at the same time we don't want to feel we've abandoned those we love. Transiting Uranus at least enables us to stir up our unexamined security issues in order to uncover the truth behind our subconscious patterns. See Chapter Nine for more on this.

OUR FIFTH HOUSE

We've plumbed our depths regarding the true nature of our emotional bonds with our family (maybe most so with our mom). As a result, we've learned to pull away from unhealthy entanglements. We realize we're not to encourage too much dependency in our ongoing relationships. Now we're ready to come up for some fresh air and enjoy the sunshine and the blue sky—Uranus loves to see plenty of sky! Our Fifth House happily awaits us with balloons and a welcoming party banner on its door. Crazy, fun people are coming out of nowhere to put a big smile on our face. It's time to laugh out loud and have some fun living out our life.

By now, transiting Uranus has made sure we know ourselves quite well on our inner levels, at least better than before. We now realize that true security comes from within, just as we've always heard. Our internal roots are strong and reliable, and are quite capable of feeding us what we need more so than any outside source can (our Fourth House Uranus transit has been a time to take back and own some of what we've projected onto others near and dear). Uranus has made the pathway to our underlying, inner foundation more accessible and less mysterious. If we don't realize this, then we probably won't emerge from our Fourth House experience feeling all that alive and well.

But let's say Uranus' transit was a smashing success (even if our family now thinks we're weird and a little distant). The new mission ahead for Uranus involves triggering an accelerated development of our ego and our individuality. This time we're learning to confidently and cheerfully relate to ourselves with even more self-awareness and courage than we had during Uranus' First House transit. Uranus could now also trine our Ascendant at this point, emphasizing that this is a period of dynamic self-involvement, one that permits us plenty of untried creative options to explore.

The Fifth House is where we do those things that bring joy into our lives. We can be determined here to have a great time being ourselves and loving it—especially when we already like and respect who we are. In this house, it's time to play, and Uranus is ready to show us how exciting quality playtime can be. What Uranus really enjoys about the Fifth House is that it's an area where we benefit ourselves most when we follow our heart and learn to get real with who we are on our own terms.

If we're smart, we won't repress ourselves here or try to go unnoticed. If we want to submerge our greatness and appear humble and self-effacing instead, well, there's always the Sixth House for that kind of attitude awaiting us in our future. While in the Fifth, Uranus will insist on making us shine like never before and in ways that tell the world we're special. (Uranus supports unique self-expression and the Fifth House finds the right stage for us on which to perform.)

However, we cannot become completely self-involved in our own glory in this house, because it's also is a place where many of us must deal with raising kids. How we go about relating to them is shown here. Uranus doesn't really like kids (don't forget that old Greek myth) unless they are perfect little self-contained geniuses who don't demand much parental attention and involvement. That's not very realistic. In addition, it smacks of selfishness on Uranus' part (again, this planet gets unnerved by neediness in others, and youngsters are naturally quite dependent on us). But since this is only a transit, parental neglect is less due to flaws in our character and more a case of circumstantial pressures.

Still, dealing with children will become one of the trickier challenges of this Uranus assignment, because our real urge is to get away from all pressing duties and obligations. However, with the right attitude, we can use Uranian energy to bring out the free-spirited child inside us. We can have fun with our children while also having a real good time being adventurous with the world. Chapter Ten provides more information.

OUR SIXTH HOUSE

No party lasts forever—it typically runs out of steam shortly after all of the bean dip and salsa have been consumed, and the party-goers realize they are much too spent to carry on (plus, they want to drags themselves home before sunrise). What starts out in our Fifth House as creative exuberance could end up as a nasty case of burnout if we were to allow ourselves to go on and on with our self-indulgent ways. That's one reason why our Sixth House mercifully follows our Fifth. Even if we are not quite ready to go back to work on Monday morning, necessity says we must. It's not healthy to be without sensible order and purposeful routine for long periods of time, or so warns the Sixth.

This house is concerned with restoring our sanity on the levels of both body and of mind. The Sixth observes how being non-productive and lacking in concrete goals can literally make us sick (an idle mind is the Devil's tool, and all that). Uranus is not too thrilled with the fact that the Sixth House is hung up on life's smallest details, because there can be so many of those scattered about for us to fret over. The Sixth is also a bit too focused on minor tasks at hand and on the proper way to do everything (if you want something to work right). Uranus won't be able to easily handle the Sixth's long list of rules to follow, and instead has other plans for us involving lots of smart short-cuts that make any job easier. However, set procedures often govern the way this house wants to operate. A measure of self-discipline is therefore a must, meaning this house rejects random spontaneity. "Uh-oh, this could be the beginning of years of tedious hard labor," worries Uranus.

Actually, transiting Uranus is sharp enough to know how to convince us that we can bend a few of those time-honored rules by trying out new approaches to problem-solving. A predictable and routine life will not benefit us during this period. In part, this transit is about being inventive in how we handle all-important tasks at hand. We'll learn some amazing stuff regarding how to stay healthy as well, seeing physical fitness as our key to continued freedom of action, especially as we grow older. Illness takes away our autonomy, something Uranus wants us to cherish forever. We'll just have to be sure we make exercising an exciting thing to do. That ensures we'll stick with it longer than usual.

Our Sixth House is a life zone where we learn practical skills that help us cope with the rigors of daily living. This is supposed to keep us on our toes, alert and ready for whatever comes our way that might upset the manner in which we organize our life. Maybe it's a bit too much to ask Uranus to always come up with practical solutions. This is an experimental planet that seldom gets bent of out shape if the results of its actions aren't always workable—Uranus will simply turn around and brainstorm again using an entirely different approach. Yet the Sixth House implies that we should already be knowledgeable enough and prepared enough so that we'll make few mistakes, if any, the first time around. We can at least minimize errors by being patient and attentive to all details. But that's not Uranus. That's Virgo, the patron sign of this house!

There will be areas where this planet and this house are at odds. Too much unresolved tension due to this can result in the development of health symptoms as manifestations of our stress. But the "house rules" of the Sixth state that whatever we do here has to be something useful and realistic. It must serve a well-defined, functional purpose. Uranus cringes when it hears that word "functional," but heck, perhaps this house is one of those crazy challenges that Uranus will have to accept in the spirit of true growth. We just need to allow ourselves to not become slaves to monotonous procedures, or else Uranus will gladly take his unappreciated genius elsewhere! Before this transit is over, we'll probably learn that serving others doesn't have to be done in ways that make us feel over-obligated and over-worked. Read Chapter Eleven for more.

OUR SEVENTH HOUSE

Flexibility and cope-ability are still going to be our best assets around the time that Uranus enters our Seventh House, because we now must put all of those constructive, useful things we've just learned in our Sixth House into fuller practice. We can no longer hide behind our paperwork in the office or get lost in our hectic work schedule when the focus is on this house. Someone may be waiting for us to finally come home, get out of our work clothes, and directly relate to him or her. There are less situational distractions to be found in our Seventh, and thus fewer excuses for not having time to more intimately relate to one another. With Uranus ready to stir up a little healthy disruption, this could be a "showdown" time for us!

Our partner in marriage (or otherwise) is often the main reason we will need to test out our new Uranian coping skills (oh, and let's hope we've learned about true tolerance by now, because we'll really need tons of it during the next several years when dealing closely with people). Unlike the First House, our Seventh is where everyone who's not just like us nevertheless requires our complete cooperation. Here we learn to bend and yield to those whose temperaments contrast, but at times complement, our own. Harmony and bliss are not what marriage is really all about—regardless of what was said in that runaway bestseller Venus and Neptune collaborated on—yet learning to resolve human differences is one unconscious reason why people wed.

What we are really marrying is not just another person, but select, unexamined parts of ourselves that we unknowingly thrust onto that person with whom we then expect to spend the rest of our lives (even if we later fall short of fulfilling that dream). Uranus here signals a time of awakening to our projections, allowing them to be more consciously recognized, and then helping us to determine how they've either enhanced and expanded our self-understanding, or perhaps made us feel trapped in a small, unventilated room with no apparent exit (Uranus' own personal nightmare). Social harmony is a big deal to the Seventh House, but that's not what Uranus is all about. Such harmony sounds too settled and static-producing, even though Venus and Neptune find the thought lovely.

Yet in our Seventh, generally speaking, everything seems to work best when we get things out in the open, where no one is deliberately keeping secrets. Even personal enemies found here, traditionally, are our "open enemies"—those who are directly in our face doing battle with us. Uranus appreciates having all troubling matters exposed and laid out on the table (it apparently doesn't have a sneaky bone in its body). Uranus is, at heart, an advocate of "telling it like it is" rather than keeping anything that disturbs us buried and unaddressed. We must suddenly unload at some point when Uranus triggers a key area of our chart.

Therefore, during this transit, we are forced to be more truthful about our needs and feelings about our partner, and vice versa. We cannot afford to be vague or indirect in our mutual communication if we want our union to survive this transit. Uranus has no real desire to keep a couple together for merely practical, economic reasons if they otherwise really hate each other's guts. Thus, it sees divorce or permanent separation as the quick and obvious remedy. However, if we feel Uranus' drastic solution is the wrong one for us, and that our marriage is salvageable, then we'll have to make an extraordinary attempt to come clean and relate to our partner with an unaccustomed degree of clarity and honesty.

Even that may not be enough to keep us together, because the plain truth Uranus may want us to know is that someone in this relationship has outgrown the other—perhaps rapidly so, during this Uranian transit. Still, this is a house that believes in fairness and careful deliberation. Uranus may think it has the right answer—just let the person go and move on—but if we have natal planets in our Seventh House, we'll find ourselves weighing all sides and not making a final decision easily. Chapter Twelve discusses this topic further.

OUR EIGHTH HOUSE

Sometimes—especially in emotionally complicated, long-standing relationships—not everything that troubles a couple gets to surface and be heard, no matter how much of a ruckus transiting Uranus has previously tried to stir up in our Seventh House. Of course, Uranus takes years to move through any life zone, and when it passes through our Seventh, we and our significant other

become much better at getting things off our chest. We cease playing denial games when it comes to dealing with our biggest marital problems. Our aim is to smash through mutually defensive
barriers and confront these problems head-on. Yet if not all gets
worked out during such a truth-telling phase in our marriage, we
still have another crack at it when Uranus transits our Eighth
House. We may also be packing more psychological ammo just in
case things turn ugly!

However, at this point, such matters are critical, because
there's not too much more time allotted for resolving our differences. Discord (even sexual hostility) has reached deeper levels,
and it will take stronger measures to detoxify this union and turn
hearts around. A little psycho-surgery may be needed. Uranus
really doesn't tolerate the murky, twisted elements of this house
(the cruelty, the exploitation, and the underhanded manipulation
tactics), but it does love getting a chance to explode now and then!
Intense soul-drama is associated with our Eighth, whereby we get
to erupt and shoot jets of fire when the internal pressures are too
much. Uranus loves a good (but quick) cathartic release, and so
it's willing to stick around and further provoke matters until the
psychological lids covering our most secret feelings about each
other are blown off.

Sex and shared financial resources are two hot Eighth House
topics that can get things to suddenly change for better or worse
in our relationship. Both issues involve a level of honest, intimate
relating and the willingness to ventilate and share our sometimes
contrasting needs and feelings. Transiting Uranus will shed
needed light on such concerns if we've been operating in the dark.
It's not always easy to openly discuss Eighth House desires, since
at least one partner may feel vulnerable or defensive and thus
may resort to shutting down emotionally. Uranus itself doesn't
care all that much about sex (aside from its curiosity about potential thrill or excitement), and is not too interested in everyday
money matters. However, it hates to think that certain subjects
are forever taboo, totally off limits, and therefore not open for free
and frank discussion. Uranus is saying, "No more secrets!" in a

house that's famous for covert activity and underground transactions. It'll be tough convincing us to come clean with what we really feel deep down inside.

During this transit, therefore, life will pressure us to confront whatever's been boiling away in our psyche's basement for much too long. The sky god's visit to our private underworld can turn out to be quite interesting, although it's not something psychological weaklings would enjoy! Eighth House processes are much like grueling work-out routines, during which our personal trainer barks, "No pain, no gain!" At least Uranus is always telling us when it's time to detach from our emotions and look at the overview of our situation more dispassionately. Maybe this is a time when we realize that not everything here has to be approached as if we're playing the lead role in a heavy Greek tragedy.

One area of the Eighth House that appeals to Uranus deals with transmuting energy so that we can further open up our doors of perception. How we can contact the less-visible realms of human experience—what has been called the astral plane—is hinted at in our Eighth House. If you ever thought about talking to the dead, this is the house for you. Our interest in matters involving life after death and even past lives may be heightened when Uranus is making stimulating aspects from this house. Read more on this in Chapter Thirteen.

Our Ninth House

By the time it transits our Ninth, Uranus is more than ready to get away from the intense, pressure-cooker atmosphere of the Eighth House. "Those maniacs can get pretty warped in that swampy Eighth—I couldn't wait to blast my way outta there!" Our Ninth is a house where the petty concerns of the small world and the frustrating daily melodramas of being human are put into proper perspective, as we now raise our consciousness to broader realms of spiritual speculation. In the Ninth House, we are no longer stuck on the ground trying just to survive, but are now about to fly into outer space in our search to understand the ultimate purpose of it all—what is existence about, and why? How do we fit into the larger scheme of things?

Uranus, of course, feels right at home in outer space, a place of unlimited expanse and with a relative lack of physical obstacles to hem it in (things get much more congested on the ground level). This transit, therefore, will make us feel as if we have another chance to breathe more fully and become well-oxygenated—but this time it's inspiration that fills our metaphorical lungs. Uranus is a planet of knowing, and the Ninth is an area where we are eager to seek what's worth knowing—all those lofty concepts that smack of Universal Truth.

"Truth? Did somebody say Truth?" asks Uranus. Well, now we have this planet's complete attention! Issues belonging to our Ninth House will always fascinate Uranus, a planet with a knack for coming up with brilliant and sometimes outlandish theories to explain away life's cosmological mysteries. During this transit, we have an opportunity to question our current beliefs about God and the Universe. In fact, we may be feeling inner intellectual rumbles that suggest we're no longer satisfied with whatever we have put much of our faith into in our past. We are open to new revelations that may just clarify the Big Picture of Life for us.

It's best to just hang loose mentally and not try to replace one dogma for another. Uranus is a little nervous about the Ninth House's tendency to institutionalize collective beliefs (using thought systems to organize people on mass levels). At this time, nothing we discover about religion or metaphysics, no matter how soul-invigorating, should be seen as conclusive—the last word on the matter—because when we start sounding like that, we can be sure it's not Uranus talking. Uranus never knows where any journey will end—it doesn't use maps!

Exploring new worlds arouses Uranus' interest, which is great, because in the Ninth House, the foreign and the exotic become alluring. We feel the call of adventure. This is a time when we can do some of the most fascinating travel we will get to do in our whole life. There might be at least one very special place we get to see before this transit is over (a place that really opens our eyes to how varied and wondrous life on this planet really is).

The Ninth House gives Uranus the stimuli needed to turn on this planet's intuition and its ability to see far into the future. The Ninth itself helps us plug into the world of tomorrow.

Therefore, some of us could end up feeling like visionaries on a world mission before Uranus finally enters the Tenth. Society becomes our laboratory, where we experiment with progressive social concepts that could liberate the consciousness of many. To help this along, we'll just need to get lucky with how and when we network with others. Fortunately, it's in this positive-thinking Ninth House that our "luck" is born and then fed by unlimited optimism! See more about this in Chapter Fourteen.

Our Tenth House

After flying high during our Ninth and feeling like a state-of-the-art weather balloon circling the stratosphere (wow, what a view!), we are brought back down to earthier, worldly levels to test out our new theories about society, about spirit, and about the unfoldment of human potential. Yet it might come as a rude awakening to find that the Tenth House is not one that caters to impractical or aimless theorizing.

This is a life zone that expects to see worthwhile, tangible results from sound and fruitful speculation. Anything that is too vague or unstructured, or that results from unrealistic mental journeying in our Ninth, doesn't have much chance of surviving in our no-nonsense Tenth. Here is where the competition can get really stiff, because there's only so much room at the top. Here we test out our ambitions in earnest (as we climb the ladder of success) and, hopefully, become honored for excellence in achievement at some future point. This is where making significant social contributions—perhaps leading to fame—becomes important. This is not a house where we are content to just make a living and get by (that's our Sixth).

Once again, Uranus is in a house where we won't be allowed to go wild with our freedom impulses. The Tenth is another place where "house rules" are strictly reinforced—note that it's always in the "earth" houses where Uranus has to tone it down. Our

Tenth House is not only an area where everyone gets to see us taking responsible action in a public manner, it's also where folks get to judge our relevance as a participating member of society. Our performance level needs to be above par if we are to capture positive attention from otherwise unforgiving social critics. Already this sounds like the kind of authoritative pressure Uranus would rather live without!

Our career is perhaps the most dominant issue of our Tenth House. Uranus can suddenly put our face on the map of success like never before, because here we can advance quite quickly if we have something unique and exciting to offer the world. That world can simply be our professional field of operation—our company. Uranus wants whatever we're offering to be revolutionary, innovative, or at least trend-setting. The Tenth expects, even demands, masterful expertise—the result of years of hard work and steady application (this house is Saturn's favorite place of worship).

With Uranus, however, we're unexpectedly electrified with a futuristic vision that's probably never been tested. It takes a lot of guts to convert our vision into a marketable Tenth House format and convince others that this baby has wings to fly! Some of us will succeed beyond anyone's wildest dreams, while others— already in it way over our heads—will fizzle out in the wink of an eye (or perhaps first we'll have a meteoric rise and then inexplicably disappear forever into the night sky). All of this is the gamble we take when we hitch a wild ride on the tail of dazzling but unpredictable Uranus.

Our Tenth is another parental house (like our Fourth). Uranus has trouble with parenting even when it's not transiting this house. It fears getting caught up in the thick of human dependency with all of the emotional demands that get triggered as a result. However, one thing that is helpful about Uranus in this house is that we can gain greater objectivity when dealing with a parent—usually it's our father—and look at this person in a new light. Uranus can help us break up parental fixations that rob us of our true autonomy. We can suddenly view our dad quite differently as we start to pull back to reclaim parts of our adult identity. A certain degree of detachment helps us here.

During this transit, we may suddenly stop laying blame on an authority figure for any weaknesses we find in our nature. Uranus would rather we take back the power we unconsciously once gave to such people of our past, and instead begin to take full charge of ourselves and our direction in life (without still seeking parental approval or anticipating parental disapproval and judgment). Read more on this in Chapter Fifteen.

OUR ELEVENTH HOUSE

Finally, Uranus enters a house where it can really kick back, ignore house rules, and do whatever it likes whenever it pleases. This planet's Tenth House transit assignment, which has just ended, had its moments of glory for us and its periods of demanding that we work hard to manifest our uncommon ambitions. At all times, dedication to our vision was essential. Still, Uranus always feels it's being watched and evaluated a little too closely when in the Tenth—one wrong move and it could be all over for us as we topple off the mountain peak of fame and acclaim. Of course, by then we could be ready to kiss off the experience anyway and move on to something entirely new. (Uranus never wants us to feel stuck at any stage of the game.) However, all that is behind us now and, assuming we've done well with taking advantage of unexpected Tenth House opportunities, the next transiting phase for this planet is one in which we get to work on experimenting with our own broader potential as a human being (someone more than just a well-known public figure).

The Eleventh House observes how often outstanding worldly success in our Tenth House—after a certain point—requires too much artificial role-playing, whereby we feel we must live up to a certain cultivated "power image" that keeps society interested in us and in our achievements. This does little to enhance the real individual within. Sometimes in the Tenth, at least for some of us, image and appearance unfortunately mean everything. Achieving elevated status becomes too important a goal.

Therefore, if we wish to come back home to our real self and truly grow, our friendly Eleventh House is ready to greet us with open arms. However, we're not going to be able to function alone in relative peace and quiet for long here (that's what our Twelfth House promises us). Part of our continued journey in self-awakening will involve our direct exposure to a wide and colorful range of people who are also developing their greater human potential. We still have an ego that needs attention from the world, but now we also find we can get excited about how others show off their special talents and genius ability. With Uranus, we are learning to enjoy such individualists without feeling ego-threatened or overly competitive. Many of the new people we will be meeting and befriending are actually catalysts for our own accelerated development. If they weren't who they already are, we couldn't become what the future says we will be. Approaching these relationships with fear, jealousy, and insecurity makes no sense at all, because Uranus is trying to convince us that everyone involved is in a potential win-win situation.

Uranus passing through the house it naturally rules (in the Aries Rising Wheel) suggests that it's less concerned with appearing in ways that satisfy the System, the Old Guard, or the Silent Majority of supposedly conservative and easily offended citizens. Uranus feels it has no reason to appeal to orthodox viewpoints of any sort, and certainly not at this special time in its cycle. Its urge to shock people out of their complacency might be even stronger than usual, egged on by radical Eleventh House social experimentalism.

However, it's unlikely that many of us will go off the deep end and gamble with drastic non-conformity in our lives. (If we've done our Tenth House transit right, we don't feel as oppressed or held back by society's structures as perhaps we once did, and therefore we are less inclined to defiantly rebel against a life we've already learned to redesign for the better.) The fascinating potential of the future is of greater interest to us than ever. Social idealism that may have subsided within us years ago is now ready to suddenly emerge and give us hope once again about humanity's progressive unfoldment. These can be very special years for us, during which we get to view life and our world from an unconventional angle, but one that helps to truly set us free. Read Chapter Sixteen for more.

OUR TWELFTH HOUSE

A planet as excitement-filled and action-packed as Uranus eventually needs to learn the wisdom of winding down and deactivating itself, without viewing this as some kind of senseless punishment. Sometimes there's not a whole lot to do in the Twelfth House (it's pretty much all been done in our eleven other busier, worldlier houses). Actually, the thought of not always having to manifest itself in concrete terms appeals to Uranus, a planet that now is free to explore more abstract realms of mind and spirit. Transpersonal Uranus is actually more in its true element when in our Twelfth, a life zone where structures and boundaries easily dissolve.

This is the house of our underlying soul-connection to the One Source of All Life. It's a house of transcendental reality, where invisible currents and pathways allow various subtle energy-linkages to happen, especially on a less obvious emotional plane (this is also were the Collective Unconscious taps into our being to either inspire or torment us). We are less slowed down by the laws of the physical universe here, something Uranus should love to hear, and thus we can travel at the speed of thought (even at the speed of love) and reach our destinations in a flash (astral projection...soul travel?), all by tuning in to our divine power. At least, that's what those adept at mind-altering spiritual techniques will claim regarding the Twelfth.

Many of our Twelfth House experiences cannot be proven in rational, material terms. That doesn't bother any of the Outer Planets, because they constantly live in relatively dreamlike worlds that seem alien to most of what happens on this planet. Besides, Uranus does feel it's blessed with its own special kind of logic at work, although it's not that Saturn/Mercury kind of slower reasoning power that has been hailed as indispensable during the past three hundred years or so. Uranus likes things to happen instantaneously and even simultaneously. In this house, we may have experiences that demonstrate this (but shucks, we're usually alone when such phenomena occur, and thus we can't prove to others that any of it really did happen—this is a typical Twelfth House frustration). We'll need to accept that in this house, it's sometimes going to be only the Cosmos, the angels, and us having a private form of communication that no one else shares.

Actually, there is a bit more work to be done in this house before we can close a megacycle of inner growth, and Uranus may be just the right one for the job. The Twelfth becomes our psyche's storage bin for the many unwanted and neglected parts of ourselves that we've abandoned—often because they've been too painful in the past for us to deal with and confront. As in the Eighth House, this can be sticky, unpleasant territory for Uranus to cover. There can be some pretty terrifying thought-forms locked away in our personal unconscious as a result of unprocessed phobias and traumas experienced in the past. Yet with so much trapped psychological energy involved in this suppressed and repressed material, Uranus feels it must take on the idealistic role of our soul's liberator.

Thus, during this transit, we may experience breakthrough after breakthrough, and may exorcise the ghosts and demons that otherwise haunt us and keep us feeling anything but whole. The outer world may not realize all the inner explosions in consciousness we are going through, since the results of all this may not be apparent to others for quite some time. Still, some personal revelations at this time can be radically life-altering. It's little wonder that some of us are more that ready to undergo a major overhaul in behavior by the time Uranus begins its cycle again in our First House. For more on this read Chapter Seventeen.

There will be more to say about Uranus' transit activity in all our natal houses soon. However, as we begin Part Two, let's first get a better feel for the basic temperament of Uranus itself and what we can generally expect when it transits various factors in our birth chart.

PART TWO

OUR AMAZING
URANUS TRANSITS

CHAPTER FIVE

THE THRILLS AND SPILLS OF FREEDOM

COSMIC THUMBNOSER

Of all the planets and other celestial bodies currently applied to the birthchart in modern astrology, none seem to compare to the wild and woolly ways of galvanizing Uranus. This planet is like a high-octane version of Jupiter, except with more brains and will than heart and soul. It's not patient in seeking knowledge and meaning—it would rather smash through the mind's Saturnian barriers, by storm if need be, to uncover the real Truth behind *anything*. It wants a shot at having an exhilarating *"Eureka!"* experience.

Uranus seems to have little respect for our human need for consistency and permanence. It acts as the joker or wild card in our psyche's make-up, providing an element of surprise. It's also a Cosmic Thumbnoser, ignoring established applications of reason in favor of its own unique brand of "logic." However, its actions are typically inexplicable to those who abide by the old format of linear Mercury mentality. Uranus also has little desire to first go through standard protocols of behavior or normal channels of conduct set up by regulatory Saturn.

This maverick planet feels under no obligation to apologize for ignoring the rules and abruptly shifting its focus when the urge hits. Its nature is to dart from point to point in an unpredictable manner. This may seem wayward and "just plain crazy" to convention-bound observers. Nevertheless, Uranus has its own unique sense of *knowing* that enables it to rip through the veils of ambiguity and illusion in order to see and understand things with unmatched clarity. I call it the "flash and zap" factor. Perhaps it is this one feature of Uranus that proves disturbing and disruptive to our ordinary waking consciousness. We don't always appreciate being shown such sharp-edged, uncompromising takes on reality, especially when they abruptly intrude on our long-held assumptions. To top it off, Uranus flashes and zaps with lightning speed and brazen force, which also proves unsettling.

MAD SCIENTIST

Instead of dissolving or annihilating form, Uranus would rather break things up and play with rearranging the pieces into new, original patterns (it excitedly reforms old, tired structures). The end results can seem weird and bewildering. Even mythological Uranus had offspring that were freakish looking, so much so that this repulsed sky god banished them. Of course, times have changed: Uranians are more apt to be turned off by people or things that look and act too ordinary. Astrological Uranus can show a marvelous sense of inventiveness in how it reassembles whatever becomes a target of its intensified interest. The spirit of innovation is alive and well with this daring "I'll-try-anything-once" planet.

For Uranus, life is one big experiment where we get to play "mad scientist" at times. We just need to give ourselves permission to adopt this planet's unusual perspectives more often in our lives and with a greater sense of inner commitment and fortitude. But don't expect society to applaud our offbeat Uranian efforts or even comprehend them, at least at first. True Uranians stand apart from those who uphold moderate mainstream approaches to living. It takes courage not to always graze with the herd and thus forfeit the benefit of safety in numbers; yet to persistently stand apart from the crowd invites unwanted social problems.

However, thank goodness Uranus is loaded with such needed courage, made stronger by its ability to mentally detach at will and remain fearless.

So far, Uranus sounds quite interesting—a real character! Who *wouldn't* want at least a small dose of its high-charged energy running through his or her veins to liven things up in an otherwise humdrum day-to-day world? The problem is that we humans typically want to control the flow of the unexpected in life much too much; such an attempt only robs this planetary principle of its special electric vitality. Similarly, mythological Cronus (Saturn) castrated his uncaring father Uranus, robbing him of his (pro)creative potency. Cronus then went on to restore order by ushering in a "Golden Age" of stability and peace. However, peacefulness and predictability are usually not what transiting Uranus brings to our current situations.

CREATIVE CHAOS?

The "human" within us is baffled by Uranus' "off-the-wall" principle. What is the possible benefit of periodic disruptions of our carefully laid plans? Why should anyone have to suffer personal "earthquakes" in the name of growth? What's so wrong with sticking to past tried-and-true ways of handling life? They worked then, so why quit on them now? What's so great about flirting with the fickle fates of the Great Unknown? Isn't chaos usually destructive in nature rather than creative?

Inquiring minds want to know but, unfortunately, Uranus' transits do not waste time answering such questions for us. They instead get right to work scanning the deadbeat or dormant factors of our inner make-up and our outer situational affairs—and in no particular order of importance (which makes these transits all the more perplexing). Once a sudden opportunity has popped up for this planet to strike, Uranus shakes up any outworn pattern we've become mired in, seemingly at random. Expect Uranus to work swiftly and unambivalently.

If you really want to experience a more robust thrust of Uranian energy, go out of your way to stifle change and progress in your life. That's right. Refuse to experiment with new avenues of self-expression. Shun all state-of-the-art technologies and at all times be mulish and do things the slow, hard way. However, then you'd better be prepared to put on your seat belt and let the roller coaster ride begin—for you have thus set yourself up for a major vulnerability that will unpredictably activate the Uranus principle. This planet will be compelled to unstabilize things for you in no uncertain terms during its next major transit.

It may do so even during a significant synastric contact (when a notable Uranus aspect forms between two people's natal charts). That other person may be the perfect Uranian trigger we've needed for so long. Still, we may be quite unprepared for the directness of our unexpected behavioral reactions (while we likely claim, somewhat apologetically, to be acting "out-of-character"). It's probably a good idea to expose ourselves to a little Uranian energy on a more regular basis as we age, so it can eventually feel somewhat more familiar to us and thus less alienating later during this planet's pivotal moments in our life.

Flying Machine

We also can project much of our own transiting Uranian needs onto others, especially strangers, who then present us with this planet's provocative energies at full strength. That can prove to be either fascinating or shocking for us, depending on how much and for how long we have repressed our own Uranian potential to live a liberated life thus far. What is our basic attitude toward people who are different from us and from our family background? How do we feel about trying new things out of the blue? In general, it's good to have inner dialogues regarding such issues long before Uranus whirls an erratic path into our life, stirring up everything thought to be solid, stable, and everlasting, and bringing in fresh new faces and circumstances. Some amazing changes may be in the air that could leave us feeling displaced on some level should we completely be caught off guard.

Such a Uranus transit can be a time for us to feel stunned and jolted out of our complacency. And for once, we are not allowed to manipulate much of the outcomes in store. If we were, we'd get in the way of our own best chance for true progressive freedom. We'd let insecurities and self-doubts take over and stop us from making new moves and letting go of old and needlessly heavy baggage. Therefore, Uranus grabs that dumb black control box from us and reuses its parts to invent a mind-powered flying machine instead!

My point is that Uranus will opt for a better, more ingenious way to help us out of our stale ruts and heavily mismanaged Saturn scenarios. Actually, for some of us, our intuitive self is probably now allowed a chance to be in the driver's seat for a while. We may not even need to buckle our seat belt this time. In fact, we'll soon be doing some mental hang-gliding!

RISKY BUSINESS

To go willingly with Uranus' challenging agenda is to open ourselves to greater levels of spontaneity. Like Jupiter, Uranus thrives on risk (although Jupiter is more conservative regarding the chances it will take, because it always wants to come out on top as a winner). Uranus values the thrill of sudden change for its own sake, even if the end results bomb. Uranus' energy is paradoxical in that it can be fired up and electrified in its driven urges one moment, then detached and totally disinterested the next. It plugs into the "peak" stage of any action or condition but may then quickly withdraw its focus shortly thereafter. However, real life is not one continuous string of exciting, charged-up peak experiences. That would eventually prove to be exhausting. We'd always be too sleepless, high-strung, and wired-up, not to mention having that shocked look on our faces and our hair standing on end. Folks would think we just pulled our finger out of an electric socket!

Expect Uranus' transits to stimulate intense but short-lived interests in people and situations that offer us a momentary opportunity to break up our routine patterns and introduce fresh social stimulation. We need to seize the moment and try out a few different approaches to handling our existing circumstances before the opportunity peaks out and Uranus abruptly demands another (unrelated) avenue of release. We can't afford to sit on our Uranus potential too long and mull things over or procrastinate until we think we feel absolutely sure about things. There are no guarantees of security with Uranus and it's doubtful if our timing under its influence ever feels completely right. Usually our actions feel premature and a little half-baked.

WHEN LIGHTNING STRIKES!

Part of any dynamic Uranus experience seems to be the unsettled feeling in the pit of our stomach along with the rousing adrenaline rush this transit pumps—even when we feel positive and excited about making changes. Uranus needs us to be sufficiently revved up and ready for the break-ups and breakthroughs in our personal life that its transits help instigate. This doesn't have to be a weird or bizarre experience for us, but we *will* need to approach life with a gutsy straightforward attitude that we seldom have shown in our past.

Uranus is like Saturn on steroids, with a bit of an attitude problem at times. Like Saturn, it signals our need to let go of the unworkable, stagnant features of our lives. It also speeds up this process and urges us to adopt an "I don't care, it's gotta go...*now!*" stance when confronting the blocks and barriers we have created or have unconsciously allowed our personal environment to erect. Saturn carefully eliminates what is no longer needed, while Uranus triggers swifter removals. Remember, it's the planet of dramatic lightning strikes. Sudden turnabouts in our affairs are common during Uranus transits, as the old and the obsolete are quickly shattered and left behind, especially if they've had a stranglehold on our freedom urges for too long.

What might take their place depends on our attempt to know the "truth" of our individual selfhood. Living an authentic life free of socially conditioned self-distortions seems to be an ultimate goal for this planet. Achieving this requires a lot of courage and the willingness to go against the grain. Uranus will always go the opposite way of wherever mass consciousness wants to go, and thereby experiences exclusion from status quo endorsement. Of course, a Uranus transit doesn't have to mean a drastic change of status for us. In less vivid ways, we can free up our energies and psychologically rearrange our priorities in order to support our individualism and further experiment with our potential.

CALL OF FREEDOM

Uranus suggests that we learn to wing it alone, without our former dependencies, at least until we better reorient ourselves. Uranus/Venus transits pose a problem here, because Venus (our urge to merge) has needs contrary to what Uranus strives for (our will to separate and be free). We could fall hard for someone during such a transit and then later realize we'd do better being alone. Perhaps the partner we choose soon gives us that old Uranus "gotta be free" routine—that distancing act—and quickly drops out of our (emotional) life, leaving our heart shell-shocked. Somehow, we are to learn not to lean on others for the wrong reasons. We are instead to discover what we can do for ourselves on our terms. There will be more about this later. Regarding the progressive potential of any Uranus transit, we have a special opportunity to rid ourselves of repressive or oppressive elements in our lives and thereby open ourselves to a new, exciting world of promise.

URANUS TRANSITING A NATAL HOUSE

Although Uranus stays in a sign for about seven years, its passage through any of our natal houses depends on the size of those houses (and thus the house system we use). In my Placidus-calculated chart, because of intercepted signs, Uranus took only five years to whiz through my high-flying Ninth House, but it took eleven years to explore the nooks and crannies of my Sixth. In my case, Sixth House matters apparently need a long time to properly develop and unfold (the energies of three signs are contained within this house—just like its opposite, the Twelfth—adding to its complexity). All transiting planets take their sweet time cruising through this situational zone of my chart. Pluto took a whopping nineteen years, a long time considering it was already moving through signs associated with the speeding up of its orbital cycle. Anyway, it's good to establish the length of each natal house, especially if you plan to make it to age eighty-four (symbolizing a Uranus Return) and even beyond.

The *transiting sign* of Uranus is a secondary influence of less personal importance. It can give us clues about exciting dynamic themes that are trying to burst forth from within the collective unconscious. That means, in theory, everyone is tapping into a little of this sign's influence according to the dictates of Uranus. This typically becomes more noticeable when given a larger societal focus—a major humanitarian goal, an important social cause, a cultural revolution, a technological breakthrough, and so on. Sometimes this sign will play into the dynamics of the house Uranus is transiting. Most times, it doesn't.

In general, our transited house takes on a new orientation marked by a greater willingness to let go of the obsolete and to try out avenues of expression that offer more freedom of action. Not all of us, however, will rise to the occasion, because it takes guts to live out Uranian energy. Should we take a passive approach to these house matters, and we often do, life will make sure that Uranus' energy plays out through a wide assortment of people and predicaments that challenge us to let go of our increasingly monotonous behaviors. Actually, Uranus seems to specifically attack those factors of the transited house where we have resisted change and adaptation the most. Yet with this planet, it's best to expect the unexpected at all times.

Remember that Uranus takes quite a while to explore any house. That suggests we don't have to revolutionize everything about our current lifestyle overnight regarding the house affairs in question. But neither should we just sit on this energy with a tentative wait-and-see attitude. At some point, we will have to make a conscious commitment to embrace the future and the unaccustomed changes implied. This doesn't have to be frightening. In fact, it's often illuminating (as flashing light bulbs of insight start to turn on inside our head, dispelling many of life's troublesome, dark shadows).

Typically, after the Uranus transit has come and gone, we look back a few years later and think, "Gee, why didn't I make such changes long before this transit? Why was I so insecure? I'm like a new person now with some exciting things in my future. Nobody is ever going to hold me and my potential back again!" Of course, that's what we innocently say while we're still feeling high on life (that "oh wow!" phase), at least until our next major Pluto transit stalks us—ha!

Still, we can't deny the amazing feeling of release many of us experience once we have had a successful sweep of Uranus through a natal house (and perhaps even over a few planets in that house). Of course, with such liberation come several new responsibilities brought about by our major readjustments—Saturn is always keeping score it seems, taking meticulous notes and restoring a workable balance. (Doesn't that old geezer ever take a break?)

URANUS TRANSITING A NATAL PLANET

House affairs are circumstantially limited by our actual current environment and the historical times in which we live. In this respect, there is an element of fate that creeps into the picture (situational factors are not under our direct control or subject to our manipulation). We are apt to feel like a part of our destiny is in the hands of a complex but capricious Universe. That's a bit unsettling for us. When and where will lightning strike next? Zap! We usually don't feel totally responsible for any personal Uranian turn of events, knowing that others have played a catalytic role in their outcome. And sometimes, startling things just happen without any rhyme or reason. Logical frameworks collapse.

However, when it comes to our natal planets, look out! Here is a symbolic part of ourselves, operating very much on the level of character, not circumstance. When transiting Uranus triggers one of our natal planets, an inner part of our psyche is stimulated to react according to its current degree of development (not an easy thing to figure out just by looking at the chart). What's important to realize is that a planet is a more potent factor in the chart than either a sign or a house. Although I love the Zodiac and its rich and colorful symbolism, I can still accept that planets are even more powerful indicators of attitudes and behavioral response.

Therefore, Uranus contacting one of our natal planets is potentially a big deal. It all depends on how we register Uranian energy and attempt to introduce it to our internal drives and urges as symbolized by the natal planet in question. Uranus will try to challenge the status quo of existing patterns we have allowed ourselves to develop according to our (limited?) understanding of this natal planet. This doesn't have to result in a major seismic shock, but it at least suggests that a few inner rumbles may be felt.

The interpretations to follow demonstrate the influence of transiting Uranus interacting with each of the twelve primary principles of astrology (Principle One = Mars, Aries, the First House; Principle Two = earth Venus, Taurus, the Second House, and so on). I will cover, for example, Uranus going through the First House and Uranus aspecting Mars in the same chapter (but not Uranus going through Aries) and continue in that way for the transits of the remaining houses and their associated planets. Remember, reading about Uranus transiting a planet may sometimes shed additional light on Uranus transiting its correlating house in the Natural Wheel (the prototype with Mars ruling Aries on the First, Venus ruling Taurus on the Second, and so forth) and vice versa. Also consider the actual house or houses ruled by that planet in your birthchart.

A planet and its corresponding house (Moon and the Fourth, Sun and the Fifth, and so on) do not share *identical* traits or functions. Planets basically represent inner states of consciousness or components of our psyche (only in an individual's birthchart, since they wouldn't mean such things in a corporate business chart, for instance). Houses are highly situational, mostly external, and dependent on the development of our cultural environment and the historical times in which we live. We can talk about certain issues in each house *now* that we could not have discussed in the 1950s; they didn't exist then. Who knows what the next century will provide? No doubt, there will be more circumstantial complexity. Houses give us an objective look at our inner planetary processes.

Because planets symbolize things closer to home—facets of our interior self—we feel them more directly and personally. When they are triggered by transit, we first internalize the results even if we then immediately project the energy onto people and outer situations. Uranus crossing over a natal planet is often more dramatic in emphasis than when it passes through a house and apparently aspects nothing (although if planetary midpoints are used, Uranus gets to aspect even more sensitive degrees). The themes of the Uranus-planet combination in question are more pinpointed and are often in need of resolution, urgently so if squares and oppositions are involved.

Undergoing a major Uranus transit means we cannot rely on well-established, social guidelines to figure out what to do or how to respond to matters at hand. We are very much on our own to come up with "creative" solutions to our existing situations. But Uranus isn't always so intuitive about things. Often, we simply get aggravated with the spirit-squelching routines of mundane living, and we react by venting Uranus in rash, volatile ways. To stimulate this planet's more enlightened features, we'll first need to take a look within and determine if we are obstructing our growth due to narrow or false self-concepts. Also, the more we let other people construct an image of who we are supposed to be, the less in touch with Uranus we become, and the more out of sorts we feel with ourselves.

The good news is that our urge to improve our circumstances is strong with this planet. We will likely have an eventful turning point that rearranges our priorities and clears the path for living more authentically and, for once, according to our own rules.

Uranus is not big on cooperation if that really means giving in to the will of others just to keep the peace and assure their approval. We will need to realize that, as we further individualize during our Uranus transits, we run the risk of breaking up our existing patterns of relating. That means some people leave the scene (sometimes in a huff) due to their inability to handle our changes. At other times, we are the ones who bolt out the door and never look back. It can feel like a period in our life when the only support we can count on is self-support. Later, however, we'll reap the rewards of living an independent lifestyle that's more suited to who we really are.

CHAPTER SIX

URANUS TRANSITING THE FIRST HOUSE

A FREER IDENTITY

Transiting Uranus in the First House is not something everyone will experience, especially if natal Uranus is already in our First or Second. We have to be at least eighty-four years old to be able to claim that Uranus has done its job on every house and planet in our chart (in every aspect formation possible). The First deals with identity issues and the quasi-conscious development of our self-image. What we manage to assemble regarding our self-presentation is here continuously put forth for others to directly encounter. The general response from others is symbolized in part by both our Descendant sign and planets in our Seventh House, especially planets nearest this angle. Much of what we automatically act out at the Ascendant occurs when we're actively engaged in our day-to-day environment, most so when we must immediately respond to a situation—similar to the way our Mars behaves.

Uranus crossing the Ascendant would qualify as the beginning of a needed surface identity make-over; this could feel like a personal breakdown of sorts for those who've been in deep ruts or long-term holding patterns of inhibition and stagnation. Our repetitive Ascendant responses to life are very subjective and are tied to survival defenses we've learned to use early in our lives. For example, Libra Rising defends itself by becoming "the other" for the moment in order to keep a dynamic harmony alive between people. Leo Rising defends itself by putting on a warm and entertaining show to defuse possible enemies and the threat of rejection. Pisces Rising's self-defense is to change personalities and blend into the immediate environment—it socially survives by using appealing camouflage to charm others into lending a helping hand rather than raising a fist. The problem with our Ascendant is that we get too good at turning on these readily available personas; because of this, we become reluctant to modify them as we grow older. They can become stale and even regressive should they conflict with the objectives of our Sun sign (symbolizing our ongoing conscious self-unfoldment).

Wherever we become too predictable in behavior is exactly where Uranus will want to generate a few frictional sparks. Our Ascendant self-image is thus a perfect candidate for this planet's experimental energies. Uranus crossing back and forth over our Ascendant (taking up to twenty-one months in some cases when applying a 1° orb) can be a time of remarkable self-discovery. It can also feel as if our whole world is suddenly crashing down around us and making a lot of noise!

Those of us with weak identity structures or rigidly defined personas seem to provoke Uranus' volatile energy the most, triggering this planet to play havoc with our lives. The more insecurely we have handled ourselves up until this point, the greater the pounding we may get when Uranus shakes, rattles, and rolls us. Uranus is a natural fog-destroyer, helping us to emerge from spells of muddled confusion. It pushes for clarity and self-knowledge. However, we can't expect to take advantage of this planet's best assets if we continue to remain timid, cowardly, or riddled with self-doubt. Uranus is now asking us to set our sights on a brilliant future that we can create for ourselves with a measure of courage and complete self-honesty. Those of us with fixed rising signs need to be especially aware that our stubbornly persistent

ways of emerging into the world will probably *not* be further supported during this transit (that even goes for Aquarius, which
often advocates progressive yet *inflexible* ideals for society to
rebuild itself on).

ELBOW ROOM

If we knew more about ourselves at this time—if we could flash to
an overview of our life and where we're going—we would have little trouble recognizing that Uranus is trying to introduce us to a
vibrant, new strategy for coping with life, one that permits ample
freedom of thought and spontaneity of action. We'd also clearly
see that, in the past, we have unwittingly held ourselves back
from expressing our greater potential. This sounds similar to
transiting Jupiter conjuncting our Ascendant, except that Jupiter
evokes a warm, optimistic glow that encourages us to become an
eager participant in the bigger, expanding world around us.
Gregarious Jupiter needs its joy-filled people-connections to help
it expand its level of social influence.

However, Uranus is not a planet associated with such warmth
or sociability. Its evolutionary process enables us to separate from
any direct identification with the masses (who collectively function on a Saturn/Moon level). Uranus' growth process can make
us appear at times cold and aloof, although highly individualized.
We may not endear ourselves to others during this transit,
although we can fascinate people while we also keep our distance.
We may wish to be involved in important collective projects, but
only as long as we can retain our autonomy and our newly awakened style of operating. If not, we probably will pull away from
any group or network that tries to lay down set standards or rules
of expected participation. We can be in a self-enthused, spear-
heading mood and are eager to do things never tried before (the
First House supports the pioneering, innovative actions of
Uranus). Tightly organized setups run by others who are sticklers
for following traditional procedures appear to only slow us down
and frustrate our vision. Why mess with them? We need to be
around people with big ideas about tomorrow!

INDEPENDENT STREAK

We may feel driven to go out on a limb and be a trailblazer. At all times, however, Uranian energy seeks to remain free and unfettered. This sounds great for us, but our existing circumstances and relationships may have a problem with that. Their problem, in the beginning, soon becomes *our* problem, until we reach a point of utter rebellion or indifference. It's unlikely that we'll want to give in to the will of another at this time. The challenge is to not be discouraged from taking independent action, but to also be fairly reasonable in the pursuit of our new and exciting ventures.

Uranus is paradoxical in that it can be intensely interested in furthering social progress, but not all that interested in personally bonding with the individuals who make up any loose social unit. It doesn't want to relate closely to people and later find that the intimacy that results gets in the way of freedom. We can, therefore, give mixed messages. One moment we can be passionately involved with human issues, and the next we're strangely detached and inaccessible. Some of us find ourselves wanting a lot of space in all relationships and are uneasy dealing with the dependency needs of others. We may have little inclination to fulfill such needs, and anyone who tries to cling to us now might be subject to a rude awakening.

BROKEN BOND

Because of the separative tendencies of this transit—Uranus initially is also opposing our Descendant—in some of our existing unions we could feel mentally disconnected, which especially affects our partner's emotional security. Restless, we feel an urge to divorce ourselves from those people and situations that have either pigeon-holed us in the past or that still continue to box us in. The rest of the chart can describe whether or not we have the overall temperament to quickly bolt out of any committed but dull relationship, or whether we will anxiously wait for the right time to make our getaway. Meanwhile, we may feel like climbing the walls. Because Uranus often provides us with an urgent need to alter our patterns, attending to our normal, daily routines becomes an unwanted distraction. We may find ourselves, as a result, running around like the "absent-minded professor."

THE BURSTING POINT

With lots of natal *fire-air* placements in our chart, this Uranus transit suggests a high degree of impatience, fed by an inner storminess regarding the status quo of our lives. This evokes much willfulness on our part and a demand for sudden, unequivocal change. We cannot take one more day of boredom and pointless conventional activity. Some of us, at this juncture, are willing to dump our past with little hesitancy and start a new life as a completely different person, perhaps with an experimental new look and a spunkier attitude about living life *our* way.

Those of us who are less bold and openly defiant (heavy *earth-water* types) will opt for gradual, maybe sporadic, attempts at freedom, at least until we have built up enough nerve to break away from dependencies we've finally outgrown. It takes earth-water folks longer to feel secure about making any kind of change. In some cases, we could start becoming very intuitive about our personal needs. This may seem a little uncomfortable if we've never before listened to our inner voice or allowed ourselves to follow our hunches (in this regard, our contrasting impulses pull us in opposite directions). Both elements are heavily caught up in security issues; therefore, Uranus can feel like quite an unsettling force. But once we have determined that the time is right to fly the coop, nothing can hold us back. Acts of rebellion can become very deliberate, backed by unwavering self-will.

OUR LITTLE PARADOX

It is important to realize that here we have a transpersonal planet operating in the most personal and self-contained of all houses. While we're learning to identify with a force that furthers the impersonal goals of collective evolution, we're also realizing the strength of our own individuality and how it makes us separate and different from anyone else. We're homing in on qualities that allow us to stand apart from the rest and to refuse to lose our identity in another. This transit doesn't necessarily "spiritualize" us, if that term only means making us more altruistically concerned with universal human issues (although this transit can and does suggest this for those who already love being on crusading social missions). Instead, Uranus puts its focus on how we can relate to ourselves with more clarity and authenticity in an often uncompromising quest for our life's truth.

A degree of self-infatuation can be noted as we unfold in ways we never thought possible—ways that make us more interesting than ever. This time period signals a very personal revolution, during which we will want to shake off the undue influence of people and situations that stifle our self-expression in the long run. Both the First House and Uranus can be self-preoccupied when taking action; bending over backward for the sake of others is not part of the psychology of transiting Uranus in the First. We'll need to make sure we aren't too willful and uncooperative, or else people will gladly avoid relating to us any more than is absolutely necessary. We can appear too temperamentally difficult to be around. We can become high-strung and even exasperating, especially during those times when Uranus squares a natal planet and we find ourselves rubbing others the wrong way with our unyielding attitudes. On the plus side, we can appear magnetically vital to others, who then find themselves excited and motivated in our presence to also act vigorously independent and willingly adventurous.

MIND EXPLOSION

Uranus passing through the First can mean we're ready to turn on our intuitive abilities, which in turn can encourage us to make fresh starts in life. Some of us, already pros at doing this intuition thing, will simply use our Uranus transit to really "bust loose" as if on an extrasensory perception (ESP) frenzy. We could feel especially telepathic and be filled with an uncanny knowing. Many things suddenly seem transparently clear to us. This could be a time when we'll need to let our flashes of insight guide us toward our newly emerging objectives. Uranian intuition by its own nature is interruptive, not gently flowing. It typically comes at awkward times, usually when we're trying to concentrate on something else, often something quite mundane. Uranian lightning strikes without warning—zap! Some of us unfamiliar with this way of receiving information will need to adjust to its urgent abruptness. If we prefer a smoother, meandering stream of intuition, we'll have to wait instead for a major Neptune transit. Uranus is more a wild-eyed god whose intuitive power crackles with electrical vitality.

SHARP SHOOTER

After several years of this transit, we may find we're not easily fooled by anything or anybody. Uranus is idealistic, but seldom gullible. We can quickly sense that which is false. We just need to "obey" our intuitive hunches and not let rational analysis detour us; we know what we know even if we can't sensibly explain why! If we have natal planets in our First, Uranus will want to give each one it conjuncts a dose of truth serum. That means we'll have to come clean about our real motives regarding issues symbolized by each planet in question. We'll need to become more honest in dealing with how we're handling what that planet internally means to us.

Uranus wants us to be relatively illusion-free by the time this transit is over, which is asking a lot of us. To eventually get to our inner truth, we'll battle against much social conditioning. None of this happens without some degree of internal disruption and the unnerving feeling that some situations are way beyond our control. It seems as if fate is taking a few potshots at us at times. Our ego is not always going to have its way when Uranus runs the show. We'll need to let life redirect our energies toward unfamiliar areas where we may surprise ourselves with unexpected abilities.

A BOLDER LOOK

How we package our general appearance can be shown by the First House. Uranus could indicate we're ready for a new and sometimes bold look. After many years of enduring the same old boring way we present ourselves to others (whether it's a three-piece suit or motorcycle gang attire), trying out different hairstyles or wardrobes can suddenly appeal to us. Maybe we have already been pretty wild and unconventional in appearance—perhaps born with a Venus/Uranus aspect squared by Neptune—but we now dare to be shockingly conservative in how we look. We could even be more interested in heavy metal—*wearing* it in the most unlikely places, that is. Unsettling others with large tattoos, multiple skin piercings, and dyed eyebrows becomes a brazen Uranian act of unconventionalism, defiant expressions that a few of us might daringly employ to show off our newfound individualism. (Others may simply and quietly switch from boxers to briefs!)

Perverse Uranus will always try to shatter other people's expectations, keeping less gutsy folks scratching their heads in amazement; it's a planet that can make at least a few jaws drop—especially when some of us leave trendy uptown salons with streaks of silvery magenta in our hair or a bleached blonde buzz cut contrasted by dark sideburns. Rock stars and radical performance artists could obviously have a blast with Uranus transiting their First. Maybe a simple tiny rainbow tattoo on our Uranus-ruled ankle is all we really need to make a social statement on behalf of free souls everywhere. Shaving our head is optional.

URANUS/MARS TRANSITS

BLOWING UP

For some of us, Uranus transiting our natal Mars could be a time akin to opening Pandora's box. The lid covering our inner issues regarding self-assertion and aggression can now fly open, and out can come those stored-up passions and even hostilities we've normally managed to keep under wraps. That's especially true if we have assumed ourselves to be the detached, cool-headed types who seldom get our feathers ruffled. If so, Uranus has a few surprising announcements for us regarding our closeted anger potential. The truth is that even a brainy rocket scientist can get steaming mad and want to throw a fit, at least once in a while. It's a very human (even animal) part of us that we see at work whenever our Mars is aroused. Since Uranus doesn't want us to hold back from expressing ourselves, it enables us to be as furious as we need to be. Getting things off our chest can be very important to us during this period.

Expect a degree of tempermentalism. We can be easily irritated, with a tendency to flare up from time to time in a manner that startles or dumbfounds others (as well as ourselves). If we already are known as a hot-head who flies off the handle a lot (we'd rather call it being "feisty"), then this Uranus transit could spell danger. It all depends on how physically impulsive or impatient we are, as well as how headstrong. Such a volatile mixture of planetary traits can cause trouble. We can have fiery outbursts that others find memorable, even though *we* may quickly brush off such scenes. For a few of us, a bigger danger here is giving in to an urge to smash things to bits and attack innocent bystanders—maybe with a high-tech laser pointing device! Senseless fury is a horrible way to relieve the internal pressures we may feel at this time; we must explore more constructive outlets to help us get our intense feelings across. Uranus/Mars can also be an accident-prone combination due to our own hasty or headstrong actions. We can get ourselves in trouble by ignoring the rules and trying to get away with stuff. We could be bold in inappropriate ways or at unsafe times, almost as if we're too detached to recognize the consequences of such behavior.

REVVING IT

We don't easily find appropriate Uranian channels to use up our Mars energy reserves, and such energy ends up spilling over into areas that do not benefit from our overdrive. Uranus renders Mars energy hard to control and direct. We might not wish to be told what to do by anybody, and therefore we may flirt with law-breaking behavior. To avoid problems, we'll need to be more conscious of all impulsiveness during this transit, as well as more aware of the details or restrictions involved before making our moves. We certainly should not push others or force issues, because the response we get could prove unexpectedly aggressive and powerful.

Mars enjoys a good Uranus transit in most cases, because it anticipates that life won't be boring and that nothing will remain still for long. This is a great time to try out new things. If the transit is a square, quincunx, or opposition, just take the proper precautions before plunging into any desired activity; remember not to go overboard and get reckless with energy expenditure. Mars wants to keep active while Uranus demands a lot of mental stimulation. We could get much done in short spurts of busyness (those famous Uranian "peak" periods). It can be a great transit for first clearing out the cluttered elements of our lives before embarking on new enterprises. Each of us personally knows those specific areas of life where we've allowed matters to pile up or remain stagnant. This transit can stir things into motion, resulting in a clean sweep of whatever might hold us back or slow us down in the future. That can also mean people who get in our way and put up roadblocks to thwart our self-expression, although most relationships typically are Venusian issues.

THOSE SEXUAL SPARKS

Excitement comes to mind with this transit, and sexuality is certainly a factor of our being that can appreciate varying degrees of excitement. However, the problem here is that Uranus demands a variety of completely different experiences to prevent boredom and restlessness. We may find ourselves less willing to uphold the virtues of monogamy (if only in our heads). Those who are married will have to brainstorm new ways to spark up their love life and keep the flame of desire going. With Uranus, whenever disinterest enters the picture, physical desire for a familiar partner can be the first to make a quick exit. Yet passion shared with a complete stranger in a chance encounter can open the flood gates of arousal like never before. It's not always easy to figure out where Uranus is coming from when it pulls little stunts like this on unsuspecting humans, catching everyone off guard and making otherwise mature people throw caution to the wind.

Sex under this transit is less complicated for those who are single and interested. That still doesn't mean it makes any more sense. This transit may prompt us to experiment with our desires if we're already not the shy type. However, even the timid can suddenly blossom. At this time, we're capable of highly irregular attractions—*erotica erratica*. Oddly enough, sexual activity itself can become a source of our boredom. We may choose to put it "on the shelf" for a while as we redirect our excitement more toward unexplored mental territory. This befits Uranus, a planet that is seldom satisfied anyway with the results of earthy, sexual expression—all that sweating, heaving, and getting elbowed! The sexual ideal Uranus often envisions rarely meets up with the primal reality experienced.

There is an urge for thrill-seeking, especially with the opposition, that can have some of us doing crazy things when it comes to powerful but momentary attractions. If we're married, we may seek out other sources of sexual stimulation, and the outlets we choose can be inappropriate and risky. Uranus and Mars together will live for the moment and pay little attention ahead of time to lust's potential fallout. This transit is not conducive to safe sex, whether from a disease or unwanted pregnancy concern. Sex becomes an irresistible energy rush we suddenly get caught up in, with little time to be sensible about things. Everything is unplanned and hurried.

In some cases, detached Uranus can cool down Mars' fever, preventing desire from having its accustomed physical release. Our sexual drive can be frozen by disinterest. A transiting square or quincunx could block normal Mars functioning (maybe we're inwardly afraid to give ourselves permission to freely gratify strong passions). We also might not be willing to share such energies with another if we feel we're being seduced or manipulated. We can get ourselves or someone else all aroused and then suddenly and inexplicably want to stop the action, which can trigger the anger of our baffled partner. Uranus perhaps will be forever ambivalent about sex, curious about it but also emotionally distant. Sky god Uranus doesn't understand sexual heat. We'll need to honestly self-review any conflictive feelings first before working up someone's passions unnecessarily.

QUICK MOVES

With sex and violence out of the way, what else is left for Mars? There is a school of thought that suggests Mars may be a key to our identity in action (meaning Mars is something more than just agitated hormones in flux). Due to its Aries connection, Mars can offer clues about how we develop a very personalized concept of ourselves. Mars isn't going to actually analyze or articulate this self-concept as would, say, Mercury; instead, its level of identity can be revealed through the things we do and our general style of action—how we directly move out into new experiences. Uranus transiting Mars could denote a "coming out" party, especially for us more reserved types, whereby we get to really feel our vitality and our zest for living—on our own terms for once. A surge of independence can hit us, and we suddenly may want to do as much for ourselves as we can without outside support or interference.

The nature of the aspect will indicate the probability of whether we experience true self-reliance or are simply being wayward. The square and the opposition are a warning for us not to become too headstrong, or we might clash with all sorts of authoritative types who certainly will try to stop us in our tracks (after all, we scare them, and they respond by retaliating with lots of Saturn and maybe even a few sneak Plutonian tactics if they're *really* terrified).

Our Mars expression during this transit can take on a hard edge that may grate on some people. We can seem brusque, abrasive, and very impatient (not every single day, but more so during our sudden flare-ups of Uranian energy). Let's count to ten before jumping all over those we find "slow and stupid." Someone's lack of speed is a Mars gripe; someone's lack of brains is Uranus' pet peeve. Again, we need to act in reasonable ways, making intelligent moves involving acceptable risks (that's really how Uranus would like it). All in all, this can be one heck of a rip-snorting transit, capable of livening up the joint if we just play by a few simple social rules and show some consideration for others.

We're not going to always have things turn out our way at this time, no matter how dynamically we thrust ourselves out there into the world and do our "mover-and-shaker" routine. There will probably be a few surprise turns and twists awaiting us. Yet that's how life on the planet is sometimes. That's also how Uranus teaches us not to ever get too settled or secure with anything belonging to the physical dimension. We're only here temporarily. Time is short, too short to not be doing what we can for ourselves rather than leaning on the wrong people. Uranus puts things into keener focus for us; we quickly realize it's okay to be completely self-directed during this time in our lives. We simply need to temper our impulsiveness with a measure of common sense. Quick moves are not always smart moves.

THE DIRECT APPROACH

What Uranus transiting Mars shares in common with its transit over our Ascendant is the push to take direct action for ourselves. We have to turn something on inside that awakens greater independence, calling for us to act alone for a while, and even to be alone. Such a period of freer self-relating can be insightful. Whatever the aspect in question, forthright (but not foolhardy) expression is to be encouraged. We're to move out into fresh experience with the determination that it will open exciting doors of opportunity for us. However, with the square, quincunx, or opposition, we might be inclined to bulldoze our way through situations, but soon find out that this "technique" doesn't work well. It even makes matters worse, triggering the surprised outrage of others and getting us caught up in weird situations that heat up too fast. We might automatically blame someone else, but we also need to share a degree of responsibility for any crazy outcomes that occur.

With the trine, sextile, or a well-managed conjunction, this Uranus transit can have an effervescent quality that helps us feel alive and well, and alert to our surroundings. Obstacles are not our concern because life seems to be giving us the green light. We don't have much reason to be overly cautious or to wait for anything we want for too long. Of course, this is not the only transit going on at this time in our chart, and thus its expression will be modified by other concurrent patterns (for example, transiting Saturn square our Mercury could certainly make us think twice and analyze matters before taking action). In general, we will be eager to put plans into decisive action when the time is right. Often, the moves we make have far-ranging benefits that we may not recognize at the moment. Uranus will surprise us later with advantageous outcomes as long as we're productive with our energies now.

WORKING FAST

Mars is more of a key to the nature of the work we do than we realize. Work is not just a matter of Virgo, the earthy side of Mercury, the Sixth House, Saturn, and the Tenth House. Work involves a focused release of energy for specific purposes, plus the intention of completing tasks. We should see if our natal Mars sign describes the nature of some of the work we do daily. It often will. Mars energy motivates us to tackle all jobs or chores at hand. Take Mars out of our charts and we won't even have the strength to get out of bed, much less wax the floor! Uranus transiting our Mars suggests we'll require a lot more stimulation from our job than usual. We won't endure a straight work load without several breaks in between. Some jobs simply can't deliver the goods, perhaps because they're too sedentary or routine. Maybe we should have stayed in bed after all, rather than waste this Uranian energy on such mind-numbing labors.

If we're professional race car drivers or have live "wild cat" stage-acts in Vegas, we're already getting more of an adrenaline rush than anyone should need. However, many of us have day jobs that cannot provide the exciting challenges we seem to want.

Are we now finding ourselves doing a lot of that nervous foot-shaking at our desks? That's often a giveaway that Uranian energy is looking for a quick and ready outlet. Are we itching for a few interesting distractions? Again, Uranus is looking for something different to do. There needs to be another area of our life where we can let loose and tackle things that highly interest us. It's best if we feel challenged mentally by our activities at this time; then we won't be as tempted to challenge others and get into crazy fights.

Due to our accelerated drive and eagerness to move forward with our objectives, we can get a lot done quickly by taking advantage of short spurts of intense but highly focused energy. Uranus is not good for anything requiring sustained efforts over long periods of time (that's Saturn or Pluto). It will demand that we take periodic breaks to avoid nervous exhaustion. However, we'll soon need to get back to our tasks to keep the momentum going, because once it's lost, it seldom returns (Uranus abandons people and things once indifference takes over). Our excitement and enthusiasm keep both planets happy and well-motivated.

FINAL ADVICE

In general, this transit pinpoints a high-energy period that can make us feel vitally involved in the here and now. We have an opportunity to show gumption and to come up with inventive ways to make our mark. Our actions can be outstanding in some manner, even trendsetting in some cases. What we need to keep in mind is that this is a time when we can be more vibrantly charged up than ever—emphatically so with the conjunction, square, or opposition. Uranus/Mars aspects have a reputation for being as volatile and explosive as dynamite. Much of the turbulence generated is usually a result of not finding suitable outlets to handle this transit's sudden swells of energy. It is imperative that we take action during this phase, but very *conscious* action as we consider the pros and cons, and then move forward with self-assurance.

CHAPTER SEVEN

URANUS TRANSITING THE SECOND HOUSE

MATERIAL WHIRL

Uranus transiting our Second House might initially worry some of us who don't like the thought of *any* potential trouble-maker entering this private sector of our chart—especially if we have safety-conscious Taurus, Cancer, Scorpio, or Capricorn on the cusp. After all, this is the house where we store and enjoy our valuable goods and where we count and spend our hard-earned cash; we expect nothing less here than total, everlasting security. It's easy, living in a materialistic culture, to believe that secured acquisitions will be of greater worth to us than personal happiness (a more abstract, hard-to-define inner state). Should happiness be a part of pure Second House fulfillment, it's mostly the result of that which we tangibly own; happiness is thus something that comes with a price tag attached to it. Knowing this, Uranus is eager to punch a few holes in such a limited worldly outlook.

One thing we might worry about is Uranus' reputation for restlessness and changeability as it enters this house. Uranus doesn't want to keep things as they are and as they've always been; it likes to tinker and experiment, which just ends up destabilizing everything the Second House wishes to keep intact. It's hard to feel grounded when all that we own is in a constant state of flux; the Second likes things to stay nailed down to the floor, as does the sign also associated with Principle Two—Taurus. Therefore, this transit may not be welcomed at first. It certainly won't be easily understood by those who have a tight grip on their possessions (often sore-loser types). Many of us are already too jittery about our financial futures in today's complex economy. Do we really need Uranus to add to our overall uncertainty? Can we endure periodic shake-ups here? And, of course, how much will this cost us in the end?

THE POSSESSED

It's true that Uranus and material stability do not mix well (after all, it's a planet associated with earthquakes and sudden turn-abouts of fortune). That doesn't mean they never can work together; however, it will take a major adjustment on our part concerning how we view our value structure. We are dealing with a sky god of the rarefied ethers, not an earth deity of the plowed and fertile fields. For starters, Uranus doesn't seem to understand this preoccupation with ownership we earthlings get so worked up about. Mythological Uranus didn't even want to own his freaky kids and made sure they didn't get to hang around! Owning things seems all so energy-trapping to this planet; having a lot of material objects may weigh us down too much, keeping us unsatisfyingly earthbound. Then we're not free to move about as we please, knowing we have all these physical *things* to maintain.

Perhaps Uranus would concede that owning exciting high-tech toys, for example, may be great fun for a while. They could be awesome to have for the moment. But why own anything or attempt to own *anybody* for a lifetime? Why risk being possessed by such possessions? Also, how can we make room for brand new, ingenious contraptions if we first are not willing to part with old stuff collecting dust and slowly devouring our surrounding space? Uranus is no junk collector or pack rat with closets crammed full of little-used items. The Moon's guilty of that. If we are especially lunar in our attitudes about what we own, we give transiting Uranus a darn good reason to break up our attachment-fixation in the Second.

HOLDING ON

We may rationalize our urge to accumulate by saying that what some call clutter, we view as our "cherished collectibles," things that have great emotional value for us. We need such items around to feel good about our link to the past. It may not be coincidental that a few of these things also have great resale value, such as rare antiques handed down for generations. "Cherished collectibles? Aha, maybe so," responds quick-thinking Uranus, "but why don't you take that same emotional energy and apply it toward *new* objects of desire? They, in turn, will become items that you'll also feel you cannot bear to part with, unless the price is right! It's your sentimentalism you're hanging on to more than anything else." We're not sure if we like the sound of this kind of blunt reasoning, but that's Uranus—always opting for an unusual viewpoint directly stated.

Uranus, although not sensitive to the emotional tie people have to their possessions, may still be pointing out a truth about this "holding on tight to what we value" phenomenon that many folks seem destined to experience. Is it truly the possession we must have or is it the sense of intimate attachment our possession triggers that is actually what we wish to cling to? There is something seductive about owning things from our past and where they take us in our heads. After all, memories can also become cherished collectibles; we can even image ourselves "storing" our recollections in our memory "bank" for safe-keeping.

LETTING IT GO

In our Second House, some of us can develop the attitude that certain attachments are never to leave us. We always want them close by and ready to be used in case we need them in the future. That's why we have attics, basements, and closets. It's hard to throw out or part with long-owned Second House things. For example, most people with self-possessed Taurus Rising have Gemini on the Second House cusp. If this non-possessive air sign is going to be acquisitive, it will be in the area of our mental world. Gemini placed here is learning to possess ideas—one's own and those of others—especially ideas presented in tangible form. Book collecting, study tapes, articles, newspaper clippings, and interesting tidbits from magazines are what we tend to accumulate.

Gemini is nosy enough to love getting e-mail or junk mail, but it hates being weighed down by heavier baggage. Paperwork of all sorts could pile up and needlessly eat up space. The trick is to choose what we wish to possess intelligently—thus satisfying the Gemini principle. Otherwise, Taurus Rising has a steady appetite for having more than it needs—some of that being nonessential information collected, making our Second House environment too congested for Gemini to properly function. Ventilation is important for this air sign; therefore, the less we stockpile the better. See if you could readily do without various items or concepts represented by the sign on your Second House cusp. With transiting Uranus passing through here, we may have to let go of some possessions without much notice or time to emotionally prepare ourselves. They can suddenly be yanked away from us.

There are people with this transit who perhaps have had to endure the horror and pain of losing an entire house and all their belongings to a hurricane, tornado, or raging wildfire (maybe even the IRS took a big chunk out of their material world). Some freak incident might have entered their lives and turned everything upside-down. The trauma triggering such shockingly quick and sometimes total loss must have been great. It is doubtful that they would have been in any mood at the time to listen to Uranus' little speech about the "here today, gone tomorrow" transient nature of all physical things. Uranus probably detaches from such loss so easily because it has never allowed itself to become emotionally bonded to anything or anyone. Yet we sense, on a gut level, that we have a lot to lose with such a planet so fundamentally disinterested in attachments and material outcomes.

Most of us will not experience a natural disaster that destroys all that we own in one fell swoop. Nevertheless, this transit will force us to rethink any future plans about amassing too much in life. Uranus pushes for material detachment, which doesn't mean we can't continue to play the happy consumer, and reward or indulge ourselves—it just suggests we are not to get so immersed that we'd fall apart at the seams should everything come to a screeching halt due to devastating events beyond our control. Maybe we could cooperate with Uranus by having a few yard sales or allowing for spontaneous giveaways, ridding ourselves of excess material build-up; then it's best to take a break from buying new items altogether for a while.

SURPRISING SKILLS

When it comes to money-making (a strong Second House pursuit), Uranus during the next several years will have some new and hopefully exciting plans for us. We'll need that much time to be convinced that we can derive a livable income by going down different roads of opportunity. Because our survival requirements must adequately be met in the Second House at all times, we feel we cannot afford to play crazy Uranian mind-games here—like simply quitting our job and taking a few months off to decide who we really are and what we really want to do in our career. Who can afford to take that route in today's high-pressured world? Uranus is nonetheless ready to surprise us with the emergence of personal resources and skills we may not have effectively tapped. It's time to get a little gutsy about unearthing and using our lesser-known talents.

Upon awakening our intuition, Uranus reveals that valuing personal freedom with a real sense of conviction can allow us to earn a living in such a way that we don't feel enslaved to tedious jobs. We don't have to be pawns in the hands of an impersonal business industry that could downsize us in a heartbeat. This still could be a hard-sell for many of us who feel pulled in two opposite directions ("stick to what you know" versus "try anything new just to get out of this rut"). Eventually, we will need to do some degree of risk-taking to test the "ethers" and prove to ourselves that *mind over matter* can help us succeed.

FUTURE SECURITY

It is important that we recognize and follow our hunches during this transit (they can act as our career guidance counselor). It also helps to have the guts and determination (Uranian values) to explore those areas in business that excite us, that make us think about the world's future, and that even make work seem like stimulating fun. Uranus could mean obtaining that dream job we always thought we'd never be lucky enough to land. Another option (and a strong one) is self-employment. Uranus is anti-boss; it defies supervision by those in authority. This transit alone is not enough to suggest that we have what it takes to be a successful entrepreneur, because that involves more than just having an independent spirit and a strong will. We'll need to find out firsthand, perhaps in small ways, if we have the working skills required. Often, Uranus means we take on several small jobs to eke out a living, something that can help fulfill our need for variety and change.

By exploring and delving into our feelings about the sense of safety our attachments provide, we'll get to examine what security really means to us—especially during those times when things feel shaky and our direction uncertain. Little should be taken for granted when it comes to material support. We'll have to be alert to any sudden opportunity to prosper. However, it is important that we shoot for high-minded goals. Hopefully, we'll eventually become less uptight about money and about paying our bills. By following Uranus' unusual game plan, we will marvel at how our financial affairs fall into place unexpectedly (money comes from the least expected sources). If we are truly stuck in a rut, we are free to drop whatever's not paying off for us and switch gears in midstream, if need be. We have more options available than we typically realize.

AWAKENING OUR SENSES

There is a sensual side to the Second House that's usually down-played in astrological texts in favor of material and financial survival needs. When do we get to slow down and smell the flowers or count the puffy white clouds above? When do we get to enrich our lives by taking time for the quiet joys of Nature's bounty? Attracting the comforts of the physical world in their most basic form is symbolized by the Second House. Uranus, although not deeply interested in physical gratification, could now awaken our appreciation of beauty in natural form. Suddenly, we may find that we want a lot of simple but lovely things around us—not just expensive, status-enhancing objects possessed solely to create a social impression. Actually, Uranus is interested in beauty's powerful ability to arouse inner states of exhilaration and astonishment. This would we a great time to witness, first-hand, an aurora borealis or a blazing comet sweeping across the nighttime sky. Nature can be very exciting, and this is a time to explore that fact!

One easy way to evoke the sense-gratifying element of the Second House is to tend a little garden or put a few terra cotta pots out on the patio. Grow flowers. Plant rose bushes or even miniature ornamental trees (the fruit-bearing variety). Bonsai arrangements might work because they're so unusual, although they're really too slow-growing to keep Uranus' interest (it's not a very patient planet). Maybe gardening sounds more fitting for earth Venus' needs than for Uranus'. Yet if we pick the most uncommon varieties of flowers, bushes, and trees to grow, Uranus will be satisfied. Regular old earth Venus would gravitate toward petunias and gardenias, the hardy, and the fragrant; but Uranus is more curious about Venus fly-traps and strangely shaped cacti, or plants that bloom only at night—anything that's different. The more exotic our plants, the more we please a sky god who relishes nature's less-popular oddities. Forget those pretty but predictable pansies; think about cultivating a giant Golden Amazon Nightcatcher!

SPENDING SPREE

When it comes to spending habits, Uranus is ready for us to try something different. Spending freely, without guilt, might be our challenge during this period. Being a tightwad or a cheapskate is not supported by Uranus. Letting go of insecurity and experimenting with sensible risks is the new trend. However, financial foolhardiness will never be condoned, because that could put us in serious debt, leaving us anything *but* free. Impulse buying can become stronger; we unexpectedly see what we like and—zap!—it's ours. Obviously, we'll need to avoid using credit cards too much for such quickie purchases, because that can invite chaos into our lives at some later point when we've become more manic with our shopping urges than we can afford.

During these years, we could also sporadically give money and other assets to groups that will fight for social causes we strongly support. Such special-interest groups have underdog status in our eyes; therefore, they become something Uranus values defending. Protecting the natural environment and the rights of wildlife (Mother Earth) would be things we could financially support. Apparently, we can part with our cash when we're angry enough to demand progressive or humane reform. This would suggest that our value system is undergoing a revolution based on altruism, social activism, and new perceptions of what we deem to be right and true. Money and social consciousness are now tied together for us like never before.

If we've having a transiting Uranus square or opposition, we may be too impulsive in our monetary support of "worthy causes" and much too impractical to satisfy Second House dynamics. Life quickly brings this to our attention by having our dishwasher break down shortly after we send off a check—for more bucks than we should have donated—to the "Save the Beavers" foundation. With Uranus involved, the foundation itself might be in a state of financial upheaval and could fold at any minute, unbeknownst to us. It's wise to do a little research before funding anybody's passionate crusade.

URANUS EARTH/VENUS TRANSITS

NOTHING PERMANENT

Uranus transiting our natal Venus can take two different routes of expression, depending on whether we're dealing with this planet's earthy side (Taurus) or its airy facet (Libra). Let's look at the earthy factor first, and then get back to the airy element later, when we cover Uranus going through our Seventh House.

We know that Venus symbolizes things we like and admire, things that give us pleasure and comfort. This is the earthy, "feel-good" side of Venus that seeks solid, tangible satisfaction. Earth Venus means money, possessions, pleasant body sensations, things that are beautiful, or anything that appeals to our senses. It's Venus' need to express itself through the fulfillment of our physical appetites. Uranus by nature doesn't like to become too submerged in this for fear of being sucked in and trapped by the gravitational pull of earthbound desire. Solely from a material perspective, ensuring any sense of long-term security is not the main thrust of this transit. An early indicator of this might be that things we've owned for a long while start breaking down on us. They unexpectedly stop functioning. They could suffer sudden damage or whatever else it takes to get our attention. Life is now teaching us that nothing is meant to last forever.

The more fixed and/or earthy we are by temperament, the more likely we will periodically suffer such sudden losses. We feel we are being forced to let things go (sounds like Pluto, except Uranus operates so darn fast that, at first, we probably are stunned by any loss rather than bitterly angry). Our material structure is ready to undergo a few progressive changes, and it's best if we don't get into such a tizzy about losing anything we currently own (easy to say, harder to do). Stability as we have known it is not promised by this transit. Our emotions, if heavily invested in what we own, will be subject to many ups and downs. A shake-up of our worldly values may be the best way to awaken an awareness of the power that our attachments may have over us. Of course, how this affects us psychologically depends on the nature of the aspect. A conjunction, sextile, or trine allows us to see Uranus as stimulating and even as a source of excitement.

QUICKIE SPLURGE

Maybe we're ready for a few, significant purchases, while some older, little-used items are ready to be tossed out or given away. Perhaps we buy a super-duper computer for the first time with all of the bells and whistles. That alone is a thrill. In addition, computer literacy becomes a new asset we learn to value and maybe later capitalize on—a thought that appeals to profit-minded earth Venus. It's the square, quincunx, or opposition that can make Uranus energy such a pain in the neck at first, because our carefully organized plans often take quick detours or sometimes even unforeseen nose-dives. Earth Venus represents fixed wants and desires that do not appreciate life's complicating disruptions. It psychologically equates steadiness with well-being.

However, even reversals of fortune are not always as bad as they feel at the time—especially if we've been living too long in the foggy land of Neptune, wasting or mismanaging our resources left and right. Uranus is no Saturn, a planet that really likes to tighten the budget, but still it hates to see us get dumb and dumber about things—even our finances. It could be that this transit will teach us how to make a few smart financial moves (intuitive investments) that open doors of future opportunity. It will be impossible to predict how we might use money during this period, except to say erratically so, and based on our hunches at times. Uranus is susceptible to splurges, but less so to extravagant or needless ones (that's Jupiter). It's just that many of those state-of-the-art, high-tech Uranian toys that catch our eye usually cost a lot.

SUDDEN WEALTH?

This transit could make us a prime candidate for financial surprises of an extraordinary nature—like windfalls, money dropped in our laps, seemingly from out of the blue. However, we won't benefit as much when we try to force the urge to win big bucks (like going to Las Vegas with a determination to leave that town wealthy). Uranus demands a degree of surprise, so the less we plot and scheme, the better. Even if we did win top prize in a major lottery or a national sweepstakes, instant-millionaire status would probably become a very unsettling Uranian experience. We'd naively anticipate it to be tremendous Jupiter abundance ("Thank you, Jesus!"), but the reality could throw some of us into an unanticipated and premature life crisis: Do I keep my job or not, my lover/spouse or not, even my current physical features or not? Should I relocate to get away from jealous neighbors or money-mooching relatives? What should I do about tax strategies and future investments? What about protecting myself from "the bad guys" (scam artists and such) who illegally want a chunk of my loot, or even from the well-meaning but endless stream of charity groups begging for donations?

Our life suddenly can be turned upside-down. We surely weren't expecting this. Be careful about wanting a quick ticket to "the finer things in life" when Uranus is calling the shots. It rarely means we can keep on living the way we've always known, making only minimal changes. Money comes into our life mainly to shatter old structures. Some of us may find, ironically, that we are less willing to share what we own as much as we did before. We can display a strong independent streak regarding our finances by wanting things to go our way without the interference or supervision of others. We won't tolerate anybody telling us how to manage our wealth (and at this time, cautious Saturnians seem to come out of the woodwork to give us their sage advice, which we are prone to flat-out resist).

BREAKING HABITS

It is likely that we will at least have a chance to appreciate new things about day-to-day living during this transit. Although much depends on the natal condition of our Venus, in general we will tend to loosen up regarding our likes and dislikes. Maybe we are learning to value tolerance and to accept things that do not fit standard norms. We even might experiment with modifying our routines, habits, or whatever it takes to break the monotony. This could be very self-educational. Another value we adopt is respect for impermanence. Flux and change are the way this world runs, even if we are temporarily lulled into long stretches of stability. Because earth Venus can have a problem with passivity, we often get stuck in our safe but dull daily patterns. Uranus abhors inertia, however, and can quickly intuit the difference between true stability and stagnation.

If it senses that we are just vegetating in our overgrown fields of non-development, Uranus will go right to work by zapping us in those personal areas that most need to be recharged. We may require a good shaking to wake us up to our unused creative potential. Earth Venus represents latent talents that need our attention and a little prodding. Maybe we are budding artists or musicians and don't realize it. Then something happens, often by chance, to awaken our talents. It might excite us to realize that skills discovered now can materially pay off later, depending on how we market them. They at least offer us much self-satisfaction. However, for talent to go anywhere requires self-discipline and hard work, which are in short supply during this transit. Uranus only introduces us to potential we didn't know we had; it will take a good Saturn transit or two to give us the staying power to further develop any potential and make something lasting of it. At least our eyes are open wider so that we see what our future could promise once greater effort is made.

BRAND NEW STYLE

We can also get this transit rolling by upgrading our surroundings and purchasing a few, innovative items, usually electronic or digital. That could be something as simple but odd as a programmable doorbell or a dishwasher with its own remote control (anything to make the mundane chores of living easier). How we decorate can be shown by Venus in both its earth and air facets. Uranus is ready for new colors and patterns. Is there a room at home to which we've been longing to do something different? This could be the time to play with fresh concepts, although it's best not to get too *avant-garde* (something we could regret later when this transit is over).

A few colorful decorative touches and an interesting new arrangement of furniture—maybe moving a bed or sofa from one side of the room to the other, and placing it at a different angle—is enough to make everything look suddenly different. This is how we can evoke Uranus. Earth Venus' first instinct is to prompt us move things back the way we first had them, because it doesn't respond well to unfamiliarity. However, to consciously choose to go with Uranus means keeping that bed, sofa, or whatever in its new spot for at least a week or so. *Then* we can evaluate the benefit of such a change. By that time, we might find out we really like it (because by then we've gotten used to things). The same could go for a new hair style, new clothes, or other changes. Our initial resistance, serving to support only inertia and inflexibility, needs to be challenged. It also helps to know that any changes made do not have to be permanent (and probably won't be).

A few of us, however, won't need much prompting when it comes to bringing Uranus into the picture. What if we are born with Venus in easily bored Gemini, a Venus that is now squared by restless Uranus? We still might need a little balance when determining how to go about ushering in the new and the different. Uranus has a wacky sense of experimentation that can upset earthy Venus' simpler, unaffected style and taste. We may go overboard with adornment (using too much of something that doesn't really coordinate with something else). We may blend things that look weird or jarring (stripes and plaids). While *we* think our look is highly original, others may find it simply tacky and ill-conceived.

Too much Uranus—like the fire element—means we can be reckless with our spending, and unwisely impulsive when buying things. Such sudden urges can be expensive, and our satisfaction with certain newly-owned goods might be fleeting. Venus alone offers us patience and the ability to wait to attract desired items at the right cost, meaning at irresistibly marked-down prices. This practical Venus does a little comparative window shopping and wants to feel the material before buying. Its ability to assess the value or worthiness of anything is excellent (although the actual sign Venus is in modifies this planet's traits; fire signs, for example, don't do their homework and usually end up paying more).

Earthy Venus likes to savor things slowly; things become more attractive after time has passed. However, a mismanaged Uranus doesn't bond well on the feeling level; therefore, we may purchase something unexpectedly, use it intensely for a short time, and then put it in our closet or in a little-used drawer as the victim of underuse. During this transit, we'll have to be careful not to thoughtlessly give in to faddish or gimmicky items—such as a lot of those things we see advertised on late night TV that "are not sold in stores anywhere." Then again, maybe a few of us really would love a fabulous set of Gonza knives—guaranteed to be strong enough to even cut through bureaucratic red tape!

NO GUARANTEES

We may even wish to use our savings in different ways ("creative investing"), although we'll need to keep in mind that Uranus doesn't much care if we lose money as long as we've allowed ourselves to try out different financial strategies. Venus, however, won't feel safe with investments that cannot guarantee security ahead of time; no wonder it rules a house that squares the risk-taking Fifth and Uranus' own, the Eleventh. If we are the sore-loser type, we'll have to approach any investment potential very gingerly. Better not experiment with money until we know that we are detached enough to not get too broken up about any unanticipated losses should things go awry. Once we truly can adopt a "take it or leave it" attitude, Uranus may then turn around and surprise us, making us big winners! We have to be caught at least somewhat off guard if an experience is to qualify as Uranian.

TIME TO UPGRADE

Transiting Uranus evoking the earthy side of Venus suggests that whatever we choose to do, we'll want the outcome to be pleasing or comforting (satisfying our natal Venus requirements). We are not to feel deprived or left wanting when this transit is completed. This facet of Venus is associated with self-appreciation issues (as is our Second House). The results of our responses and actions during this period can enhance our personal value and upgrade our sense of self-worth. Yet all of this will have to come about by consciously inviting more Uranus manifestation into our lives. How do we do that? First, we establish that change itself can be very gratifying when it releases us from having things around us that we're outgrown (including antagonistic people who continuously dump on us). Being surrounded by whatever is stale or static can feel suffocating. We need an invigorating stimulation of the senses—because we're worth it!

CHAPTER EIGHT

URANUS TRANSITING THE THIRD HOUSE

MINDBURST

Uranus transiting our Third House can be an excellent time to awaken our mind to the ever-exciting possibilities of expanded consciousness. This is not a period to get lazy with our brainpower or underestimate our intelligence. Cerebral-intuitive Uranus likes to live in its head; here, it can create amazing new concepts at will without being slowed down by the conventions of a limited, external world. On our more abstract levels of thought, we have the freedom to explore what we want without self-restriction and without censorship (although Uranus can sound like a heretic who willfully rocks the boat of conformity).

The Third House is where we toss about a variety of stimulating ideas that keep our minds well tuned and open to smarter ways to communicate. Variety and adaptability are the keys to our success here. Both the Third House and Uranus support brightness, quickness, lively expressiveness, and the speedy transportation of data. The Third House, however, mostly describes standard methods of education (books, classes, lectures, online forums and newsgroups, or even just asking smart or nosy questions). In contrast, Uranus doesn't depend on outside "authoritative" sources for answers; it comes up with its own startling revelations or else invents previously unheard-of theories on the spot!

We may currently feel mentally bored with the usual information that we use in our day-to-day dealings. We need an intellectual boost, especially if stuck in rut-bound thinking patterns. With Uranus, this boost involves being introduced to fresh and sometimes provocative data. This is not always a conscious quest. Nonetheless, we may stumble on eye-opening information that makes us sit up and think twice about our former assumptions. Not all people are interested in feeding their minds challenging or controversial material, so one option is to channel this planet into heightened physical movement. This could mean a restless phase in life when sporadic travel is accented (usually local trips rather than international tours, unless perhaps we already have natal Jupiter in our Third and/or Sagittarius on the cusp). Our short-term adventures usually don't take us too far away from our familiar environment. We're more apt to meet up against real Uranian culture shock in our Ninth House.

MY MAZDA, MYSELF

We may be more on the go than usual, so one understandable mundane manifestation of this transit would be to buy a new car. It's usually one purchased under pressure because of the unexpected malfunction of the vehicle we already own—that old workhorse we expected to run forever, the one we claimed thrived on neglect! We may stubbornly try to hang onto our clunker even though it's now undergoing a series of unrelated but exasperating Uranian symptoms, all requiring sudden repairs and stunningly high bills. We'll have to weigh all sides and determine if it might be wiser to unload this high-maintenance nuisance and choose something more reliable—this time, something with a great sound system.

Sometimes we have little choice: our car was stolen, or vandalized, or perhaps totaled in a weird accident (and we weren't even inside it at the time, thank goodness). Remember, Uranus symbolizes life's more confounding matters. Strange things just happen as fate intervenes to force a decision on us. A clue that Uranus is active and feeling mischievous would be when odd little symptoms pop up involving the electrical system of the car, most commonly the battery. We could be more absent-minded than usual and accidentally leave the headlights on, the keys in the ignition, or the door unlocked while we're off shopping. We'll need to pay extra attention to the details of what's happening in the now rather than let our mind distract us by impulsively jumping into the future.

It may seem superficial at first to be talking about mere car problems for a transpersonal planet like Uranus, especially to those who see deeper esoteric meaning in all Outer Planet activity. Yet for me, one of the more fascinating correlations astrology makes is that of the mind/car connection. Our Third House thinking patterns directly tie in with not only our driving habits but also our experiences with traffic. I can't help but wonder if the person who drives around with dents and scrapes has an angry Mars in the Third to work out; what about a Third House Neptune for someone whose car displays the message "License plate has been stolen. Applied for new one." Could transiting Pluto be involved in the chart of a victim of a terrifying carjacking?

The health of our car and the way we drive can be a tip-off regarding our general or current state of mind. Uranus suggests off-and-on-again symptoms, paralleling our erratic mental patterns at this time. Our inner nervous tension can be discharged in dramatically quick and surprising ways, then projected onto our auto or any other daily means of transportation.

STRANGE NEIGHBORS?

Neighbors will sometimes show us how Uranian energy works. Here we have a variety of possibilities available for transiting Uranus to freely express itself, assuming that our nearest neighbor is no more than a few miles away and that we're not total recluses. We'll need to consider especially the folks right next door or just down the block from us in a Third House context. Uranus, often when aspecting our natal Mercury, Venus, or the Moon, could indicate some disruptive or exciting changes are going on with these people. Maybe some folks are suddenly moving away or are noisily renovating their house. Maybe they've undergone an unexpected upset recently. Perhaps a neighbor is just being a pain because of his eccentric behavior or her inability to follow standard rules of neighborly conduct. Uranus symbolizes oddballs who don't fit into the social system well—nor do they care to. There might at least be one of these on the block or in the apartment/condo complex acting in ways that we find annoying. These manifestations alert us to possible disruptions also going on inside our head, those rebellious mental qualities generating similar friction and heat. Sometimes a neighbor's behavior gives us a chance to play the detached witness and observe a few Uranian principles in action.

Even sections of the neighborhood could undergo physical changes that some would deem to be progressive (road construction, housing developments, a new fire station or post office built, and so on). Our aspects at the time might better determine whether we will interpret such changes as true improvements or just unwelcome aggravations; they probably are both. Things will be disruptive and chaotic for a while, and we may be forced to take different routes (Uranus) to get around. The Third House deals with nearby localities; therefore, if we are dedicated homebodies, even hermits, we may not even be aware of the extensive alterations going on around us. We need to cruise around and see for ourselves what Uranus has stirred up several blocks away. The Third House tries to arouse our curiosity about the social scene around us; Uranus provides a few interesting people and things that capture our attention.

SIBLING SURPRISE

Of course, siblings are another Third House issue. Uranus can mean a brother or sister either in a rebellious stage in life or at a revved up phase of self-discovery (maybe both). Perhaps siblings are undergoing a major upheaval, bringing chaos and unexpected opportunity into their lives. The potential for bust-ups, break-throughs, and abrupt departures can be strong. This is a non-issue for those of us who are without brothers and sisters, unless Uranus means we're feeling agitated about our lonely "single child" status. Of course, maybe due to a parent remarrying, we could suddenly end up with step-siblings that we'll have to get used to. Uranus here presents us with an unusual set-up that requires flexibility and tolerance. However, our transition might not be smooth at first due to someone's lack of cooperative spirit. Then again, depending on the aspects made, we could become instant friends at first sight—an unexpectedly quick bonding.

If Uranus forms stress aspects to our natal planets, those of us with siblings can be at odds in our relationship with at least one of them. It's not a great time to engage in a battle of the wills or threaten one another with ultimatums and other strong-arm control tactics. Big blow-up scenes can result, as well as baffling falling-outs (baffling to other relatives, at least). Under difficult conditions, Uranus can symbolize alienation and estrangement. Keeping our distance may seem like the best policy at this time. However, Uranus can also suggest—under trying conditions—sudden reconciliations with a brother or sister we have avoided honestly relating to for a long time. We may warm up to one another quickly once the initial ice has been broken and our main problems have been aired. Under easier Uranian patterns, we could share something exciting with a sibling. Maybe we are thrilled for them as they embark on a more promising future. We may enjoy open and spirited communication with them as we seldom have before.

Anything goes with Uranus, so it's futile to try to pinpoint exactly what will "happen" regarding our brothers and sisters—maybe it's absolutely nothing out of the ordinary—because Uranus could be working on other, more pressing Third House themes for us. Many of my clients with this transit seem detached from the affairs of their siblings, not really knowing what to say about them sometimes. It could be that a sibling's circumstances might prove to be a big surprise (or shock) to those clients later, when the cat's finally out of the bag! Situations could change overnight, and suddenly we might become involved in their urgent affairs. Wouldn't it be great if all this transit means is that we find out a brother or sister has developed a sudden passion for learning astrology and wants to have stimulating conversations with us about chart interpretation? Just don't be too quick to lend out your favorite book on the subject; that book may not come back in the best shape, *if* it comes back at all. Uranus deals with life's "oops!" factor; therefore, unfortunate mishaps can occur when we impulsively lend our Third House items, including our cars.

SHOCK PROOF

Uranus going through the Third is typically a time for a major overhaul of our thinking process. We are to learn to trust the intuitional development of our mind, which does not mean foregoing logic and reason. Intuition may reach the same conclusions as does logic, yet it does so faster and with greater excitement. Also, after several years of undergoing this transit, we may find ourselves more open-minded, tolerant, liberal, and almost unshockable. I suspect life will especially test us regarding that last quality to see if we can remain unflappable and non-judgmental in the face of strange events.

A healthy detachment can develop, preventing us from being as touchy, defensive, or as easily offended as we may have been before this transit. We will now seldom take anything said as personally as we once did, if we ever did (especially if we have a water sign on the cusp or a water planet in our Third). We can be amazed at how people's statements just bounce off of us, whereas in the past they would have disturbed us for days on end. This could be a great transit to experience in old age; our mind will surely not dry up with boredom and non-stimulation; and we certainly will make sure that we *don't* just settle down and act our age!

ELECTRIC BRAIN

We also have opportunities to undertake studies that can blast away our former assumptions about life and about ourselves. Our need for intellectual arousal is typically heightened. Often, we unexpectedly bump into new information that can strongly alter our thinking—at times, radically so. However, it is important not to become too wired up about any new theoretical data or study. We'll need to slow down and learn to digest things at a reasonable pace (which still can seem rapid, because Uranus by nature is no slow poke). We'll also have to ask ourselves if we can apply our new insights in workable ways. Is this information relevant to our needs in the real world?

Sometimes we may have a sudden urge to go back to school (even if not formally). We will find a way to continue our education, even if what we wish to learn is deemed unconventional or futuristic. Maybe taking classes in specialized subjects is suddenly required of us (usually for the sake of career advancement). Although everything can feel very uncertain at this time, learning a new body of knowledge is a step in the right direction. Doing absolutely nothing with our brain power can leave us feeling uncomfortably up in the air and even cranky—Uranus can entertain some pretty bizarre thoughts when it's cranky. When we don't allow ourselves suitable channels for experimentation, irritability increases, our nerves can act up, and our sleep patterns become disruptively altered.

All in all, this can be a great time to enjoy being a free thinker, less influenced by the opinions of established authority. Of course, we'll have to live with the possibility that nobody may agree with us. We're on our own with ideas that may not get a lot of support from others. Still, we're able to size things up more quickly and invent new ways to problem-solve. It is important that we give ourselves a chance to challenge our minds with concepts that allow us to explore new horizons. We're ready to look at things with much farsighted understanding. If we consciously work at it, this could be an excellent time to shatter a few mental hang-ups that have kept us from living an optimal life.

Nonetheless, we should reserve any final conclusions about everything we've learned, unlearned, or relearned until Uranus transits out of this house—long after the excitement factor has peaked. It's very doubtful that we will ever go back to our former boxed-in ways of thinking. Once we've seen more with our mind's eye, we will continue to also want more to explore throughout our life. We're been zapped with a hunger for truth that can lead us to no-nonsense paths of enlightenment. Actually, indulging in a little absurd nonsense at this time is also good for the soul!

URANUS/AIR MERCURY TRANSITS
SPEED OF THOUGHT

Transiting Uranus stimulating the airy side of Mercury can be invigorating for the mind, but sometimes it's too over-energizing for the nervous system. We will need a variety of physical outlets as well to syphon off this energy (even Gemini can't sit still for long without getting fidgety). Much depends on the aspect being made: a square can be unnervingly intense, while a sextile is refreshingly stimulating. We can find ourselves mentally restless, but not in the adventuresome manner typical of Jupiter, because we don't necessarily need to relocate elsewhere geographically in order to feel freer. We can find satisfaction exploring our immediate environment. Thus, wanderlust is usually not an issue (unless Uranus is also passing through our Fourth or our Ninth). We probably are ready for many elements of our private world to change. Yet our inner ideal of a better life-script may conflict with mundane outer realities, and that can create varying degrees of unrest and agitation. The thought of being stuck in certain life patterns can aggravate us, but this could also fortify our will to alter our scenario for the better.

Some of us may become growingly distracted, unable to concentrate on the routine matters of the daily world. Boredom with the ordinary stabilities we've grown accustomed to starts to take over; one initial reaction to this can be irritability (as rebellion

hits the mental level) or even absent-mindedness (our unconscious way of not having to deal with trivial details). Others may
note that we are not paying attention to things like we used to;
we even appear to be indifferent to certain responsibilities and
commitments (less so with the sextile and trine). Our mind acts
as if it's far, far away from the here-and-now. Indeed, we may
even become suddenly disengaged in those repetitive facets of our
life that normally define our stability; we may begin to show an
"off-again, on-again" interest in such everyday matters, a take-it-
or-leave-it approach. Perhaps we are getting fed up with these
tiresome day-to-day structures and want to bail out from such
commitments.

INSIGHT FULL

We may wonder what's wrong with us—why the sudden yo-yo attitude and why the contradictory inner dialogues? Why do we have
a sense of undefined urgency to overhaul those things that now
seem to slow us down? We'll need to give ourselves permission to
feel a bit flaky for a while, until we've become more adjusted to
this transit. We need to catch our breath and get a clearer perspective on what we are trying to realize about ourselves and our
developing needs. Such insights will come in spurts, usually as
interruptions to our normal train of thought. They can be vivid
reminders of how we're only as trapped as our mind allows. Action
follows thought, and if we wish to significantly change things,
we'll need to open our mind to all future possibilities—and then
we'll need the guts to act on our hunches.

Uranus encourages Mercury to become less analytical, less
logic-bound, and more willing to use intuition and spontaneous
association as ways to input data and to learn new things (or to
perceive old things in new ways). But don't expect overnight
enlightenment. We have long been socially conditioned to place a
high value on Mercury's way of knowledge-gathering with its sensible, linear approach to the organization of thought. Mercury—a
natural list maker—likes things in orderly sequence, although
Uranus doesn't believe anything has to be so systematic to be

understood, nor should we have to work that hard to get the data we want. Also, the way that Uranus free-associates may not make sense to air Mercury, who wants to be clear and precise about the way two or more things are linked together.

We will still need to do our air Mercury process during this transit, except that Uranus will now and then alert us to smarter short-cuts and cleverly inventive techniques that can really speed things up. Airy Mercury thought it was already fast enough, but apparently not. However, Mercury should appreciate extra speed if it's natally in fire, air, or in cardinal signs (most so Aries, least so Cancer). In earth, water, or fixed signs (except for Aquarius), our Mercury doesn't always enjoy going at such a furious pace. Pent-up inner tension and mistakes can result. There also is something about being on a mental roller coaster ride, turning quick corners at sharp angles, that we probably find jarring!

GOTTA MOVE

Nervous energy is more noticeable during this transit, and we will need a variety of invigorating outlets to use it up effectively. Needlepoint or potting plants may be too sedate and too quiet, something better undertaken during a more tranquil, meditative Neptune/Mercury transit. However, not all avenues of expression have to be intellectual and verbally-oriented. Thrilling and daring physical activity (ski surfing anyone?) normally does the trick. Riding on a bucking bronco, rodeo-style, could do it if we were so inclined (Uranus enjoys jerky movement and doesn't mind the danger element). Whether our mind or our body is in motion, what we need is something challenging and uncommon to preoccupy ourselves. Nonetheless, it's important that we learn to be discriminating in what we choose to do, because we're tempted to take unneeded risks. We could also make up our own rules as we go along, only to find that we get in trouble for not following critical details or paying attention to standard procedures—whoops!

Although we may impulsively decide to forego acknowledging certain regulations, especially when dull Saturnians are in charge of operations, we may not land in the same degree of hot water as we would under a Uranus/Mars transit; that's because air Mercury intrinsically is not the willful, confrontational planet that Mars is. Rather than become loud and combative, this facet of Mercury will quickly use Uranus' brilliance to try to talk itself out of a jam. This may not always work, however, because this transit is known for unpredictable outcomes. We may not get fired for being a rule-breaker, but a supervisor could give us a good tongue-lashing (which is a Mercury manifestation). Let's not be the ones to lose it and explosively bark back at the boss, unless a new job is already lined up and waiting.

ONE TIME SHOT?

It is important to realize that any Uranus transit can feel like a "chance of a lifetime" to us. It can seem that Uranus is presenting an opportunity that may not come around again, which is why we typically feel we just cannot put things on hold any longer—especially if we are undergoing the urgency of the square or the opposition. If we are very creative types, Uranus contacting Mercury could spark a major brainstorm, a new, breakthrough concept that might excite and enlighten the public.

It's as if this new trend or progressive direction is waiting in the ethers for an independent and original thinker to reach out and grab it. This suggests that if we don't give form to this energy and do something with it, someone who's more fired up will—maybe with spectacular results. Nobody cares later if we claim, "But I had that idea first!" Procrastinators and deadbeats always lose with Uranus. Should we even hesitate because of self-doubt, the rewards of following Uranian flashes of insight are bestowed on somebody else who will dynamically put them into motion. We need to have the gumption to follow through *now* with our futuristic ideas and make that daring leap into tomorrow. Uranus doesn't benefit those who always want to play it safe, waiting around for the "perfect" time to take action. Besides, does a truly perfect moment even exist?

INTO THE UNKNOWN

Once it aspects a planet, Uranus will not aspect that *same* planet again in the same angle formation while transiting through that *same* natal house for approximately eighty-four years. Some of us may unconsciously sense that this is a rare moment, which is why we feel a pressing need to make decisive, turning-point changes (not because we feel totally self-assured). We probably can feel uncertain up to the very last minute before we make our decision and move on it. Uranus carries with it that nail-biting sense of the unknown, producing a little anxiety until pivotal missing elements suddenly come together and impel us to act decisively.

Inwardly, we may know that once we start, things will never be the same. Are we completely ready for that potentially destabilizing experience, asks our Mercury? Air Mercury enthusiastically answers, "Yes, siree!" It then adds that we might want to check with old worrywart Saturn first to make sure we've got our facts straight and that we're well aware of any pitfalls. How much we let Saturn dominate our consciousness can determine the degree of faith that we have in our Uranian opportunities at hand. Saturn doesn't care much for being uprooted and would probably give our seemingly rash plans a thumbs down. Well, that's just *too bad,* Saturn—this is *not* your transit. Mercury has the last word, at least in theory. Still, it's not wise to ignore Saturn's "look before you leap" policy.

Often, should we resist doing *anything* because of fears and insecurities, external changes will be forced on us (with the blessings of our Higher Self) in order to liberate us in the long run from stifling attitudes. Uranus wants to make us look at things differently and more courageously. We may think some matters are beyond our immediate control, and maybe they are. That's usually because the kind of control we would try to enforce, based on our limited awareness, could botch up this planet's bigger and more ingenious plans. What we'd be willing to settle for seems a bit dopey and boring to Uranus compared to what we really could have instead. Mercury likes to make plans, but Uranus seldom will abide by orderly step-by-step procedures. The anticipated timing of an outcome is thrown off, usually in a unexpected manner that benefits us. By taking advantage of Saturnian delays, Uranus has more time to reinvent its scenario for the better. Making broad but tentative plans that allow for a few alternative routes seems the smart thing to do during this transit.

TALKING TRUTH

With Uranus transiting Mercury, we may finally have a chance to think for ourselves regarding a wide range of topics. Ideas seem to come at us from nowhere, but they also move us to sit up and pay attention. Some of our thoughts seem to be crystal clear and full of certainty. We could discover that we have very strong, if unpopular, opinions. Airy Mercury is sampling a little of Uranus' truth serum at this time. One result could be that we don't wish to further engage in meaningless chit-chat, to say just what's expected of us, to communicate merely to keep social interactions smooth and harmonious, or to say what's politically correct. Instead, some of us may want to let our words fly and say exactly how we feel, shocking others with a few zingers just to keep everyone on their toes! However, the results could prove disastrous if we're freely mouthing off to the wrong people. Neither Uranus nor air Mercury are well-qualified to read others emotionally, which is very important when we're trying to get our point across to a potentially sympathetic audience. Our not-so-carefully-chosen words could backfire.

Some folks will find us moody (although that's not quite what's going on), or perhaps erratic (they may have a point here), or at least contradictory (now that sounds right on target)! We indeed can be very changeable while, at the same time, appearing to be intensely sold on our ideas of the moment, especially if ours is a fixed Mercury. The winds blow in all directions at the same time. It becomes harder to determine if we have any real mental focus, since we can operate in fits and starts. This could make us appear merely unreliable to others, even if it's solely due to unexpected events.

With the transiting sextile or trine, we have an extra amount of adaptability working for us, partially due to a more cooperative environment that facilitates our changes. That alone can make us less mentally willful or defiant, as well as more able to make the best of any unstable situation. Should we decide to alter plans in midstream because of unexpected last minute factors coming on the scene, such changes often amazingly benefit all parties involved (leaving everyone satisfied). It's a beautiful thing to watch the workings of an unfettered Uranus creating such synchronicities. People born with a Uranus/Mercury trine or sextile are probably used to this phenomenon by now. Paying attention to these transiting aspects in full swing and observing how Uranus works can be an educational eye-opener.

LIFE'S INTERRUPTIONS

Because Uranus and air Mercury can be fickle influences, most so when they join forces, it's best not to firm up long-term commitments before we've allowed ourselves sufficient time to consider the pros and cons. We're more unstructured mentally than perhaps we realize. Others could suffer annoying disruptions due to destabilized thinking on our part, and vice versa. Particularly with the opposition, we may attract people whose behavior is undependable, interrupting careful plans and wasting our time and resources. Something we weren't even thinking about can slip us up, and it might be due to another's restlessness; he or she could abruptly pull out of a mutual agreement at a most inconvenient moment, and leave us little time to get our plans back on track. Yet Uranus works well under such crazy pressure, and we surprise ourselves with our brainstorming abilities, or perhaps we benefit from the brilliance of those we attract. Still, nothing feels solidly in place for a while.

However this plays out, it's good to read the small print before signing contracts during Uranus periods. We can't afford to leave even minor details unattended if our project involves precise coordination. We need to make sure that nothing is too binding and unalterable. An escape clause is always a handy thing to have. Uranus is much happier with flexible negotiations, because it knows we will have an itch to modify things in the near future. It's quite a reviser of plans. Only the hopelessly stubborn or chronically pigheaded bring out the worst of these transits.

STUDY MANIA

It is typical during this transit that we stumble upon a body of knowledge that fascinates us. Let's say it's astrology. Our inclination is to want to learn as much as possible as soon as possible. We could buy plenty of books recommended by others and plunge into our new study with great gusto. Usually, what we learn is absorbed more quickly and easily than expected. The air side of our Mercury is excited by what it's getting into. We are able to keep up the momentum for a while, especially when we attract others with the same interest.

However, at some point, and without much warning, Uranus has had enough of putting its laser beam focus on just one topic; it's ready to move on. Unless our Mercury is in a fixed sign or in a natal aspect to Saturn, we'll probably lose interest in our new studies as quickly as we began to pursue them. We therefore shouldn't try to buy too many books or sign up for too many courses during this time, because we are in a manic stage regarding knowledge-gathering that may not be easily sustained. Other interrupting life conditions could arise and force us to put our studies aside for a while. When we return to them, the thrill is gone. We can't get as motivated to learn as intensely as we were before. So, let's not make this venture too costly. Buy one or two good books on the subject and see if we are truly as interested as we think we are before going hog-wild.

CHAPTER NINE

URANUS TRANSITING THE FOURTH HOUSE

ANGULAR SUPERIORITY?

Uranus transiting our Fourth House takes on a bit more importance because our Fourth is also an angular house (like the First, Seventh, and Tenth). Planets—especially slow-moving ones—crossing over our natal angles surface their energies more readily and expose their intentions to the outer world for better or worse. This has given the false impression that their influence is therefore stronger because of their greater visibility. I discounted that theory early in my studies—even before reading that Michel and Francoise Gauquelin's exhaustive research determined that planets in "cadent" houses (the Third, Sixth, Ninth, and Twelfth—all traditionally deemed the least potent areas) were nonetheless quite vital in determining outstanding success in certain careers according to the nature of the planets posited therein.[1] (I was born with Uranus in my Third and, during my preliminary studies, didn't appreciate reading that this planet was in a "weak" zone in my chart. It's not a planet that feels weak to me—wacky, maybe, but certainly not weak!)

Each cadent house has something to say about the workings of the human mind. Probably during the time that this concept of cadent house "weakness" came into being, ideas—especially those bombshell ideas that went against teachings of the Church and the authority of the Crown—were allowed fewer channels of open expression. The mind's potential was more suppressed, few people were allowed formal education, and the printed word was less accessible. I would bet that this contributed to astrologers' dim view of cadent natal houses in those days. Hopefully, by the next century, we'll finally get off this "weaker" versus "stronger" kick regarding any factor in our birth chart. This terminology can be highly misleading when analyzing a real person's chart; the same goes for using the biased and inaccurate terms "good" and "bad" regarding aspects.

SHELL BREAKING

Uranus is here passing through a house known for its tight emotional hold on potent security symbols—our family, our home, heirlooms, or anything that makes us nostalgically feel attached to the roots of our past; it describes our heritage on all levels. The Fourth pushes for close and enduring connections that support long-lasting stability and protection. Well, Uranus certainly has its work cut out for itself! Bonding Fourth-House style can feel more like *bondage* to this planet. Before this transit's over, there will be some hard shells to break and maybe a few tears to shed. We've kept a lot of emotional memories internalized that now need dramatic release.

Yet Uranus refuses to support much of the pitiful whining that can go on in this house—not to mention any regression technique we may seek to help us dig into the dirt of our childhood years. Parts of us can refuse to grow up in our Fourth, and yet Uranus insists that we free ourselves of any and all infantile needs that, unfortunately, still pack a punch in our psyche—especially if we've already gone through our transiting Uranus square Uranus cycle (around age twenty-one). However, impatient Uranus has little interest in delicately exploring every detail and nuance of early traumas suffered. That's Neptune's territory. Neptune, feeling no need to rush the awareness process, doesn't care how long it takes to resolve such painful matters; it's not paying attention to time anyway.

THE SKY'S FALLING?

Just because Uranus conjuncts our Fourth House cusp (the "astrological Nadir"),[2] it doesn't mean everything here starts to come apart immediately. The sky isn't falling...yet. The Nadir's degree itself can act as a trigger point. For such a so-called "up-front" angular house, the Fourth can be quite indrawn and subjective. It would seem that Uranus intends to shake us at our roots and show us in clear terms the limits of seeking complete security through others, no matter how much we need them or depend on them to need us, too. Uranus figures that nobody should hold that kind of emotional power over us. Saturn going through the Fourth usually comes up with similar conclusions, but we react to this awareness with less volatility. In the course of the next several years, we will find ourselves pulling back from such dependencies, sometimes very unexpectedly, even defiantly so.

THE GYPSY IN US

Circumstantially, a growing restlessness deep within can manifest as an urge to relocate, to get away from our current home base and try out previously unexplored places to reside—not necessarily for logical reasons. Some of us are looking, even if unconsciously, for any location that will remind us little of where we presently live. We are growing tired of the sameness of our surroundings. We just want to pack it up and move it on out. It's not something we may even be willing to mull over for a while. Short-term escape urges now override any long-term security needs, with unpredictable results; this goes against the grain of typical Fourth House motivations. How uncomfortable we are with our "inner" home is often determined by how ill at ease we are with our current domestic conditions. The impulse to transplant ourselves elsewhere is a sign that old security patterns are no longer working for us. Feeling trapped in our current home and being eager to uproot ourselves is an example of this.

Often, we have not given ourselves adequate time to do our homework and plan our getaway (particularly those impatient Martian types among us). Therefore, we could wind up in a locality that soon becomes too much of an environmental shock. Unpredictability strikes again. We probably don't even get to visit the new area we are relocating to first, at least not for long. Our move is based on a strong hunch that this is the right place, for now, to resettle. While we probably will learn to "make do" in this new place, we may not quite feel at home with our living arrangement. We might even regret the fact that we can't fall back on those former comforts and familiarities of our previous residence that we took for granted.

Of course, the transiting trine or sextile imply that we may fit in different surroundings more quickly, probably because we are not as much at odds with ourselves regarding security issues. Uranus is ready for us to enjoy anything and everything unfamiliar. However, with more frictional aspects, we may discover that moving far away does nothing to relieve our inner dissatisfaction—this could come as a big, perplexing surprise! The turmoil we may feel within does not disappear just because now we have lovely palm trees with coconuts swaying in our backyard.

All Fourth-House matters reflect sensitive, feeling-charged issues that often have to be approached delicately. We wouldn't enjoy some freak tornado sweeping through, swooping us up skyward and then dumping our psyche in the strange land of Oz. So why intentionally stir things up prematurely that will strongly impact our feelings? Of course, Uranus insists on its own unusual agenda. Maybe because of our tendency to suppress Fourth-House energy due to our insecurities (it's the house of late-bloomers), Uranus operating from our unconscious depths feels especially eager to activate a few drastic inner and outer changes. It might be better to first take a pragmatic look at our options.

DESIGNER HOME

Do we really need to live 3,000 miles away just to bring in that new Uranian energy, or should we first attempt to remodel parts of our home (along with elements our life)? It's not an easy call, since do-it-yourself home repairs can be a nightmare under the wrong Uranus transits (i.e., Uranus squaring or opposing the Moon or our Fourth House ruler). Even the most carefully laid plans would undergo interruptions left and right (heck, *especially* those). It would be better to call in the experts, because a dose of sensible Saturn usually helps Uranus work out better. It is common with Uranus in this house to have an interest in doing a few very individualized changes to the structure of our home. That alone suggests needed professional feedback from those with plenty of experience.

If everything turns out fine, we could end up with the most unusual-looking house on the block (which is also the case should everything turn out horribly). Besides, moving thousands of miles away or even replanting ourselves in a new country means we might have to let go of many of our possessions in order to lighten the load. Yet under a Uranus transit, we could be too quick to get rid of too much, only to regret our hastiness later. I would think this deserves thoughtful planning (of course, my Moon in Taurus—a sign easily attached to possessions—rules my Fourth House cusp).

Only you will intimately know the details of your circumstances and what has led you to this "once in a lifetime" crossroads. Remember, even the best decisions under a Uranus transit don't always seem like they make a lot of sense at the time. In general, we need to rearrange the furniture in our "inner home" even if we do not relocate anywhere. This transit will last a while, giving us ample time to experiment with letting go of unworkable attitudes toward what we emotionally cling to. We can embrace new versions of security that free us inside. We learn to be gutsy, and it can feel good. From this point on, we become less rooted to things of our past. The Fourth represents how we handle our "old age" (the closing years of our life); therefore, Uranian insights gained now can even prove very valuable later during our actual retirement years.

Maternal Freedom

The Fourth House describes mothering themes as well as the family dynamics we were exposed to while growing up, particularly during our tender years. It's the house that depicts whomever symbolizes for us a protective matrix, specifically that parent more likely to envelop us in loving care. This is not true of all moms—what about mothers who are crack addicts, mentally ill, or child abusers? Also, more dads are trying out "Mr. Mom" roles than they ever have in the past. Fathers with a strong maternal streak are not as rare as they once were. Still, the Fourth is most likely our mother, or, more precisely, our *impression* of our mother and the qualities we draw out of her as a result of our personal security needs. However, we should remain open to the possibility that our Fourth House actually depicts the more emotionally nurturing parent of either gender, plus our response to such comforting treatment is also implied.

Whatever the case, Uranus passing through this house marks a time when clarifying our dependency issues becomes critical. How emotionally connected have we become to our mom (or dad), and is this connection unhealthy at this point for either of us? How much does our family depend on us for psychological support, and are we finding this burdensome? What parts of this relationship make us feel trapped or uncomfortably obligated? Wherever Uranus is, we can become claustrophobic when unable to have the wide, open spaces we need to breathe. Too much closeness becomes problematic over time.

Uranus helps us to become more objective about our situation. There is an urge at this time to pull away from too much family commitment. If this urge remains unconscious, it gets projected onto our mom or perhaps the family at large. Disruptions can occur. Sudden and unanticipated conditions can alter the rhythms of our usual emotional patterns, catching us off guard. Uranus implies that there could be a parent in a state of agitation, restlessness, defiance, or internal chaos. It can also suggest a parent ready for an inner awakening, a breakthrough in self-awareness, or a strong urge for freer independent expression (such as a mom who goes on strike and refuses to further cater to the needs of other family members, suggesting a mix of rebellion and liberation). All of this becomes symbolic of how we subconsciously feel regarding ourselves.

Because Uranus stays in this house for many years, changes may not be all that drastic. Our mother could start to seek freedom in small ways in the beginning, leading eventually to more courageous efforts on her part to become her true self. Her unfoldment can be delightful to watch if she has formerly been inhibited or suppressed. Maybe Uranus hints at a change in mom's marital status that could be self-affirming. Certain limitations are removed. Her ability to strike out on her own in some way can surprise us. We'll have to back off and not throw our unresolved insecurities onto her while she's in this special transition state (although her interactions with us will change in ways that can unsettle us, at least in the beginning). A truth we may have to accept is that our mother needs to separate from us and from anyone else who has made constant emotional demands on her. This is now her turn to find out what she's all about, on her own terms. We can say bye-bye to those limiting labels that have neatly described her and her role in the past.

If we have been blinded by the depth and power of our own attachment to this parent, meaning we still cling to a maturity-defying part of ourselves, Uranus is ready to shake the very foundation of our relationship, if need be. Life will remove mom's influence from us in some way, by force in some cases, making us feel suddenly abandoned. Our reactions to this can be very chaotic and subject to rage (the fury of the infant denied); or we could turn cold and become emotionally distant. Gaining a clearer perspective is healthy at this time, but being remote and out of touch with our feelings is not a good sign. Still, it's hard to tell anyone how they are supposed to behave during this period. We can be filled with psychological insight one day, and be shaken up at our roots the next.

FLASHING ON FEELINGS

The Fourth is also where we get a chance to develop our own maternal energies. Uranus can here awaken hidden strengths involving our ability to nurture and nourish ourselves and, eventually, others. Although Uranus itself has no interest in tenderly mothering anything or anyone, it is very concerned with releasing any human potential that we otherwise hold back due to fear, doubt, insecurity, or ignorance. Uranus sees Fourth–House issues as challenges that allow us to have remarkable insights about people's emotional subtleties. We can learn to read others better than we usually do. Our instincts prompt us to respond quickly to unspoken needs. We can't remain unaffected by our environment for too long. Therefore, somewhat paradoxically, Uranus triggers human warmth and caring if we have been too cool and uninvolved in the past, but it also does whatever it takes to thwart or turn off our hungry feelings of neediness if we have failed to develop a sufficient degree of separate identity. Uranus spurs flashes of insight into the nature of our deeper feelings in order to make necessary improvements. As a result, we learn to interact in closely bonded relationships better than we have in a long time.

URANUS/MOON TRANSITS

ROOM TO BREATHE

Uranus aspecting our natal Moon heralds a time of needed, honest self-encounter regarding how much we either feel or do not feel at home with ourselves and others. Uranus is no empathetic hand-holder (that's Neptune), so we may experience a few rude awakenings regarding our existing emotional attachments (of course, with the sextile or trine, emotional insights could prove clarifying and even exhilarating). One realization that may result is that sufficient detachment is imperative in order to discover how to best "mother" ourselves at this point in time. We don't need total detachment from our dependencies—just more breathing space from others than we've allowed for in the past.

We also need to anchor ourselves in a reality that allows for the freedom of greater self-sufficiency. That should not preclude close, warm human involvement of a healthy nature, but it would suggest that we stop latching on to people we think we need (or those who need us) for all the wrong reasons. The Moon reflects habitual, subjective responses that are hard to break, especially involving people we subconsciously attract for security purposes. Such folks reinforce our emotional expectations, for better or worse. Uranus wants this process to become more conscious than it usually is. However, it's doubtful whether it will ever become completely conscious. Then it would no longer resemble a true lunar experience, which often requires us to go inward to reach those parts of ourselves that defy intellectual analysis or articulation. That doesn't mean we can't get a better understanding of at least some aspects of how our Moon's dynamics work. Uranus will always try to shed light wherever it can, particularly where it is most needed.

FEELING CRAZED?

Uranus is famous for its reversals, so those of us who are already a tad too remote and emotionally distant may find, during this transit, that Uranus smashes open a dam of buried feelings. We've already been quite good at detachment in relationships. This transit becomes a time to experiment with intimacy. Uranus won't stick around for the long-term results, but works as a catalyst during this transit. Uranus' patterns have a touch-and-go quality; there are no guarantees that its varied human experiments will prove successful. Still, it's worth a try to uncover our capacity to feel deeply for others so that we willingly seek closer human involvement.

A lot of powerful feelings we never suspected we had can come gushing forth, surprising us and others. We have to watch out for that transiting square, quincunx, or opposition! Each can describe a temporary period when we feel crazed by our mood swings, at least until we've had enough time to subjectively adjust to Uranus' newer and more unusual rhythms. We can appear hyper and high strung, or have ups and downs that unnerve us and sometimes agitate others. Emotional outbursts can occur, although some of this can help clear the air with those we feel have stifled us. Uranus is trying to get us to push off people's oppressive influences and whatever else has weighed us down in our relationships. Our Moon could use a little creative instability at this time to help us shake off habitual ways of responding in close partnerships.

TURNED UPSIDE-DOWN

To effectively use this transit, emotional objectivity is a must (and that's no easy feat). Maybe we realize that if we are to make any sense of our current life situation, we will have to address exactly how independent and self-reliant we really may be deep down inside. Few may appreciate what we are internally going through at this time. We alone are to discover what we are about at the roots of our less-conscious being, as we go through possible periods of fitful and turbulent self-revelation. We've been harboring inner self-impressions for a long time that we adopted due to early family conditioning. Some of these are what we needed for our growth and still need; but other imprints from childhood have turned out to be unnecessarily heavy weight.

This becomes a time to turn our emotions upside-down, to shake things up and see what falls out. Whatever hits the ground no longer needs to be kept. The feelings that do remain intact are those natural to our being rather than programmed by others. They don't need to be uprooted and changed, but simply recognized for the strengths they provide us. Uranus is trying to get our emotions to support and feel comforted by a greater reliance on ourselves for true security.

JUST SAY NO

Uranus, a loner at heart, advocates our singular self-focus as a less complicated way to explore ourselves without the inhibiting influence of others intruding. However, this does not have to be a lonely experience; we can still have people around us, as long as our links are not too emotionally sticky. If we have a lot of water or air in our chart, we might not like feeling temporarily cut off from the familiar impact of others, even though we are excited about redirecting attention back to ourselves and our newly discovered, subjective awarenesses. Social isolation is not the intention of this transit, but neither is giving in to the manipulative neediness of others. We are learning how to say no to people's demands, maybe while also showing them how they can independently fulfill their own needs.

Fire and earth are more self-reliant to begin with, and may not encourage a lot of heavy-duty emotional interchange at this time. Freedom from the influence of others could be quite welcome. Finally, we're taking a breather from giving attention to our committed relationships. Of course, we won't feel like this every day during this transit, because Uranus is sporadic in its operation. We still will have moments of contentment in our partnerships, but usually not for long. Freedom calls, and that can get in the way of our or someone else's beautiful dreams of undisturbed intimacy. Uranus is a particularly restless energy when blended with the changeable Moon.

NOT SO DEEP

This is not a deep or quietly introspective "going within" period, unlike Neptune aspecting our Moon. Uranus itself is not that reflective and it doesn't really ponder experience; it doesn't slowly mull things over. Instead, this can be quite an active and vigorous time, although occasionally erratic and nerve-racking: we're up, we're down, we're feeling settled one minute and displaced the next. Emotional sparks fly every which way, yet the overall result is externalizing and enlivening. Life may even seem faster than usual to us, with its day-to-day tempo quickening. Still, we may feel eager to go at this peppier pace, avoiding or abandoning whatever seems to drag on; our instincts are telling us to keep the excitement level up and to bypass the boredom of routine whenever possible.

When we stubbornly refuse to make needed inner alterations and instead hang on to our worn-out security blankets, we merely goad Uranus to turn on its higher voltage (that special juice it normally reserves for those electrified experiences that catapult major changes). One result is that life suddenly yanks the controls from us (as if disgusted with our predictable responses), and sets off dramatic, attention-grabbing fireworks. Typical target areas become those "people" attachments we have invested with a lot of emotional energy, seemingly forever. A few of our unions may threaten to break down. Interruptions occur to alter our accustomed patterns. This could prove exciting or threatening to someone in the relationship. No matter what the transiting aspect is, we'd do better not to try to keep everything in its habitual place. Why tempt Uranus to jack up the voltage? Here is one time when just going with the electrical flow makes the most sense, although this may not be apparent at the moment.

MIXED FEELINGS

The Moon, one of our parental planets, usually tells us plenty about how we relate to nourishing, maternal energy. Perhaps some of the symbolism of this transit gets projected onto our actual mother (or maybe father, if he's the only parent left in our life). Observing our mom and the changes she may be going through could offer us a few indirect clues about our own internal challenges (maybe *we* are playing out a "mom" role with someone else). What we cannot afford to do now is to get too caught up in her or any other family member's dilemma, because this goes against the purpose of this transit (it's a time of emotional detachment enabling a broader self-perspective). The same advice would also apply to Uranus going through our Fourth. We need to reconsider our nurturing urges. Some of us may flat out go on strike and refuse to cater to others, especially those who have been expert emotional manipulators in the past.

Still, we could feel torn between wanting to involve ourselves in parenting a parent or other loved one in turmoil and unrest (the result of our old lunar conditioning) versus wanting to pull back or just get away from it all and let others fend for themselves (due to our newly emerging autonomous urges). If we make the wrong moves at this juncture (due to nagging feelings of Saturnian guilt), other people's lives, with their crazy, disruptive issues, can send ours into an uproar. Our feelings become agitated and our normal rhythms upset. We can get ourselves off track psychologically and suffer a major developmental detour. Any freshly formulated goals could then be put on hold or even aborted, while everything in our life feels chaotically up in the air. It can happen so suddenly. It's good to think this out before becoming embroiled in any interpersonal involvement that we already have mixed feelings about on a gut level. We can't afford to ignore our inner voice now.

A BIT FLAKY

Uranus contacting our Moon is a time when we find ourselves getting out of emotional slumps or black moods, because something in our world is now invigorating our feelings. Our emotional batteries are having a good recharge. We may instinctively magnetize events and people that put us in touch with novelty and originality. Our world can look offbeat to us, but in a satisfying way. Creative unpredictability is part of the atmosphere we attract. We are not to securely ground ourselves at this time and make firm commitments or promises that we are in no condition to keep. We are allowed to be fickle and a bit flaky in our moods and impulses. We should just enjoy the freshness of experiences that come our way. Obviously, only very short-terms obligations involving others will do (and we do them because them seem like fun to us, like something different).

RESTLESS HEART

Home-improvement projects may appeal to us at this time and are reflective of our internal renovations. We are trying to give birth to an ideal form of shelter. However, we'll have to make sure that we don't try to do very complex jobs by ourselves that take a lot of time or many separate steps. That's because we could lose steam and fizzle out in the middle of our undertakings. Once the thrill is gone, we might want to disconnect from the project entirely. Things could come to an abrupt halt. Still, if we can pace ourselves and try not to do too much too soon (no urgent deadlines, or five things going on at once), we can replace an old, tired look we've lived with for a long time with one that shows touches of innovation and individuality.

Of course, wanting to leave our home and instead try out a different living environment could be another option (especially with the transiting square, the quincunx, or the opposition). Much of what was said about Uranus going through our Fourth House also applies here. Our restlessness is driven by a feeling of being trapped, and that alone can push our panic button. We can perceive our condition as terminal unless we do something drastic to enforce change. Practicality doesn't have to fit into the picture, but immediate relief does. We can suddenly transplant ourselves in order to achieve quick gratification, only to find out later that a little sensible planning would have made our move much smoother and more satisfying. In addition, we might bring any inner emotional turmoil with us to the new place, so that our unsettled situation continues. It's probably wise to stay put until we have had a thorough inner dialogue about our real needs. Otherwise, we're just running away from unresolved parts of ourselves—doing that can be a big waste of energy, time, and maybe money.

SUPPORTING OURSELVES

By the time this transit is over, we may learn to avoid relationships that are built on too much dependency (especially the one-sided kind). Uranus has taught us not to get sucked into the lives of needy people. We still can get close and become warm in our intimacies, but now we may have developed the internal resources needed to alert us when to back off and create or demand healthy space for ourselves. We may feel more secure, knowing that we won't always be as ready to support others while depriving ourselves of healthy independence—and we are also learning to not impulsively seek their support as well.

Notes

1   Jamie Binder, *Planets in Work*, ACS Publications, Inc., San Diego, CA, 1988, Chapter Four.

2   A technical note: Steven Forrest uses the term "astrological nadir" in his books (see *The Changing Sky* in the Bibliography list at the end of this book), probably to differentiate it from that point on the celestial sphere directly beneath the observer—also called the Nadir—used in the *horizonal* coordinate system. There, the Nadir is always found opposite the Zenith. In contrast, the astrological Nadir belongs to the *ecliptic* coordinate system and is exactly the same, northernmost point also known as the Imum Coeli, which astrologers refer to as the IC. The astrological Nadir is always opposite the Midheaven and is not to be confused with the other Nadir. Whenever I use the term "Nadir," I'm solely referring to the "astrological" definition.

CHAPTER TEN

URANUS TRANSITING THE FIFTH HOUSE

FREE TO RISK

Uranus should have a lot of fun passing through the Fifth. This house is linked to those recreational options that allow us to let our hair down and literally "re-create" ourselves for the moment. It's a risk-taking zone; of course, venturesome Uranus loves hearing that. Hopefully, we learned to accept emotional risks in the Fourth that have led us to the discovery of greater, inner security. We are now to undergo an extended period in our life during which individualism (a Uranus and Fifth House issue) is given freer reign of expression. This is an opportune time to experiment with our talents and to display ourselves in highly personalized ways. We could surprise ourselves by how and where we unexpectedly shine. Even others are amazed. Our creative drives are in need of a big boost, and Uranus can provide the right voltage of electricity to turn on the bright stage lights.

However, much depends on how comfortable we are about revolving around ourselves, acknowledging that we are the star of our show, and the main act. We're also the director and the producer. Many people do not allow themselves full permission to strut their stuff, as least with the level of confidence required to be convincing. Others may believe it is wrong to be so darn full of unabashed ego. Uranus, however, simply thumbs its nose at party-poopers who would dare squelch the human capacity to become larger-than-life. Now's our time to wow our audience with whatever we've got that's so great, even if the reviews are mixed. Hopefully, if we've emerged from our Fourth House transit feeling like a different person on a new footing with the future, with most of our old insecurities wiped out, we are now ready to explore this new and fertile creative territory.

LIVE WIRES IN LOVE

One area where we can test Uranus' electricity is in our love life. The Seventh House is actually where two people attempt to merge as one committed couple. In the Fifth, we initially are looking for sparks to fly with others who make it easier for us to show off *our* pizzazz; we also want to enjoy *their* special personal assets as well. Some people, however, are perennial deadbeats, dull and lifeless no matter who they're attracted to or what's happening in their Fifth House. Uranus works better with those of us who already are live wires, spirited and eager to further discover more of our colorful potential. The thought of losing or even altering our identity in the process of becoming more intimate with another would be very unappealing at this time. Uranus has absolutely no plans to give up its identity to anybody else, period! During this phase, it's more thrilling to appear uncommonly special to another and even to the world itself.

We will probably find ourselves wanting to call the shots in our romantic affairs—not so much controlling them (that's unlikely to occur) but attempting to steer things in the direction that most stimulates us. We may want to make exciting things happen; we can also attract a dynamic partner who has no trouble doing this for us (if we constructively project our Uranian needs onto our lover). That other person becomes a terrific turn-on, allowing us to feel alive and well, and ready to let our hair down. The electricity crackles between us. Both parties recognize the strong magnetism involved and want to do nothing but encourage it.

However, infatuation can be a pitfall unless we are really clear that we're just "playing the field" and indulging in bold flirtation. Uranus is not prepping us for serious long-term involvement. This can be a problem; we may not realize that we are play-acting at this time, and that we are more dramatically in love with the new, improved version of ourselves than with our suitors. We're not really ready to settle down with someone yet; our lovers may not feel ready for full commitment, either. Therefore, it's best to be up-front about our intentions in these relationships and let the chips fall where they may. Hopefully, we learned the value of such open feelings when Uranus was transiting our Fourth. Our lover may decide that we're much too intense, and thus may get nervous and suddenly abort the affair. Although such a quick ending can jolt us, we need to realize we are not to become too attached too soon to any love partner, and that perhaps we need to take periodic intermissions in the pursuit of romance now and then. Other creative areas of our Fifth also need to be explored.

HANKY PANKY

Yet what if we are already married? We'll need to infuse our committed partnership with a measure of romance and adventure to keep things from growing stale. If our marriage is on the rocks, we might give ourselves license to graze in stranger pastures. Any new person we pick to relieve us of our marital monotony—usually it's someone who's quite emotionally unavailable in the long run—could turn out to be an individual who can trigger both our powerful erotic impulses and our emotional volatility. We can feel thrilled and yet agitated about this, which is one of those typically paradoxical Uranian reactions. Sometimes Uranus acts like an edgier, keyed-up Mars—the energy it arouses becomes aggravating or tempestuous.

The trouble is that we may forget that we're only in this extra-marital affair for the exciting rush of experiencing somebody new who knows little about our psychological baggage (unless we are the blabbermouth types who advocate full disclosure of ourselves as the best policy). We can plunge headlong and headstrong into this affair without really knowing whom we are dealing with. All we know is that we have a lot of passion to burn for the right stranger. However, it's typically a short-lived passion. We or our lover can eventually end up feeling shell-shocked after one of us—becoming less emotionally accessible—pulls away or becomes harder to reach. We may get anxious when our phone messages are not promptly returned, or when our lover claims to have to go out of town on "business" a lot. We may also allege that neither of us had planned to get this much emotionally involved in the affair; things "just happened!" Actually, very little in our Fifth "just happens" romantically to us unless it's going to put us smack-dab in the center of special attention that we normally can't get elsewhere. The feeling is anything but casual on our part.

Those of us who are married or single and are attracted to someone who is still married should start preparing for what may come. Uranus does not engage in steady emotional commitment, perhaps especially so in torrid affairs. It prefers that heady whirlwind kind of stuff, with nothing too predictable. So tread carefully here, and avoid trying to emotionally manipulate someone into becoming what they really are not (after all, this *ain't* the Eighth House). Things here tend to backfire when we operate under ulterior motives and show false fronts. We cannot get away with trying to seductively rope someone into a sexual liaison that we hope will later provide us stability. Sometimes we are the one to suddenly bail out of this crazy up-and-down relationship, even if going back to our spouse does not seem to be a real solution. It is better for us to face our marital issues squarely to determine the cure for our emotional restlessness. If this marriage is not to work out, we at least can sidetrack a chaotic ending by first getting divorced; a third party only emotionally complicates matters at this time.

CRAZY MIXED UP KIDS?

We are alive and well with Uranus even when things get intense. Drama is part of the Fifth House package. Anything bland is to be avoided. We may witness much of Uranus operating through our children (a very common projection), some of whom may now start acting out of character. Rebelliousness on their part tends to symbolize hidden unrest on our part. Their open agitation represents our inner, unresolved friction (maybe involving our dilemma concerning creative channeling). Our kids may break a few rules here and there, and even get in trouble with authority (which includes us as parents). However, we'll also need to examine our *own* desire to have things our way and to not allow for the interference of others to change our plans. Maybe we're suppressing the need for a tantrum, especially if our lifestyle is colorless and no one's noticing who we are. Our child may be mirroring our "inner child," who is ready to break out on some level and get its need for attention met.

Rather than suffer a stunning clash of wills, we had best pull back and give our offspring needed space (basically because *we* require a break from parenting). Unless our children are getting into deep trouble due to severe internal turmoil and outer social defiance, we can afford to let them self-discover on their own terms, just as we're trying to do. Otherwise, we'll alienate them even further at this time. The key is to not give them so much space (Uranus can go to extremes) that our strategy is misread as indifference, lack of caring, or cold aloofness. Remember, some degree of Saturn (the rule-enforcer) helps an otherwise defiant Uranus work better. Our children may fight against our authority as an unconscious way to test out the state of independence they will need to exercise in adulthood; yet they will also require appropriate structures to follow. Otherwise, they can run wild with a favored Uranian group of peers and create much havoc and heartbreak.

Uranus could also suggest that we may witness a child ready to emerge with unsuspected, special talents. It is important to keep in mind that he or she represents a part of ourselves that we'll understand more clearly when we willingly own up to our Uranus experience. What our child may undergo is rapid acceleration on some level. We would likely see our kid at this time as precocious, brilliant, advanced, and even touched by genius. He or she may not look or behave like a child. One trade-off is that this kid could also be too different from other children to be accepted, and thus could be unpopular. He or she could be left out of social activities that require peer conformity, but maybe gratefully so (whereas transiting Saturn here could mean our child feels distressed about being shunned). Our kid may instead become preoccupied with an inventive world of his or her own design and therefore claim to not miss wasting time hanging out at the mall or yakking for hours on the phone to friends about nothing. As long as there is a powerful computer hooked up to the Internet in the house, this kid is probably most happy electronically exploring the world, and maybe is even designing a web page! Still, all of this could make us uneasy.

One last thing about kids: we can have one when least expected if we overlook a few simple protective measures. Uranus doesn't want to feel confined, and birth control devices may seem like obstacles to spontaneity and to pleasure. But in actuality, raising an unwanted child could be a confinement that we'd regret for many years to come (although with Uranus, a surprising change of heart is possible and we could suddenly adore being a parent to this very special little boy or girl). Still, unplanned pregnancies are to be avoided, because Uranus will be in this house for years, suggesting that we'd feel tempted to shirk our parental responsibilities (especially during those demanding and energy-taxing pre-school years). Abortion may be an option, but it's a decision that can catapult some of us into disruptive states of internal or external conflict.

Gambling Fever

What about gambling? Uranus will sometimes take risky chances just because they are risky. However, the Fifth House likes to end up as a winner; its aim is to reap rewards, honors, and whatever else glorifies the ego. First-class treatment is sought. Uranus has a more detached "win some, lose some" attitude. It's a planet that knows and accepts how unpredictable life can get—and why not, since Uranus itself is the main instigator of the unexpected? Should we allow ourselves to invest at this time, it's best to do so with an awareness that the pattern of our speculative ventures will not be a smooth one. However, that can be good in the long run, because detours and turnabouts here can lead to sudden benefits. Our transiting aspects can give us clues about whether it's even worth getting involved in all of this. This can be a nail-biting time for those of us who are rigid, poor-loser types. We'll need to be as adaptable as possible, free of fixed expectations, and willing to not always hit the jackpot. Uranus will make it clear that these matters cannot be controlled by our will.

Ego Boost

Speaking of our will, Uranus is stimulating us to apply more will-power in ways that can raise our self-esteem. Uranus can demolish an over-blown ego, but it also attempts to breathe new life into a deflated or underfed one. We may awaken to ourselves and recognize how important our individuality is during this transit—plus how necessary it is to keep experimenting with further developing that individuality. This means having the guts to stand out as being special in some way, and to be anything but mediocre. False humility is not supported. Uranus can't understand what could possibly be wrong with being in awe of one's surprising talents; such self-excitement is not unlike the exuberance of children who realize that they are spontaneously creative.

FUN HOUSE

One final thing: when as the last time we took on a new hobby? It's probably been a while since we've allowed ourselves quality play time. We now need to loosen up and do whatever feels good, whatever brings us joy, or whatever reclaims the fun part of being young again. We need to find time to go hiking, or play the casinos (without going too wild here), or take up belly-dancing. Maybe we can learn to fly planes or go hot-air ballooning. Whatever our interest, Uranus is in a perfect position to trigger unexpected skills never before displayed. This alone expands our perception of ourselves. Let's pretend that we only live once and take that creative plunge. Life sure looks better when we are free to "do our thing" and feel a little joyous doing it.

What we cannot afford to do during this period is to bore ourselves with an anemic presentation of who we are. Hobbies offer opportunities to explore our unexpressed dimensions. If you don't have a hobby that excites you, get one! Better yet, try to figure out why you've deprived yourself of such sources of pleasurable recreation for so long. Those of us who are without hobbies probably have low expectations of ourselves, and are probably unwilling to test out our talents and demonstrate our skills. Uranus can enlighten us about how much inner enrichment we could receive if we'd just radiate from our center and throw ourselves into an assortment of sparkling activities that can make living our life a true pleasure.

URANUS/SUN TRANSITS

TRUE COLORS

Any transit to our Sun by one of the Outer Planets is considered to be a big deal. For one thing, these aspects are not a dime a dozen. When Uranus, Neptune, and Pluto have a dialogue with our Sun, we have a chance to make meaningful alterations involving our core being—our life-essence (which the Sun symbolizes). This affects our overall, operating consciousness, influencing everything we do. With Uranus, depending on the aspect, we can undergo breakthroughs regarding how we demonstrate our individuality. We can shake up parts of ourselves in ways that enlighten us, even if this disturbs others. Uranian changes are seldom understood and appreciated by all interacting parties involved.

Uranus urges us to emphatically put our unique thumbprint on all that we do, reinforced by the Sun's need to stand apart from the crowd. Self-recognition at this time can lead to unexpected detours on our life-path, surprising both ourselves and others. A sextile or trine may not be the dramatic turning point that the square or the conjunction can be, but even in this instance we can feel the itch to show our true colors to the world by waving our flag of freedom. This is all to be done in an open, conscious, and unambiguous manner. Subtle reflections are not part of this planetary combo, but bolder attempts to grab the world's attention *are* often to be found here, although sometimes inconsistently so.

However, an issue of willfulness may need to be addressed, especially with frictional aspects in operation. Heightened self-will—further fed by a sometimes too self-confident, bossy Sun—can override caution, common sense, and consideration for others. People may find our actions abrasive and jarring, especially if we are at war with these energies internally (a lot of what Uranus stands for threatens the ego-glorifying Sun). However, successful Uranian transits seem to depend on effective, inner, Saturnian development. Since both join forces as co-rulers of enlightened Aquarius, each planet's strengths need to come into play. Saturn lays the groundwork that provides us with built-in safeguards that keep Uranus more manageable, if possible. However, even a mishandled Saturn can sometimes resist the set rules and regulations of outer established authority, although more from a point of resentment than active rebellion.

WAYWARD WIND

When we're out of touch with our Saturn, any Uranus transit can add an extra dose of defiance (even when these two planets are not aspecting one another at the time). Those of us who already lack self-control and maturity usually don't make the best of our Uranus transits. We are too unstructured and undisciplined to harness this planet's energy, so things can quickly go haywire for us (maybe dramatically so, with the Sun involved). We could also be too insecure to accept this transit's push for major repatterning in our lives. However, if we already feel solid and sure of ourselves from within—signs of a healthy solar ego—we are less likely to defy society with our obstinate, wayward, or bizarre behavior. Instead, we are driven to awaken our real self in a manner that can affect the progressive development of the world, even if only in small ways. Uranus can act as a light-bearer, while the Sun helps us to shine and radiate our strengths. Together, they suggest an ability to dispel those darker elements of society that are born of ignorance and of fear.

In a less idealistic manner, this transit stirs an inner urge to function separately from the influence of others. Peer pressure and conventional standards have less of an appeal for us. We can act as if immune to certain types of social programming. We ignore any pressure to behave conventionally if we deem that doing so will be confining. We won't tolerate any circumstances or people around us whose energies feel tight or binding. This becomes a life period when we might bend a few rules in order to do things in original, inventive ways—especially with the trine or sextile. It can feel like a time to have flash insights regarding stimulating new objectives and future goals. How can we free our energies and direct them in conscious, self-affirming ways? Uranus keeps us feeling positive about our future and ready to take charge of our direction. With these aspects, we are less in a state of conflict regarding our Uranian challenges (since most of the challenges presented are seldom imposing or monumental).

Yet with the conjunction, square, quincunx or opposition, our freedom drives can become amplified in ways that create a sudden and seemingly unnecessary change of course. Such a life interruption might at times prove fortuitous (accidental happenings somehow turn out lucky for us). At other times, they can seem like freak accidents. It's that old fickle finger of fate at work again. However, life can become more interesting when we wander off the beaten path. Often, we are too impatient to wait for the best time to make changes (especially with the square, unless fixed signs are involved), or maybe we're too filled with inner discontent to objectively weigh all sides before taking action involving others (a problem with the transiting opposition). Therefore, we may make our moves too boldly and haphazardly—although less so if we were initially born with a Uranus/Sun sextile or trine.

ONE LAST ZAP!

It's always smart to refer to our natal chart first to size up our basic Uranian profile before jumping to rash conclusions regarding this planet's transits. Perhaps shy but very talented people finally break out of their introspective shells to offer the world their stunning potential when Uranus squares their Sun (a time when only the dynamics of the pushy, insistent square can get the ball rolling). Again, a lot of Saturnian inner preparation in the past may now allow us success at this momentous turning-point. In contrast, those of us born with a Uranus/Sun square may feel overstimulated even with a transiting Uranus sextile to our Sun. As inner tension runs higher, impulsiveness can become problematic, leading to changes that are too abrupt or inappropriate for others to accommodate.

Sometimes Uranian aspects appear to come and go, and we don't think we're really getting any mileage out of them. Life still seems to be sadly the same for us. Nothing startling or out of the ordinary happens, at least not when the aspect makes its first exact contact. This is typically due to Uranus' retrograde cycle. Using a 1° orb, any degree of the Zodiac—and thus any planet—will be contacted three times: first going direct, then retrograde and then direct again within a span of almost two years. Things could certainly be different by the time Uranus makes its last "direct" transit to our natal Sun. Perhaps at this time, unexpectedly, our life circumstances change in major, unforeseen ways, and yet we realize we are somehow more prepared to embrace that moment. We may not have been so ready before.

Our newly developed daring at this time is a result of those previous Uranus/Sun transits. We may not have needed to show these traits then, but we do now, almost urgently so. Perhaps we get to apply our fresh insights only when the Cosmos finally gives us the go-ahead. We shouldn't think something's wrong with us if Uranus aspects our Sun a few times and we don't pierce our nose or quit our tedious government job on the spot! Either we were not psychologically ready, or the environment didn't provide strong enough signals to initiate such radical change.

Nevertheless, Uranus throughout this aspect cycle has touched us in ways that are central to our consciousness (the Sun). Any self-insight gained can eventually help us take off in dynamic new directions—and after twenty-one months or so, we should be rarin' to revolutionize at least a few important areas that are personal to us. The first place to look for change is in the house where our Sun is natally found. This is normally a house where primary life lessons take a while to unfold. We usually become more attentive to this house's issues as we get older, as we become more realistic with ourselves. Uranus helps us to discover such authenticity. Although Uranus will not worship the ego, or tolerate its prima donna tendencies, it is acutely aware of our ego's right to creatively exist for purposes other than its accustomed self-focus. It does not share Neptune's interest in ultimately obliterating ego-awareness.

ELECTRIFIED

Actually, the transpersonal benefits of any Outer Planet are not to be had until we recognize the ego's existence, its potency, and its need to evolve along with the rest of our being. In fact, the Outer Planets can wreak havoc on ego-damaged individuals who have little sense of their center and who don't know how to establish an identity separate from others or from the collective unconscious. Such people can become victims of turbulent currents in an impersonal sea of mass-fed energy. The results of such ego-fragmentation can be disastrous, manifesting as varying degrees of mental deterioration.

How can we consciously and constructively work with a Uranus/Sun transit? One way is to make a decision to improve all aspects of our life that can no longer be contained by our "safe" structures. We'll need to muster more courage than usual to let go of certain ego-attachments and thereby look at things differently. The unexpected enters our life to help us to break away from our past; willingly allowing for this minimizes undue disruption. Uranus enables us to reinvent ourselves in ways that liberate qualities we've trapped within us for too long. This transit signals a time to release what we have formerly held back due to either suppression or self-ignorance. Uranus is determined to shatter elements of our past that have never once looked out for our best wishes, a past sometimes filled with powerful people trying to "keep us in our place" (whatever that means to them)!

The Uranian spirit of experimentation can be rousing. We are filled with confidence about a brighter future when we engage in unusual activities that help us embrace our otherwise dormant potential. Uranus electrifies both our will and our intuition in ways that facilitate insightful self-discovery. For once, we're alive and well, and wide awake. We can be thrilled with the individualist we turn out to be. The square and the opposition, although aggravatingly unpredictable at times, can open our eyes to our truer self versus the phonier roles society has convinced us to perform (an issue we meet again when Uranus passes through our Tenth). Once we actualize who we are with typical Uranian certitude and verve, we often upset the status quo of our close relationships. People may initially have a harder time relating to us, because now we are presenting ourselves in unfamiliar ways, and dynamically so. We can't understand why doing something so real and so valuable to us could, in turn, threaten others; maybe they fear we'll outgrow them and leave for good (which does sometimes happen).

With such tensional aspects, a measure of internal eruption is suggested. We not only are pushing away from control factors found in the environment or in personal relationships, but we may inwardly battle warring forces in a futile struggle to keep our ego intact (the ego loves to tamper with everything, and yet it unreasonably expects a "hands-off" policy regarding itself).

Uranus won't allow any planet it touches to function in its accustomed fashion, so here the solar ego gets a uncommon lesson in flexibility. One outer manifestation of this inner conflict is social malcontentment, but it can become a stormy attitude towards established structures that can backfire on us. Our self-will at this time is its own worst enemy. In addition, our refusal to cooperate sparks the unexpected resistance of others. The more we kick and scream for freedom, the more we invite authorities to come down on us. Dramatic clashes and uprooting confrontations can result—especially with the opposition, an aspect already poised for showdowns if need be.

EXTREME MEASURES

Extremism can plague mishandled Uranus/Sun transits. Before we've even given ourselves a healthy chance to assimilate the newly emerging awareness within us, some of us may feel compelled to allow Uranus to overwhelm us. In wild abandon, we can give in to this energy without regard for our normal securities. We rush things that would have turned out better by adopting a less hurried approach. We also may rush people into making major decisions that they're not ready to make, and sometimes they'll break off the relationship rather than undergo this kind of pressure. That's not what we had in mind. We could also wholeheartedly support social causes with more determination than needed and with more drive than anyone else can tolerate. A mishandled square, opposition, or even a conjunction can be guilty of this; we could come across as a loose cannon. This would typically be a sign that our ego hasn't undergone any stage of true reformation, but has just learned how to make more noise than ever.

However, at times, the quick way we attempt to do things becomes the ingenious way, because opportunities may only be available for a short period. Uranus alerts us to this fact, and we act swiftly. Usually, we're not so lucky, especially when our actions are fueled by obstinate defiance rather than objective truth-seeking. With the trine and the sextile, we have a better chance of getting the environment to work with our freedom needs without stirring discord in others. Our leadership potential (the Sun) becomes more attractive and less threatening; we're not as annoyingly single-minded in our objectives and instead may show greater adaptability. We're willing to make last-minute changes that appeal to all involved.

After this transit, life will probably not look the same, which is something we may not be grateful for until the dust has settled. Our perspective has gone through a major reorientation. We become increasingly clear about our individuality, and therefore more willing to separate ourselves from any standard identification with the masses (we know we are not "just like everyone else" at this point). We have awakened to ourselves. This is especially the case if Uranus has squared or has opposed our Sun, and we have survived the internal clashes and outer chaos these aspects sometimes generate. However, even chaotic situations can help us to deal with ourselves more directly and honestly. We just need to realize that we don't have to always experience such states of disorganization in order to reform ourselves.

INTUITION IN ACTION

Uranus can open doors of perception that let us know what's going on before it happens. It sounds like a very exciting talent to acquire and certainly one that fits into any New Age agenda: tune in today and be aware of tomorrow! Yet this transit is not as likely to suggest a mental awakening—where we sit around and suddenly flash on the future—as it does intuition stimulating full-fledged, timely activity (it's like Uranus/Mars in this respect). Fire planets are less happy with "thinking" about things when they can instead be busy "doing" things. We may discover we gravitate toward exciting, futuristic projects that require a keenly intuitive grasp of overall conditions. We can flash on the whole picture rather than only on the details, and we quickly realize that it is a perfect opportunity to act in some executive or leadership capacity.

As long as others are willing, this could be the right time to show off our talent and bring a progressive vision to the world. It will take much self-confidence to think in the big ways that we need to; but if we are undergoing a trine or a sextile, doors of sudden opportunity fling open. They can do so even more with a square or an opposition, with fantastic results, but we'll need to curb both planets' willful behavior. Big projects require enormous cooperation; we'll have to know when to give in to the equally brilliant insights of others. They have legitimate ego needs, too!

URANUS TRANSITING THE SIXTH HOUSE

OUT OF ORDER

The Sixth House gives us clues about the "job" of living a fully functional, productive life. It represents the details often involved in maintaining our Tenth-House career, indicating our basic approach to day-to-day work. This may not seem like an exciting place for Uranus to hang out, because the focus here is on chores and other minor daily rituals that nevertheless provide a semblance of organization and continuity in our lives. Necessity motivates us to act the way we do in our Sixth. Here we are to attend daily to those minor mundane affairs that keep everything else humming along like a finely-tuned engine. We'll also suffer consequences here when we neglect to do any needed routine upkeep—usually in the form of illness on some level. When a machine is not working and is in need of repair, we commonly say it's "out of order." This phrase underscores much of what concerns the Sixth. Here we learn to either fix what is broken or prevent unnecessary damage in the first place.

So what could an often *dis*orderly planet like Uranus offer to enhance this industrious life department? Wouldn't it seem better if Uranus just took a sharp turn and went elsewhere? After all, it seems to trigger a little chaos wherever it goes, which is just what the Sixth House would rather avoid. Uranus is not a tidy planet, while the Sixth can be quite a fastidious house. However, we do a lot of mundane things in this life department monotonously—robot-like at times. This seems boring from a soul perspective, and seemingly does little to demonstrate our true essence, or life spark. Uranus wants to break up a few such patterns to prevent us from further operating like a computer chip. Mindless repetition is the kiss of death for Uranus, especially if done at a snail's pace and without heightened mental stimulation.

Uranus can instead offer the Sixth some nifty short-cuts to run things more efficiently, which is great. Although the Sixth House hates to waste precious time and energy, it ironically ends up doing so by painstakingly following standard procedures without deviating—Virgo has this problem, too. Uranus here can apply its brilliance toward faster problem-solving in ways that satisfy this productivity-valuing house. We'll have to reconsider our time-honored ways of doing things and instead allow for more experimental techniques.

JOB QUEST

It's not uncommon to hear clients state how much they hate their jobs, but dare not jeopardize their employment status. Instead, they force themselves to cope with boredom, just so they can receive a steady paycheck and the security of company benefits. Transiting Uranus lends no sympathetic ear at this time. Its mission is to uncover the unvarnished truth behind any matter, and the truth here is that things are getting mighty dull for us. Being bored silly with anything is one way we evoke this planet to suddenly spring into action, according to its own unpredictable timetable. There is apparently a lot about our current Sixth House experience that doesn't thrill us, so transiting Uranus goes on a campaign to reform our working goals and objectives. This increases the level of excitement, while also raising the risks that such alterations imply. To cope with changing events here, we'll need utmost flexibility as Uranus rolls the dice.

This transit lasts for several years; therefore, any fireworks in store for us won't necessarily take place as soon as Uranus crosses our house cusp, unless a natal planet happens to be there or at the cusp of our Twelfth (thus opposing transiting Uranus). Due to its recent passage through our Fifth House, we probably have learned that life will now allow us more fun and creative self-extension. That alone should help us to feel we can tackle more interesting, individualized work in the Sixth. We need to know that it's worth doing tasks that further awaken our self-esteem. More of our true self needs to sparkle on the job, and it can, if we allow ourselves room to stretch. Any job situation that discourages individualism and independent thought will prove to be a dead-end road for us.

It sounds like we just might have to quit that tedious stint at "Burgers R Us," and seek new avenues of vibrant but workable self-expression. That oddball but seemingly ideal job set-up that we've been toying with now deserves serious consideration. Many a Uranian path can seem impractical or too radical at first, only to prove to be a stroke of genius later, as everything suddenly comes together. However, we'll have to make sure that we are completely sold on our talents before making any big leaps. At least, we'll need to realize the therapeutic effect that a job that allows for extra freedom can have on us. Some of us have felt straightjacketed in the way that we serve the world. We're now ready to follow our special vision. We also have to know when to allow for unexpected changes to occur in our future game plan.

SMART CHANGES

We will need to assess our working skills, taking stock of ourselves and clearly appraising our more versatile assets—but not in the slow, analytical manner typical of the Sixth House. Uranus instead quickens this experience by providing us with flashes of insight and spurts of inventiveness. We are to brainstorm our way to vocational independence, being careful not to repeat any self-limiting attitudes we've adopted regarding employment in the past. Turmoil felt at this time is often due to inner clashes between our daring urge to try out something different, versus our fear of abandoning all that we have carefully developed for ourselves in our career. Perhaps not *all* needs to be abandoned, just those parts that feel stagnant. Still, in the Sixth we worry that it doesn't make a lot of sense to suddenly change our direction according to Uranus' impulses.

Some of us will resist making needed changes here to the nth degree (while armed with plenty of rationalizations about why we can't afford to alter our pattern). If transiting Uranus aspects our planets—maybe setting off a natal T-Square—and we still refuse to dislodge ourselves from a going-nowhere job, life will toss a few, hot lightning bolts our way to shock us into making a no-nonsense change: we are suddenly downsized, or maybe unexpectedly our company goes out of business, or perhaps we've come in late a little too often and now get fired for this and other seemingly minor infractions. Not being punctual or careful on the job is one early-warning sign that we no longer want to be employed where we are; it's just that we haven't let our conscious self know this yet. Maybe we quit in disgust, realizing that we are wasting our talents in an unappreciative work atmosphere.

However it happens, our removal from a work environment that no longer offers growth for us should be seen as a step in the right direction. It is true that we won't feel very secure and surefooted while things are shaking and shifting, but we'll need to suspend expectations of security for the moment and live in the now. Instability is normal and necessary for the transition we need. Why fight the process and flirt with a nervous breakdown or even attract a major accident that can render us unemployable for a long time? It's best not to unduly frustrate the Uranus principle by refusing to modify our unsatisfying situation (besides, this can be an unhealthy exercise in futility).

HEALTH SHAKE

Speaking of health, this Uranus transit can throw a few surprises our way if we've been neglecting the body maintenance ideals that the Sixth House supports. Through our aches and pains, life lets us know that elements of our inner and outer selves are also "out of order." The Sixth House works hard to ensure that things operate flawlessly through the development of efficient systems, methods, and techniques. Here we are motivated to take practical action regarding whatever malfunctions or impedes smooth performance. Uranus knows that being sick or physically unfit can rob us of many simple freedoms that we normally take for granted. Anyone who has broken a leg has found this out firsthand. We simply invite limitations when we are not well or able-bodied. Uranus will use sudden illness or physical mishaps (accidents) as a wake-up call, showing us the results of building up unchanneled inner tension (all of it tied to our resistance to accept change). Physical symptoms, of course, also are the consequences of not taking good care of ourselves in terms of nutrition, exercise, and overall physical fitness (in our willful neglect of the laws of health).

Such symptoms tend to pop up out of the blue, demanding sudden adjustments. Sometimes we're the ones who demand instant relief. We can develop tics or twitches overnight, nerve disorders involving strange tingles, or sudden, inexplicable flare-ups of whatever has bothered us in the past. I don't wish to provide neurotics with a long list of ailments to further worry about (besides, with Uranus, the symptom we'd probably fixate on most is rarely the one that actually manifests). However, rigidity is one of Uranus' most despised enemies—paradoxically, Uranus itself can display unbending self-will when it meets with opposition. Doing whatever we can to release our body's stiffness also helps to relax the psychological tensions that create such symptoms.

We need to find some lively technique involving body movement that can loosen us up; aerobic exercises have enough get-up-and-go to appeal to Uranus, as long as we throw in a lot of variety. Even "jazzercise" classes or tap-dancing lessons may do—anything peppy that keeps us on our toes or in continuous motion. Maybe we'll have to mix a few techniques that interest us and come up with our own formula for staying healthy and fit. Of course, since Uranus loves gimmicks, some of us may be tempted to buy those amazing-sounding health gadgets advertised on late-night TV, especially those that promise quick results.

Before this transit is over, we may have made some sweeping reforms regarding body/mind well-being. We can become more conscious of how we deal with our basic physical requirements. Learning to value good health as our most direct source of vitality is important to Uranus, because it does its best when we feel alive and well, and in tiptop shape. Still, Uranus can go to extremes. We'll need to make our positive changes without resorting to ideological fanaticism or to a strident revolutionary spirit (that's when we obnoxiously try to forcefully awaken the whole world to our exciting health discoveries). We shouldn't be so quick to jump on the bandwagon, and attempt to suddenly reform everyone else's health habits. After all, that tactic would not have worked on us before we became so "enlightened," and we certainly would have resented the pushy approach involved. So why indiscriminately pitch it to others? In general, Uranus moving through the Sixth is a time to clear away blockages on all levels that have kept us functioning inadequately. Think of it as a well-needed, cosmically timed tune-up.

IDEAL WORKPLACE

Another Sixth House theme involves our dealings with co-workers. Uranus would love for us to be independent, free-spirited types who are non-salaried and not under any binding contract—we're happy freelancers instead. Total self-employment becomes highly attractive to this planet. In Uranus' vision, we'd preferably work from a home office equipped with state-of-the-art goodies (computers, high-tech toys, amazing stereo systems, voice-activated machines, and other futuristic touches). Lots of window space, opening up to a breathtaking panoramic scene, is also important (we need as much natural light as we can get). What we wouldn't do in Uranus' ultimate fantasy is rub shoulders with co-workers on the job. Uranus can't handle the hive mentality of busy working bees that obediently follow orders from the higher ranks. In real life, however, most of us do have to relate to other employees in a work structure that is often anything but relaxed and casual. The impersonal roles we perform can feel artificial after a while—and we may not even have a room with a view!

REBEL CO-WORKER?

Transiting Uranus can easily be projected onto a co-worker who suddenly doesn't want to play by the rules. Small acts of rebellion can make this person a pain in the neck in our opinion, even annoying to other workers. Uranus represents the one who is a nuisance in any group, the disruptive one who doesn't agree with social protocols. This person simply ignores whatever he or she doesn't want to do. Perhaps this individual engages in inappropriate activity, or merely attempts to violate something as basic as the dress code. Uranus can't understand what's so important about dress codes anyway, and sees them an just another excuse to enforce totalitarian rule (an extreme reaction typical of Uranus). Our relationship with any singled-out co-worker can be uneven, choppy, abrupt, or unnerving. However, if the aspects formed at this time are less tensional, we can view this worker as a breath of fresh air, as somebody novel, fascinating, brilliant, or even multi-talented. He or she is certainly not stodgy.

Such co-workers become symbols of our unrecognized inner rebelliousness and our readiness to be liberated from conformity. We wonder how they dare to get away with such behavior, and yet their spirited independent streak holds an appeal for us. We probably need to show more of our real self on the job as well, at least in measured amounts over time (let's keep old Saturn happy). However, if our environment prohibits this, we can soon begin to feel that we are definitely not in the right job atmosphere. Unexpected clashes with other employees are warning signs that unrest is building within and is looking for ways to erupt. We can get flat-out fired for being confrontational, or perhaps we feel so alienated and disgusted that we quit in a huff. Uranus here means there will be no beating around the bush when it comes to making timely or untimely exits. Once we break away from our intolerable condition, we may feel disoriented for a while until we land a new and hopefully improved job situation.

Uranus/Earth Mercury Transits
Not So Airy

While the air side of Mercury deals with our mental development in terms of Gemini themes, earth Mercury becomes an indicator of how we apply ourselves when organizing tasks or tackling jobs through analysis and practical evaluation, Virgo-style. It deals with mental activity brought down to a functional, efficient level. If our Mercury is aspected by transiting Uranus, there is no sure-fire way to indicate if the earthy side of Mercury will be triggered into action versus its airy facet. However, if natal Mercury is in our Sixth House, rules that cusp, or is in Virgo itself, we probably will handle this transit along more pragmatic lines of thought. There may be less of an urge to concoct brilliant abstractions that sound fascinating but have little chance of being useful and marketable. Earth Mercury will also operate better at a slower pace; it cannot afford the needless distractions that air Mercury looks forward to during the course of a day.

Going on Strike?

How we feel about doing the daily routine services that we provide (either in the home or in the office) may now undergo radical change. Uranus is urging us to halt operations on some level, often unexpectedly so. This is especially the case when we realize how totally unstimulating or slavish many of our little rituals have become for us—maybe it's the endless cooking and cleaning chores that bug us, or those other required tasks that do little to enhance our sense of individualism. In the home, we may rebel and disrupt our family patterns, causing temporary conflict. At least this remains a relatively self-contained protest, a minor revolt, and usually not the kind of issue that leads to divorce—just a lot of unmade beds.

However, when we attempt to struggle with our internalized unrest on the job, perhaps with our co-workers or with our work itself, we can stir up trouble for ourselves. It all depends on how

Saturnian our job environment is. Can we be flexible with rules and experiment with our own procedures, or is everything done by the book? Obviously, the options of a freelance graphic artist with this transit would be different than those of a federal government employee. Still, whatever our job is, we may find ourselves easily distracted. Our mind may be elsewhere much of the time, even though our body appears firmly planted at our desk. (Earth Mercury hates when that happens!)

Absent-mindedness on the job never looks good. Coming in late or leaving too early to go home for the day is usually frowned on (and resented by other workers who play by the rules). Frequent, unexpected sickness could make an employer reconsider our long-term value to the company. Yet sometimes this is how Uranus makes its presence known. We start to rebel against orderly habits in small and seemingly inconsequential ways at first, even if unconsciously so. But things can soon escalate.

ALL SHOOK UP

When Uranus works on this earthier part of our Mercury awareness, we can experience a disturbing kind of nervous energy that interferes with our normal capacity to tolerate repetition, routine, and minor details—all of which can lead to making stupid mistakes, the kind that drive earth Mercury crazy. With the square or the opposition, we can respond by having temperamental outbursts, and criticisms of our work environment might be sharper and more vocal than ever. We'll need to be careful about how we mouth off—and to whom. The reactions of our co-workers can be unpredictable (most so with the opposition). We find we become agitated too easily; we're more keyed-up than usual, with our nerves on edge. This could be a good time to look into foods and supplements high in nerve-soothing elements (such as B-complex vitamins, calcium, and magnesium). We also should ease up on our caffeine intake.

Keeping Our Cool

Uranus can be as hot as lightning one minute but as cold as steel the next. It's hard to figure out how it's able to switch its temperature so radically, but it can. For some of us, this Uranus transit becomes an opportunity to coolly detach from whatever might typically upset us on the job. Maybe we're learning to diffuse tensions at the workplace by turning off our emotions, and by reacting to less of what's going on around us, especially regarding the intrigues of office politics. It's best not to get too riled up about anything that could cost us our job or alienate co-workers. Feeling a bit removed from our surroundings, we may focus dispassionately on only our immediate tasks at hand, and pray that potential trouble-makers leave us alone.

This may seem to work for a while, but in the long run it's a response that is too sedate and mild-mannered for Uranus. It could also be that such detachment reflects our unwillingness to arouse the spirit of risk-taking, thus keeping us in a rutbound mindset or work situation. Eventually, some minor incident triggers a major Uranian hotflash. We discover that we can no longer keep up the cool facade. As stormier feelings enter the picture, we realize that we're not all that happy with circumstances. It's then that we can pop our cork and explode with restless thoughts of radical change. If appearing calm and collected only suppresses our more fiery feelings, Uranus will find a way to blow our cover wide open.

Not So Fast

Uranus likes speedy interactions, but many activities associated with earth Mercury can bog us down with step-by-step processes. Certain things in life simply must be done according to correct and sometimes painstaking procedures. Yet Uranus has little time for this, urging us to come up with short-cuts and quick workarounds. Sometimes these prove successful, but others times things go awry due to poor planning, carelessness, and overall impatience on our part. If this really bothers us, we'll have to slow down and use Uranus' energy to promote our alertness rather than simply to overstimulate our urge to go faster. Still, earth Mercury often makes simple chores more laborious than need be, something to which Uranus is calling our attention.

Perhaps a lesson here is learning to be less uptight about outcomes. Life is trying to discourage excessively orderly behavior in our response to the many, little details that daily living demands. It would be good to allow for original or experimental thinking to help us to come up with alternatives. Here's where a Uranus sextile comes in handy; it's an aspect that shows us the value of intelligent adaptability. Both planets can breath easier under such a pattern and can come up with a variety of solutions. We can be more accepting of different outcomes than those originally planned. With the trine, the environment usually presents us a clever opportunity to handle matters, and one that instantly appeal to us; we don't waste much time mulling things over. Uranus teaches earth Mercury that there are many ways to tackle any problem, not just one "right" way.

LEARNING CURVE

Maybe we need to consider new training that could help us to work in different areas of our current vocational field. Both the airy and earthy sides of Mercury are keen on continued education. Earth Mercury simply is more likely to study things at length that can be applied in the everyday world. However, Uranus will only get excited about short-term studies. Specialized training that lasts a few weeks or a couple of months is acceptable, but anything involving several years will probably fail (unless we have a lot of encouraging transiting or natal Saturn and Pluto support—two planets that can handle grueling schedules and that can make enduring efforts).

It could be that we receive our education while on the job via hands-on experience. Our company may suddenly require its workers to learn new procedures or perhaps deal with upgraded equipment. Maybe new business software is being introduced, involving a learning curve for everyone who wants to keep their

position. With the wrong attitude, we could convince ourselves
that this new material will be overwhelmingly complex, confusing, or too taxing for our brain. Yet with Uranus, it's worth the
effort to take on any mental challenges offered. We could surprise
ourselves with how well we progress (because there usually is a
readiness for fresh mental stimulation). Any new skills developed
now can help us achieve future aspirations.

BOTH SIDES NOW

In reality, when Uranus transits our Mercury, we probably will
alternate between its air qualities and its earth components;
we're always doing both so that we can handle a wide range of
varying day-to-day conditions. Uranus certainly doesn't want to
get stuck in only one mode when it can instead have the stimulation of variety. Air Mercury provides needed flexibility; earth
Mercury provides needed workability. Together, they benefit from
Uranus' ability to clarify matters and to allow for the pursuit of
our own inner truths. We may find we are less willing to accept
other people's thoughts or follow other people's methods of doing
things. Mercury is a persuasive rather than a combative planet,
so maybe we can present fresh concepts and insights in a way
that enthuses others to try out our novel approach. Quick-thinking Mercury wants to be ready for anything that life tosses its
way, especially those interesting things that Uranus likes to fling
at us when we're not looking.

HEALTH MINDED

Earth Mercury is not as strong of an indicator of health problems
per se as the Sixth House. Still, Uranus may spark our interest in
reading about alternative health measures or other unconventional ways of integrating body, mind, and soul. This can be quite
educational, fulfilling earth Mercury's need for helpful, useful
information that can be put into immediate practice. Even starting new exercise programs and proceeding sensibly (nothing too
crazy) can help us to mobilize our energies in different ways that

enliven us. It's important that we don't try to do healthy things just because we *should* do them (that's Saturn talking). Push-ups will not work for us if we truly despise doing them. However, if racquetball has always seemed like a really fun way to get one's needed exercise (and let off lots of steam), then racquetball is the way to go. We need to find activities that don't feel like a chore or a punishment to do. Otherwise, our Uranian restlessness will prompt us to quit sooner than expected (this can apply to Uranus going through the Sixth as well). In addition, our Mercury—less interested in purely physical activities—will claim it's utterly exhausted.

Again, whatever we put into practice may not last a lifetime. We will probably get bored and look elsewhere for new solutions. But for right now, getting charged up about our health potential and optimum body functioning is an excellent way to partake of this Uranus transit. We just need to be careful of those activities that have a dare-devil quality to them (maybe this is not the time to take up bungee jumping). However, there's nothing wrong with a little lively tennis, or at least with reading a few "how-to" books about improving our bowling technique.

CHAPTER TWELVE

URANUS TRANSITING THE SEVENTH HOUSE

OFF TRACK

Transiting Uranus entering the Seventh House actually triggers two transits. As it first crosses the Seventh, putting emphasis on changes in our more intimate relationships, it simultaneously opposes our Ascendant. The Ascendant is that specific part of the First House experience where we show to the outer world an image that helps us move out into that world, hopefully with a measure of courage and self-reliance. It is that part of ourselves that we semi-consciously activate in order to initially confront the immediate experiences we attract. Others who don't know us very well usually identify us according to surface personality traits emerging from our Ascendant. Transiting Uranus in opposition aspect suggests a sudden lack of stability in how we present ourselves. We may feel thrown off our normal track, which can be disorienting. However, this could be just what the we now need most for accelerated inner growth.

This is especially the case if we have a fixed sign rising, because then our Ascendant identity is probably well-entrenched and less responsive to outside alterations. Fixed signs usually only follow their inner promptings while ignoring or resisting external pressures. In contrast, mutable signs rising—Gemini, Sagittarius, and Pisces (but less so Virgo)—may not, in the beginning, differentiate the initial Uranus shake-up from the typical personality fluctuations they experience. However, Uranus is not superficial in its influence; therefore, it will try to effect change on deeper and more crystallized levels of our being, because here is where we can unwittingly trap our fuller potentials. Uranus doesn't waver in its actions as much as can the mutable energy of these signs.

Electrical Charge

Whatever our rising sign, now we suddenly feel as if we have outgrown many of our familiar projections and are eager to explore different patterns in close relationships. Remember that this is an opposition at work, and therefore a balanced response on our part is not easily achieved. There's a tendency at first to overreact to our new Uranian surges. The people we may accidentally involve ourselves with—our chance encounters—are quite different from us in ways we may not know or fully appreciate at the moment. We just see them as exciting, animated, very friendly, and refreshingly up-front. These folks won't allow us to get away with much psychological game-playing; they insist on our self-honesty, and even know how to bring that out of us. There is a special electricity in the air when we get together, a high degree of mental rapport that puts this relationship in a category all of its own. It's partly the newness of this energizing but often off-beat association that helps us to come alive.

Meanwhile, the person with whom we may already be involved (spouse, live-in lover) also appears quite different from us in ways we are beginning to find alienating. How interesting that at this moment *everyone* appears to be different from us! Actually, this can be a time for us to breathe fresh energy into an existing marriage, or its equivalent, if there is plenty to support in our union. We could be undergoing a temporary slump that can be remedied by the quick energy pick-me-up only Uranus can provide. However, this is not accomplished by magic or wishful thinking. Uranus demands honest confrontation and insists that we air any pent-up feelings. Things can get intense at times and periodically explosive. Sporadic emotional flare-ups force uncomfortable dialogues. We get to see a few of our own shortcomings as a partner. Still, for the survival of any relationship that's worthwhile, we'll need to simply weather the storm as we continue to get real with ourselves and our partner.

PARTNER SHIFT

Sometimes it's our partner who stirs up much of this Uranian energy on behalf of the ongoing growth of our committed union. This becomes a mate/spouse willing and able to let go of stale ways of interacting, one who instead adopts fresh approaches to intimate relating. He or she may even take on a different look to go along with his or her progressive attitudes. Maybe our partner feels very headstrong and determined about embarking on a new life direction—such as quitting the corporate world and working in a plant nursery instead, plus finding time to do a little free lance photography on the side. "To thine own self be true" becomes our partner's motto. We may fear that our significant other is pulling away from us while affirming his or her significance. Even if this is true, it could nevertheless be a good thing for the evolution of our union rather than something to be dreaded. Time will tell.

During the beginning of this transit (when Uranus opposes our Ascendant), we may be taken aback by these unexpected and disruptive changes—perhaps the biggest threat being that we may have to alter ourselves in order to readjust to our mate. We may wonder, "Why does our partner have to change on us now? Where will that change lead?" Until we can better assimilate all of this, we'll need to back off and give our mate a lot of space to act out his or her awakening urges. Maybe that's our biggest challenge. Uranus will make it clear that we do not own our partner and never did, which could come as a shock to those with Venus or the Moon in possessive or controlling signs such as Taurus, Cancer, Scorpio, or Capricorn.

If our committed relationship is beyond hope, with nothing worth salvaging (perhaps it has been toxic for years and has stifled our freedom or right to be ourselves), something is likely to come about to bust things up once and for all. It would help if we participated in this experience by calling it quits and moving on with our lives. This takes courage, and a strong feeling that a better future awaits us. However, real life shows that some people's circumstances, even at this time, can be exceedingly complicated. What if our unbeloved is sick and dying? Do we just bolt without looking back, or do we dutifully play a wait-and-see game? It never looks good when we abandon a partner who's gravely ill, even though Uranus itself would easily do so without batting an eyelash—emotional detachment is no problem for this planet, and to hell with what others think!

UNCERTAINTIES

Sometimes what keeps a miserable, estranged union going long after its expiration date is economic reality. At least one of us, and sometimes both, literally cannot afford to break away and be free of this tie. Don't blame karma—insecurity is the issue. We feel that we must establish greater financial independence before we can make any radical moves. We also may want to break away without regrets, which could create further hesitancy on our part. We may even be intimidated by the thought of the high cost of a turbulent divorce proceeding and the monetary or property settlements to be determined.

Of course, Uranus says these are not good enough reasons to do nothing and, instead, to stay trapped and angry. We certainly can't remain in such a pressure cooker situation for "the sake of the children," considering the damaging emotional instabilities of our marriage. We'll have to come clean with ourselves psychologically, address our fears, and then review all viable options. Any reluctance to change this familiar scenario, dissatisfying as it is, could be based more on a distrust of the unknown that awaits us than on any valid financial excuse.

SHOCKING PROJECTIONS

Our urge to merge is strongly at work in the Seventh House even when times are rough, and it can overpower our better judgment.

No wonder many of us are indecisive whenever heavy-duty planets are crossing this area of our chart—especially Uranus, a planet that refuses to give us clear-cut guarantees of secure outcomes. Some of us may sit on this frictional energy for a while and do nothing. That usually backfires, provoking Uranus to set up unexpected external solutions via our unconscious projections. The more external, the less controllable the outcome. Yet trying to fully control things when Uranus is active is nearly impossible anyway.

And so, instead of having the guts to divorce our spouse once and for all, he suddenly dies in a plane crash. Instead of not leaving our partner because she is "so dependent" on us, she willfully embarks on an extra-marital affair (leaving us stunned!) and tells us she's moving out. Perhaps we hang in there and cope with a loveless marriage because of compensating material comforts, only to find our spouse is now filing for bankruptcy due to bad business deals or gambling addictions we knew absolutely nothing about. Uranus can work in ironic ways. Maybe the stress of a constant clash of wills becomes a strain on our health, and one of us in our relationship develops acute symptoms that put us through a health scare. Is ignoring our Uranus challenge to change worth doing this to our bodies? No way.

STRANGE AFFAIR

Similar to Uranus in the Fifth, this Seventh House transit could create for us an opportunity to develop another relationship that is in strong contrast to our existing, dissatisfying partnership. Uranus is not particularly comfortable with the tradition of marriage or monogamy, probably because it's not comfortable with social traditions of any sort. At this time we could almost literally bump into some sparkling, effervescent individualist who operates on a fascinating wave-length that we'd love to share. And guess what? They're infatuated with us, too, which comes as a big surprise! We may feel we haven't had such special attention in a long, long time—maybe never before. Now this facet of our life gets a shot in the arm of a special kind of emotional vitality that we freely want to share.

What do we really know about this energizing lover who's popped into our life? Probably just enough to make things interesting, yet not enough to prove disillusioning. Uranus provides an ideal package in the beginning that, unfortunately, may not be able to sustain itself (time is Saturn's faithful servant, and Saturn can be a harsh reality-tester). If we think that we've found someone who will save us from the pain of a failing marriage (which is actually Neptune's fantasy), our new "friend" may see things differently. He or she may view his or her role more as a facilitator, one who helps us plug into a gutsy self-determinism that in itself clears our head of confusion about our current marital condition. The Uranian lover won't take responsibility for our actions and their outcomes, which is a startling realization that plays into our self-awakening process. We're the ones who are to independently decide our course regarding our marriage, perhaps with encouragement from our new lover. At no time, however, can we resort to clinging tactics and helpless routines, for that's when we abruptly hear this special someone we're infatuated with saying, *"Hasta la vista,* baby!"

Uranus can be a crazed planet, capable of erratic behavior in matters of the heart. We'll need to first look at our chart for natal clues regarding how much resilience and emotional fortitude we have before we start to boldly experiment in love's laboratory. Let's not forget the value of Uranian detachment and objectivity. Otherwise, some of us can fall hard for a person we just "know" is perfect for us, only to find later that nobody gets to hang on to such perfection for long. The truth is, no one's ever going to be perfect.

Eventually we'd wake up very disappointed to discover that our lover is much too different from us or too free-spirited to handle. It seems such a disruptive experience in the long run. Therefore, if we are inclined to have an extra-martial affair, realize that the person we meet does not want or expect to be with us "happily ever after." This is a transitional stage for both parties involved—which could be a healthy thing. Such a transition may even be crucial for us if our marriage is "the pits" and we need a boost of self-confidence to break things up for good. Our Uranus lover becomes the catalyst needed. The wise advice is: don't try to turn around and then marry this person! If we must marry, let's not be foolish enough to think we will possessively have that person all to ourselves. Nobody owns them!

SWINGING SINGLE

In many cases, we are single and available when Uranus enters our Seventh (typically because not everyone chooses to get married early or stay married these days). Uranus might fare better in this scenario, although it seems to have its own set of problems. Many of them revolve around the intimacy issue. Will we be attracted to someone warm and cuddly, someone obviously affectionate and caring? Nah, this doesn't really sound like Uranus. We could instead go head over heels for a charismatic, highly vibrant, and brainy gal or guy (fitting the Uranian profile), who, unfortunately, is passionate with us one minute and then strangely aloof the next.

The fact that we attract those strangely detached types says something about us—that we are not really ready for total commitment or that, perhaps, we are not open to true individualism in others. We don't recognize that this is what's going on inside us. Therefore, we could end up blaming our partner later for being too cold, too remote, too unreliable, too inconsistent, or just too plain weird! Maybe our partner thinks that *we're* also pretty strange ("too insecure," "too possessive," or "too paranoid"). He or she can wake up one day and suddenly decide to leave us (the bad types won't even let us know we've been dumped—they'll just disappear). Alive and well may not describe us very accurately at this point—dumbstruck is more like it!

Again, we will have several years to observe Uranus in motion and to take mental notes of our reactions to this planet. It's probably not a great time to get married, but it can work out fine if both temperaments are not the clinging-vine type and if a great friendship was allowed to flourish first. Unions can be real eye-openers for us, because at all times they urge the development of our greater individuality and self-reliance (even after Uranus moves away from opposing our Ascendant). Some of what's been said can also apply to business partnerships, minus the need for emotional excitement in most instances. However, Uranus is capable of stirring up the most unexpected reactions in all of our closest relationships.

URANUS/AIR VENUS TRANSITS

RUDE AWAKENING?

The airy side of Venus deals with the sharing principle as it is manifested through human relationships. It is very sensitive to give-and-take issues in partnerships, especially when there's more selfish taking going on than genuine giving. This Venus pays attention to how any couple's loving feelings are exchanged. Fairness is of key importance to this planet. Libra-oriented Venus also wants things to remain nice and pleasant between people (with lots of mutual consideration and kindness thrown into the picture). Although it can handle more dynamism and complementary differences than Venus' earthier Taurean side, it still enjoys temperaments that are composed, unruffled, perhaps even stimulatingly assertive at times, but never aggressive. It's also a friendly, cordial Venus who's able to put a positive spin on most social encounters. It tries to see something attractive in everyone.

Uranus is somewhat open to the air Venus experience; after all, it's own social straightforwardness can be interpreted at times as being friendly, although Uranus is not really an easy-going planet. Ultimately, Uranus will not be lured by air Venus' subtler manipulations (i.e., its occasional "help me" act, its passive approach to magnetizing what it desires, or the seductive psychology it uses to get others to act on its behalf). "No thank you," says Uranus! If we do possess such questionable characteristics and have applied them in many of our former relationships (or even in a current one), Uranus is setting us up for a rude awakening. These tactics won't work for us within our new arrangement; Uranus won't permit us to use such subtle strategies. Nonetheless, things may start off "nicely" as we attract a sparkling romantic partner with an appealing personality who quickly shows a clear interest in us.

TUNED IN

Venus eagerly looks for similarities in its relationships; it wants easy compatibility with no hassles. We may appreciate the fact that, with Uranus, we appear to be attracting someone who speaks our emotional language, or who is highly attuned and attentive to us in other ways. Our new Uranian partner seems to intuitively read us like a book. In fact, we both may feel telepathic at times regarding each other's unspoken thoughts and needs. Just this alone may make us sit up and claim, "This relationship is different—it's going to be really special!" Yet this union should not be confused with the empathetic bonding of Neptune/Venus affairs. Uranus doesn't want to merge, but expects to simply enjoy the stimulating exchange of two separate, well-defined egos. Uranus also doesn't anticipate that everything will run smoothly, and is not bothered by periodic flare-ups. Flare-ups keep things exciting and vibrant.

We may initially warm up to one another quickly. However, if we have been possessive with our past partners, this transit will pose quite a challenge for us. Why so? Because sky god Uranus doesn't like being trapped in earthy experiences and will only make physical contact sporadically and briefly (as demonstrated by lightning strikes, where the high voltage force of such electrical phenomena can damage form and structure). Holding tightly onto something or someone is not part of the Uranus experience. Uranus won't tolerate static conditions for long, no matter how harmonious things appear to be on the surface.

The air side of Venus can appreciate the freedom urges of Uranus, because air needs its space, too. Earth Venus probably has more of a problem here (especially if we have in the past inappropriately turned our partners into our possessions). Thus, the best advice during any Uranus/Venus transit is to live in the moment and enjoy our highly energized relationship, but don't quickly start building emotional structures intended to support long-range security needs. That could send us into an unanticipated tailspin, should this union suddenly stop working for our lover (the one who usually plays out the transiting Uranus role for us). However, sometimes it's our *lover* who wants to cling to us (we knock his/her socks off and he/she is in awe of us). That's around the time *we* start looking for a quick exit. Uranus cannot tolerate being pursued by anyone filled with too much emotional intensity and a determination to have a settled union.

Social Sparkler

One benefit of this transit, whether in romance or in our social life, is that we loosen up our approach toward people, and thereby open ourselves to a wider range of interesting personalities. This could be good for those of us with a natal earth or water Venus who typically are less outgoing and self-assured in forming relationships. We now seem to be more willing to reach out and explore different social interactions. Here's a case where birds of a feather do not flock together. A transiting sextile or trine gives us a gentle nudge in this direction, and we at first may simply experiment with changing our appearance and our attitude (before even considering new romantic or social involvements). In contrast, the square and the opposition carry with them the possibility of energies colliding plus a sense of urgency, which translates as quirky fate entering the picture and thrusting certain magnetic, irresistible types into our path. We can become intimate very quickly (due to an excited Venus in overdrive, because Uranus alone has no strong need or desire for intimacy).

In fact, these transits are notorious for infatuation or "love at first sight" reactions. If we are single and lonely, a Uranus transit can change our status really fast, as someone out of the blue is practically dropped in our lap. Are we ready for this to happen? Yes and no. Yes, because we will be learning new things about our emotional make-up and our ability to relate to another while in a state of high rapport. But also no, if we are unwilling or too scared to change our emotional patterns and express ourselves in brand new ways. This aspect requires greater tolerance of people's behavior (we are going to be attracting individualistic types). We had best not try to handle such love encounters the same way we've mismanaged some of our previous unions; that wouldn't be good for our growth (and our new partner won't play by those old rules of conduct, anyway). This realization alone—that we are not to supervise this relationship—could make us feel a bit crazy and uncertain about such an obviously hard-to-manipulate Uranian affair.

ON THE ROCKS?

If we are married, Uranus/Venus transits disrupt the more stagnant patterns of our relationship. This doesn't have to pose a crisis (although often it does), but it could signal a turning-point for us, where we get a chance to address a different set of inner needs and expectations. This could be an opportunity to clear the air of mounting tensions due to poor emotional communication and to superficial interaction. Feelings now come to the surface, causing minor explosions that shake up our accustomed stability; thereafter, we can begin to again operate based on new understandings and awarenesses—that is, *if* we value this union at such a point. Some of us discover we do not, and will therefore use this Uranus transit to find out if we can connect with someone else "out there" who might be more suitable for us. For Uranus, the unknown is preferable to a existing union of ongoing emotional sterility or of lifeless, and maybe meaningless, marital routines.

Uranus is a vow-breaker when freedom is at stake, so we can expect to feel distractingly restless in our need to establish space in partnerships. A sextile can find small ways to introduce a little zest and unfamiliarity to our relationship. There doesn't have to be a divorce in the making or an illicit affair on the side to satisfy a sextile or a trine. However, with a conjunction, a square, or an opposition, we may feel that the intimacy we once experienced with our spouse is no longer a reality. We don't feel that special, close connection we once had. Outside relationships at this point (especially if prompted by a Uranus opposition) could prove momentarily manic if we have previously been in the slump of an emotional depression for too long (maybe our spouse has been cold and indifferent to us for quite a while). Yet our new Uranian love interest makes us come alive and apparently validates who we are. For us, it feels therapeutic to seek such adventures of the heart.

What this also smacks of is a short-term fix (Uranus makes sure of that) for a long-term ache for real love and companionship, a yearning that ultimately will not be satisfied by temporary romantic diversions. Thus, quick ups also have their quick downs, which is not surprising, considering that we're not acting level-headedly in our love choices. In addition, our lover typically can lose interest in us if we try to get too cozy and settled, or should we desire a permanent state of spark and excitement. Suddenly our whirlwind affair can be over, leaving us dazed and more frustrated than ever. Some of us risked losing long-established stability for all of this, and now we feel jilted. Others, having little to gain from staying in a miserable marriage, now feel confused and shaky regarding self-worth. Just why our new love decided to split the scene usually makes little sense to us, if we can even figure out why this happened. Baffling Uranus often makes sure such people drop out just when things are seemingly going well, which makes it all even more maddening!

If married, it's better to use Uranus for furthering a shared exploration of honest feelings and needs first, before getting a third party of unknown quality involved. We'd be caught up in an idealized vision of the new lover we'd attract anyway, thus abandoning any practical assessment of our situation. What we're quick to call "love" during this transit is often just experimentation, something better left to those who truly are detached and less emotionally triggered. Imagine someone with natal Venus in Cancer square Saturn in Libra falling head over heels for another with Sun in Aquarius opposing Uranus in Leo—just because transiting Uranus is opposing that Venus. It's a recipe for volatility and eventual disillusionment, especially if this insecure Venus tries to emotionally engulf the Sun/Uranus lover. Such a disastrous experiment can blow up in one's face, with the Sun/Uranus individual soon walking away unfazed, leaving Venus shattered. We'll need to know ourselves better before we throw away an existing union that still can be renovated, in favor of a fresh and exciting new affair that could come to a screeching halt without warning.

BUSINESS AS UNUSUAL

Business partnerships may also demand our sudden attention. An associate may want "out" of a current professional relationship, and we'll probably need to go with this abrupt decision. However, it's how this partner chooses to leave us that can be a bone of contention. Sometimes our partner has no choice but to leave. More often we feel that we haven't been given fair notice. This sudden crisis has been sprung on us without the benefit of an honest, face-to-face dialogue regarding unresolved issues. Uranus (and even Jupiter) triggering the airy side of Venus can suggest irresponsible action on someone's part, in which running away from problems provides a quick and less confrontational solution. This issue is not handled in a professional or timely manner.

However, it may be good to let this person go without too much regret (like a no-fault divorce). Often, another partner, even more qualified to fulfill our future objectives, may be just around the corner. We may connect with that new individual under surprisingly "lucky" circumstances, seemingly by accident, usually by being in the right place at the right time. Many of the issues mentioned in earlier paragraphs still apply (minus the emotional/sexual themes in most cases). Of course, if the aspect at the time is a tensional one, we're very wired up about our business disruption, and may pick a replacement too impulsively. We might panic and, as a result, make a hasty decision that later could prove destabilizing as contrasting temperaments begin to surface and clash.

With a less frictional transit, we may find that we connect with someone who's brilliant in an area of our business where we often have to struggle to make things happen. Maybe he or she has sparkling social skills that we lack, and can captivate clients or the public. This person has what it takes to turn things around for us in ways that can invite an exciting level of success.

However, we'll also need to realize we're dealing with an individualist who doesn't care to be told what to do. We can't afford to be bossy with this freedom-loving partner. Protective or authoritative Saturn is not what this person wants or needs at this time. Our Uranus partner will think nothing of going out on a limb based on a strong hunch, but that could turn some of us into nervous wrecks (if suddenly we revert to our more self-protective earth Venus instincts). Still, life is teaching us to take a few chances by attracting someone who finds taking certain business risks thrilling.

WILD ARTIST

Air Venus, like earth Venus, also has an interest in artistic expression. Aesthetics, refined taste, exquisite design, fashion, and architecture are all appealing to Venus' airy facet. Uranus has its own unique way of interpreting beauty, which might prove jarring to this Venus. Harmony of line and form are not critical to the Uranian artist. Balance of composition is also less important. Uranian artistry can look or sound outrageous, be very experimental in technique, and even anger-producing (due to its ideological content). Sometimes humor is cleverly employed. Uranus can use art for revolutionary purposes that make big, bold statements about the human condition. Subtlety is not the style of this planet.

Collage seems Uranian to me, as does multi-media art (such as combinations of paint, fabric, metal, and maybe even things that light up). If a work of art disturbs most viewers and triggers revulsion or the urge to censor such art, Uranus is usually at work (sometimes it's Pluto). If we're thinking some "crazy person" did the "so-called art," again we're looking at misunderstood Uranian creativity. For those of us who are freer with our artistic expression, a Uranus transit to air Venus can open the doors even wider to a world of weirdness. The transit itself can encourage such freedom. The strange and the bizarre begin to look beautiful to Uranus/Venus (the way classical portraiture and realistic landscapes do to tamer, conventional Saturn/Venus patterns). Maybe

we have an urge to dabble in trendy forms of interior design, perhaps using unusual prints and loud colors. We're looking for a dramatic touch in our decor—out with the bland and the practical,
and in with the exotic and the attention-grabbing! Sometimes,
however, our aesthetic sensibilities may go out the window as
Uranus has a field day mixing and mismatching things. It can
seem an assault to the senses of some people, although we, the
inspired artist, may think our creative efforts are pretty cool. In
short, we are seeking, first and foremost, to please ourselves.

SHARING INDIVIDUALITY

In general, this transit helps us to appreciate people and things
that don't always blend together well. We'll learn to take a new
look at life's contrasting features, seeing them as something that
adds drama and interest to our social landscape. We may learn to
realize that, while conservative behavior can be a safe way to
function, it's much too bland for Uranus/Venus expression. This
transit wants us to emerge as a more vibrant personality, one that
would be interesting for others to get to know. It could be exactly
what wallflowers and social late-bloomers need in order to
sparkle. The key to success here is learning to value tolerance for
all that is different. This invites a broader range of people into our
sphere, who can show us how exciting shared individualism can
be. We may be surprised to find how much we love this new,
dynamic approach to our relationships.

CHAPTER THIRTEEN

URANUS TRANSITING THE EIGHTH HOUSE

TIME BOMB

Transiting Uranus entering the Eighth House meets with powerful resistance by attempting to do what it does best: uncovering the truth behind those matters that, in this case, we prefer to keep secret. In theory, the Eighth is where we have experiences that help us to probe and penetrate hidden levels of our inner self—levels typically exposed whenever life forces us to confront utmost, psychological pressure. This alone should provide us a few, truth-revealing moments. It does, except that this house takes its own cryptic, complex route in getting to the heart of any issue. Uranus may detour at times, but typically it opts for the quickest way to get from "A" to "Z". The Eighth House is more like a darkened labyrinth offering occasional multi-leveled clues as to how to safely navigate its tricky obstacle course. We don't seem to take easy ways out in this house. It's typically a twisted path of complications that we follow.

The Eighth House is also where we deal with guarded emotional realities that seldom see the light of day, at least those regarding our most private relationships. Some astrologers view the Eighth as descriptive of the darker underbelly of marriage, where a whole different set of dynamics may underlie any surface appearance of matrimonial harmony. Now that we have established a committed partnership (in our Seventh House), what are we to do when this relationship doesn't immediately cure what ails us, or when it doesn't automatically provide the state of inner balance and self-completion we anticipated? Well, the wise and the brave just roll up their sleeves and get to work by facing their thornier psychodynamic Eighth House issues!

In this regard, transiting Uranus acts as a blessed messenger undisguised (it won't abide by the usual Eighth House cover-up routine). We may be suddenly alerted in our closest relationships to the sound of inner rumbles that have been operating silently underground for quite a while. If we are married, then some vital truth about our union needs to surface and make itself known. It's usually something we've kept under wraps for good reasons (maybe it's hurtful material that could damage the stability of our partnership, emotionally injuring ourselves or our significant other). Often, it is a big secret made more powerful and ultimately destructive when not revealed and shared.

Therefore, clever Uranus—in its brilliantly mischievous ways—sets up a scenario in which we or our partner seemingly stumble by accident on this hidden stash of thoughts and feelings deliberately kept from being revealed—until now. Some minor issue can become the catalyst for a major confrontational showdown or a long-overdue blow-up scene. This is most likely when Uranus aspects a planet that's already dealing with its own natal frictional patterns (e.g., if transiting Uranus in our Eighth squares our natal Libra Moon, which in turn also squares our natal Mars in Cancer; if suppressed, a natal Moon/Mars square can have problems with allowing anger to inappropriately build to the point that dangerous levels of hostility accumulate and eventually discharge in very irrational ways).

Here, explosive Uranus is touring an emotional house of active volcanoes. What potentially erupts can be quite an eye-opener and a soul-wrencher. When the turmoil is over, if we have come clean with our initial grievances, we can renew our commitment based on deeper mutual understanding. If not, we terminate this union and learn to courageously move on. Things either die for very good reasons in the Eighth or else come alive again with greater gusto and vibrancy.

THE NAKED TRUTH

One lava-hot topic to confront at this time is sexual intimacy. Is someone in this union a taker rather than a giver? Has sex become a manipulative bargaining chip, a source of emotional blackmail, or what somebody selfishly uses to punish another through guilt, shame, or insecurity? Uranus says enough of these sicko games—let everything come out into the open (including all ulterior motives). This planet is not really concerned with sexual matters (they're too earthy), but it hates subterfuge and those nasty, sneaky little schemes we use to get back at others or to get "the goods" on them (in true Eighth-House style). Uranus would prefer that all hell break loose now, allowing everyone to show their true colors—fangs, claws, and all.

So far, this interpretation doesn't sound too user-friendly, does it? But remember, it was mostly meant as a word to the wise and the brave, not clueless cowards like some of the rest of us! We the chicken-hearted will probably not want to get this heavy ball rolling, and instead will seek to detach ourselves from or completely ignore any distress signals sent out by our wounded relationship. This avoidance act won't work in the long run. Something has to give way in order to ease the build-up of steamy pressures. Therefore, it's best to come clean and allow for a needed purge of formerly unventilated attitudes and feelings. Sex appears to be the issue, and yet there's something more covert, which involves power drives that need to rise from the underground. Transiting Uranus will operate here for many years, so we shouldn't feel too bad if it takes us a while to address all of the complexities to be found. These sensitive issues didn't crop up overnight, and they're not going away any time soon. We probably first saw signs of them in our early childhood, when we observed the dynamics of give-and-take between our parents, and within our family in general.

Sudden Arousal

Some of the issues mentioned above may still arise within new relationships in which we involve ourselves, even when we are single but are steadily seeing someone. Here, we may get to feel highly charged both emotionally and sexually in a partnership that unveils our hidden depths. We (our partner included) know that we can get out of this union quickly if it becomes too much in terms of psychological wear and tear. At least that's what we innocently tell ourselves at the moment. However, it would be better to first explore this relationship as deeply as we can stand to before quitting on it. By doing so, we could emerge feeling more empowered.

Of course, in some cases we also might feel devastatingly shattered once we realize we can't control people and make them stick around forever. Still, something potent within us, a source of hidden, inner strength, can now surge forth. Our new relationship attempts to rapidly pull an emotional resilience out of us that we need to own up to more consciously. Manipulative neediness on anyone's part is not to survive this transit (nobody gets away with pulling any melodramatic acts of desperation).

Things can get dramatically erotic as deeper passions that we didn't know existed spill out and intensify our bond. Curious Uranus arouses our instincts to delve further into our sensual needs. Maybe physical sex is what we think will most quickly awaken those needs. However, the Eighth is Pluto's natural domain, so at first we may not feel safe openly revealing our sexual hungers and our vivid fantasies. Strongly conflicted desires are more the norm in this house. Still, transiting Uranus works to destroy taboos and to break down resistances to a fuller release of intense desire. Why frustrate such needs, asks Uranus, only to have them turn into discomforting obsessions that imprison us to some degree? Instead, we are to unearth this dimension of ourselves and liberate who we really are deep down inside.

The Eighth is a house where our emotional, physical, and even material appetites can get on the greedy, insatiable side. It's

often not that obvious to others just how single-mindedly possessive and fixated we can be here, because Eighth House planets
seem more self-controlled and disciplined on the surface.
However, with transiting Uranus, we are not going to get away
with owning anything belonging to another person. People
(including their psyches and bodies) are not our property for
exclusive use. Uranus won't allow for seductive schemes designed
to ensnare others into the sticky web some of us have spun. A
burning desire for another person is okay as long as we are clean
and up-front about fulfilling that desire—but we are not to devi-
ously get what we want out of somebody who's not even aware of
what's happening. We are not to secretly take over this relationship by using unfair tactics. Transiting Uranus will make sure
this will not work for us. In fact, such tactics may dramatically
backfire. In this house, we experience the consequences of uneth-
ical behavior (which I believe first becomes an issue here, even
before we deal with it more philosophically in our Ninth House).

FOOTING THE BILL

Sometimes the pressing issue is a power struggle over money as a
mutual resource. We could attract someone undergoing a (hopefully brief) Uranian period of financial upheaval or of some other
form of material instability. However, our partner's personality is
so exciting that we figure it's worth paying for more of otherwise-
shared expenses than we normally would. If we are married, we
probably at this point are feuding about long-time financial patterns, and how they interfere with our own freedom to currently
spend money as we wish. When we're dating someone new, we
may be more sensitive to the issue of taking turns regarding who
pays for what and how often. This is so even if we are already living with a lover, and are not too sure where this union is headed
(a likely state of mind for us with Uranus here).

Whatever the case, the money issue can suddenly open up a
Pandora's box of woes, disrupting the whole works. It also puts
someone's character and integrity through the acid test. Are we

truly being generous in spirit, or are we indirectly trying to buy love? Is our partner someone in a temporary state of financial shakiness, or we are dealing with a chronic moocher? Uranus insists that we see things in the clearest possible light, free of ambiguous interpretation. It is very important that we are forthright in our attitudes about money and spending—especially if we are married and going through a rocky period concerning economic survival.

Uranus in the Eighth means that we can view our partner as either a risky, impulsive spender or as a financial genius who can pump exciting, new energy into our material structure. Whatever the case, it is important that we do not lean on a partner completely for our material support, because Uranus cannot provide the predictable security we seek. This can become painfully apparent during divorce proceedings, when another's financial status is revealed to be more of a mess than we may have realized. Part of the reason that this occurs is that our partner has not felt comfortable sharing the details of his or her management of money. We'll need to look within to see what role we may have played in this. Have we been intimidating or too controlling in this area, prompting our partner to spend freely or perhaps to make unwise investments behind our back? Why does our partner pull away from us in this area in an attempt to circumvent our participation?

RISKY BUSINESS

This would not be the best time to begin a business enterprise where the start-up capital is expected to come from several different sources (i.e., three or more business associates pooling their funds to make a dream happen), even if this feels like an exciting and novel thing to do. Transiting Uranus is the tip-off that someone may bow out of this arrangement at an inconvenient time, and leave others holding the bag. Neptune transiting here offers a similar warning. Of course, if Uranus happens to aspect Venus or Jupiter at the time, a partner may enter the scene and open doors of sudden opportunity for us, only to quit unexpectedly once the excitement dies down. Maybe by then we've taken off running and no longer need that particular partner's energy. Still, rather than depending on Uranian types, it's better to have some established Saturnian money hanging around just waiting for the right opportunity to be used. Even then, Uranian-inspired Eighth House business ventures may not have the longevity needed to satisfy either Saturnian money or us.

We'll need to do our homework and study the details of new enterprises—while also brainstorming—before taking big financial risks. We could attract amazing windfalls, but sometimes things could blow up in our face and leave us in a very unstable fiscal condition. In that case, we would feel we have no real freedom at all, because by then we may owe a lot to others who are eager to swoop down (vultures are ruled by the Eighth) and collect their money while we are in this chaotic state of affairs. This could include tax collectors. Therefore, we need to tread carefully. The bigger the deal we wheel, the greater the shock should things surprisingly fall apart at the last minute. The best advice is to keep our financial negotiations as uncomplicated as possible, yet take a few smart risks every so often when our intuition runs particularly strong. Nothing is hopeless with Uranus. Solutions may come from unexpected sources that we have yet to tap.

QUICK DEATH

It's hard to discuss the Eighth House without using the "D" word at some point. Physical death may seem a bitter ending to an unfulfilled life for some. To the Eighth House, it's just another meaningful—even momentous—transition to a more profound state of being. Like Pluto, the Eighth is interested in life's biggest mysteries, which certainly includes our death experience, and whatever part of us survives beyond that point. Uranus is not an emotional planet, so we're challenged to learn to approach experiences of death with greater objectivity and detachment. This helps us to accept the soul-liberating potential of dying, to see it as more than merely a sad loss of life. While detachment is often misinterpreted as indifference and even coldness, here it means we get to explore such matters without undue fears and insecurities getting in the way. Truth-seeking Uranus wants us to get down to the bottom of it all: what really happens to our consciousness when our body dies? Where do we stand with the overall concept of an afterlife?

Uranus believes in getting real, even about those things that disturb us. When someone dear to us dies suddenly and unexpectedly (which could happen during this transit), we are encouraged to grieve as freely as we must. However, Uranus urges us not to prolong our period of mourning in order to satisfy social customs or the expectations of others. We don't want to follow any protocol, especially in this situation. Maybe we don't wish to wear dark colors at the funeral or do whatever else emphasizes the wound of death. Uranus has little patience for melodrama and romanticized pain—that's Neptune. During this transit, quick deaths can also occur on different levels other than the physical. Seldom will this transit pinpoint our own demise. It will at least take many supportive indicators in the chart before we can reach that final conclusion; even then, it's debatable whether we could unquestionably foresee this ahead of time.

What is possible is the death of negative attitudes that have had a stranglehold on us, and crippling fears that have held back our potential for years. We can experience the release of hatreds that, in and of themselves, could prove fatal at some point, the end of chronic habits that smack of self-loathing, and the death of feelings that only serve to isolate us from rich, emotional participation in the world. All of these deprive us of the vitality required to feel alive and well. None of these conditions will magically melt away. Transformation often is not a gentle process in the Eighth, but each stumbling block is to be faced squarely and willingly shattered. Uranus will make sure that no distortions are left to take hold again and do their silent damage. We will need to want this with all of our might. If we do things right, we can clear out a lot of toxic debris and prepare ourselves for the upcoming years of wonder awaiting us in our Ninth House.

URANUS/PLUTO TRANSITS

DEEP SHIFTS

Transiting Uranus aspecting Pluto is a contact fraught with tensions that are not going to help us to feel that we are on solid ground, at least not at first. How we register these patterns depends much on our age and on our emotional flexibility. The less in charge of our lives we are (maybe we're only pushing seven), and the less we know ourselves, the more these energies operate through others who control our movements and choices in life (our family, the community, or the government). Both planets, being transpersonal, can represent societal activity beyond our conscious control, especially according to the themes of these planet's sign placements. Such signs may describe background elements that play into our current situation, but they usually do not define our main and most pressing issues. Uranus and Pluto each deal with overthrowing something inside us that has blocked our growth for too long. They also can be extremists, forcing us to get rid of people and things in dramatic ways that completely alter our life structure and our overall perspective.

SMALL DARK STRANGER

The intention of this book is to show what we can do to help ourselves feel alive and well with the Outer Planet, Uranus; therefore, our natal Pluto will here be considered as more than just some generational marker or impersonal collective force. Let's treat Pluto as just another member of our inner, planetary family, not as some eccentric out-of-town guest who likes to wear black a lot and stare piercingly at everyone. Also, realize that Pluto has a notoriously standoffish, loner temperament and doesn't mingle or blend well with others. (Pluto will rarely attend holiday dinners or large weddings. Operating rooms and funerals?— that's a different story.) This planet represents darker, more remote parts of our psyche that are seldom invited to participate in life's superficialities (besides, life already knows that Pluto won't show up). Dare-devil Uranus is now bent on stirring within us strong and vital Plutonian responses to our current situational reality. Uranus will make certain that reclusive Pluto sees the light of day.

Actually, people are becoming more conscious of having and needing the energies that Uranus, Neptune, and Pluto symbolize, and they're even acting out these forces on a more continuous basis in the external world. Hordes of individuals are doing their Outer Planet dramas daily, although often miserably so, as suggested by those disturbing newspaper headlines we read. Activating such drives provokes a stronger tendency to complicate our lives in ways history has never before witnessed. Oh, what our brain circuitry must be going through! The social fabric carefully woven by centuries of conservative Jupiter/Saturn values is now beginning to rip apart, for better or worse.

DEMOLITION DUO

What both Uranus and Pluto have in common is a fearless approach to altering the status quo of anything, sometimes drastically so. They both have the guts to revolt when they've *had it* with tiresome, useless structures. They refuse to be trapped by that which deadens the fires of our spirit. Nothing is to stop them in their tracks (our ego's resistance only unwittingly eggs them on to break things up further). Uranus shatters while Pluto pulverizes—please, keep these planets away from all glass objects! Together, they are a forceful combo designed to wake us up and shake us out of any deeply entrenched rut. They symbolize our ability to entirely remove long-standing blockages. The difference is that Pluto can be exceedingly slow, while a Uranian overhaul can be ultra-fast. However, once stimulated into action, they will not lose sight of their intended target. These planets can be cosmic sharp-shooters.

THESE REVOLUTIONARY TIMES

Does our natal Pluto trine our Venus, or square our Mars, or perhaps quincunx our Moon? Whatever our natal aspects are, we'll need to review our current level of Plutonian awareness in light of our accumulated experiences. This means we need to honestly evaluate how transformation has played a role in our lives. Can we consciously see the positive results of our previous Plutonian trials-by-fire? How well have we dealt thus far with Plutonian issues of power, deep pain, dark secrets, lust, rage, primal passions, and other primitive instincts that we rarely vent—in fact, all of that top secret stuff we don't want anybody to see in us? Pluto symbolizes qualities that are difficult to claim as part of our makeup. It's much easier to talk of Plutonian transformation or regeneration than to actually do what it takes to accomplish it.

Nevertheless, Uranus is now ready to crash through Pluto's steel and barbed wire door—the one we've learned to reinforce with many deadbolt locks. Why does Uranus dare to do this? It probably senses that, on the other side of that door, trapped energy has been festering due to a lack of fresh air (ventilation) and blue sky (greater vision). As long as we hide behind our oppressive little secrets, we rob our life of a piece of the Truth. That's one potential danger that gets Uranus roused. Pluto, a planet that is as willful as Uranus, won't allow itself to be so easily invaded. A power struggle is likely to manifest, one that often comes from tensional forces in our environment.

LOSING CONTROL

We can expect, especially with more demanding aspects, a time of psychological battles involving the breakdown of long-standing, internal control complexes that now can spill out freely into the world; they become externalized as power-plays in relationships or as do-or-die survival issues—even involving the mighty forces of Nature. Our hyper-defensive Plutonian parts can put up a good fight to resist being overtaken until, at some pivotal point, these parts begin to turn the tables and use Uranian energy to quicken our self-destruction. We can use our inner turbulence to relentlessly attack and destroy ourselves on some level, although not necessarily physically so. That becomes one not-so-great way to handle this enormous tension. Pluto annihilates when frustrated; heck, it even annihilates when feeling on top of things.

Another option allows us to finally release the tight power grip that any dark, psychological elements may have had on our inner development. We are to release any urge to relentlessly control our here-and-now circumstances. This transit can be a phase of turmoil for us, but one hinting at a more creative, empowering resolution. Whatever the case, any self-created Plutonian negativity is not to get the upper hand at this momentous time. The trend of this transit is to demolish—Uranian style—what no longer deserves to remain intact and to leach us of vitality and strength. Pluto rules vampires and other nasty blood-suckers, including the manipulative human kind. The conjunction, square, quincunx, or opposition can all be expected to take the strong-arm approach to wiping out obsessive, problematic behavior.

With the trine and the sextile, things may still feel intense (activating our Pluto always is), but at least our inner make-up is more in agreement with any reformative program that ingenious Uranus is trying to put together. We may have to release a toxic load or two—which alone makes us feel stormy and malcontented before we can absorb the greater benefits that Uranus provides. Still, this can be a time of heightened awareness about central life matters that have profound meaning for us, especially if we are willing to muster the courage and determination needed to usher in a few major breakthroughs in consciousness. Uranus is most interested in zapping those parts of our Plutonian make-up that are sick or dangerous to us and to others. These are the elements that need radical reform the most. However, wherever we are already constructively empowered, Uranus only seeks to increase the voltage and greatly expand our playing field.

BIG SHOWDOWN

From 1995 until almost 2004, members of the Pluto-in-Leo generation will have a Uranus opposition, although not all at the same time. After that, it will be Pluto-in-Virgo's turn, followed by the Pluto-in-Libra group, and so on. Given that this is an opposition, it appears at first to operate only on impersonal, external levels. We perceive its influence in the vast world around us, maybe through our nation's governmental affairs. It even seems to shout at us from unsettling newspaper headlines. Altogether, this transit appears to be collective and, therefore, less a direct part of our personal world. Our culture and our society further shape this experience, especially regarding the development of trends and events that affect mass opinion and behavior.

Still, there are some scenarios in which Uranus opposing Pluto will seem very much to be our own dilemma. One area, in particular, where some of us probably have ambitiously sought Plutonian power is in our career. Are we now being downsized after years of loyalty—mixed with unrecognized resentment—to a company that has merely used us all of this time? Perhaps for the company to continue to live, our working role must die. If so, Uranus is now trying to get us to recognize how our deep discontentment— related to the way we've been unfairly or shabbily treated in the past—is something that this career crisis must now bring into sharper focus. Being mistreated or undermined by the powers-that-be alone implies a Plutonian dilemma. Sometimes we've done everything to near perfection and have no axes to grind, yet we still get fired out of the blue or are singled out for whatever strange reason as someone who's disposable or replaceable.

In this or in some other area of our life, better described by Pluto's natal house, our structure may feel like it's caving in, with the walls crashing around us. From a soul perspective, however, we will end up more alive and well than ever just by psychologically surviving this destabilizing transit. Something deep inside will have its long-needed awakening. This Uranus/Pluto showdown can have great psychological rewards for us. The benefits will not be apparent to us until a few years later, especially if we're currently out of a job or are devastated by some other "freak-out" incident in life. It's then that we are better able to see this period as a weird blessing in disguise. We could also live in times when social revolt is stronger than usual. Think of the Uranus/Pluto conjunction of the mid-1960s to get an idea of the urgency that these planets bring to the social dialogue; that conjunction in itself was a transit that presented personal upheavals for many folks at the time.

While we may seek to fight the System during this transit, why not turn Uranus' laser beams back on ourselves, so that we can burn through our own Plutonian darkness? We may be a product of a troubled society, but such a society is also a collective network of individual, unregenerated selves. This could also be a time for some of us to face up to our greed and selfishness as Uranus gives Pluto a momentary taste of powerlessness. Pluto's not used to being thrown a curve ball like this. Our reaction to this can be extreme and sometimes violent.

SCREAM

It's hard to say what's more terrifying to contemplate: wildly unleashing the forces of Uranus and Pluto, or *not* doing so and instead pretending everything is just fine inside us—with no problems whatsoever that a little self-denial can't handle. Uranus hates deception of any sort and is ready to dynamite the psyche if pushed to the limit by our ego's willful resistance to change. Uranus "knows" we're not as content as we try to appear. Pluto, meanwhile, quietly ticks away, aware that its time to belch fiery fury is near. The more we deny ourselves release, the more explosive are the consequences when such release is imposed on us in the form of external crises. Our soul wants to let out a few blood-curdling screams and to lash out at a world that's doing its darnedest to stifle the more vivid parts of our individualism. Pent-up and sometimes toxic emotions need to be released.

Actually, some of us do scream, kick, and scratch ferociously at life for a brief, shocking moment—just as a few cultural and political revolutionaries did in the 1960s. We can feel as though we are outside of ourselves, oddly detached, while looking down at parts of our consciousness that are having furious fits about the evils of society, or that are railing against human injustices perpetrated by dictatorial authorities. Both hidden paranoia and righteous indignation can leave us agitated during this time. Uranus the Liberator is trying to engage Pluto in an all-out war against the destroyers of individualism. Uranus' worst nightmare is living in a totalitarian state (something corrupt Plutonians could easily create).

However, when Pluto comes alive and takes up a cause that serves the collective good, it acts with tremendous intensity and single-mindedness. Before we feel compelled to take our rage to the streets, maybe we should first experiment with our need to scream and holler while taking a shower. Remember, at some point, we'll still need to work out an internal solution to what appears to be outside enemy attack. Uranus realizes that we need to start dealing with our Plutonian realities in a much less tense, defiant manner.

AFTER THE STORM

In general, Uranus/Pluto transits provide an opportune time to do some serious inner remodeling. Effective self-rehabilitation can especially be accomplished during the quincunx phase. Uranus is trying to free us of whatever has weighed down our soul. It knows how heavy some of our Plutonian fixations are, plus how damaging it is to allow such disturbing obsessions to go unresolved. Hidden pockets of hatred are associated with Pluto, and hatred is one deep emotion with which Uranus is very uncomfortable. Uranus associates hatred with being tremendously entrenched on some painful level. This planet will do whatever it takes to blast us out of that terrible feeling. Sometimes, we'll have to let all of our seething poisons surface and vent before we can be healed. Pluto at this point is not going to feel satisfied unless it can erupt in flames.

After the storm is over, and we've gotten a heck of a lot of garbage out of our emotional system, Uranus is ready to unfold its future plans. It will present a clear, exciting vision for us, a goal that will require that we use our Plutonian powers of concentration and stamina to the utmost. This doesn't have to mean going on some messianic mission to save the world from its corruption and darkness; that route could even lead to more unexpected power struggles. Instead, we still have plenty to do in just trying to further redeem ourselves. It's our personal existence that needs a big face-lift. Uranus seeks to transform our approach to living a boldly authentic life, free of the illusions and the hypocrisies of our past. The good news is that such a stifling past may be gone forever. Amen.

CHAPTER FOURTEEN

URANUS TRANSITING THE NINTH HOUSE

STRANGE SAFARI

Transiting Uranus entering the Ninth House is so glad to be free from the complications of the sticky Eighth that, as rumor has it, it's designing a rocket ship to take it straight to Jupiter! That's right, Uranus now has travel fever and wants to get far, far away from the scene of the grime (uninspired, survival-oriented existence). If we have dislodged and freed potent underground parts of our nature when Uranus was passing through the Eighth, we probably are ready to allow ourselves to reach out for more gusto in life and sweet "soul" adventure in a positive, open-handed manner. If so, the Ninth House makes the perfect next stop-over for hearty Uranus. There's plenty of room for at least mental, if not geographical, exploration in this area—thus satisfying Uranus' yen to wander down a few less-traveled roads of learning, paths than can seem weird to less mentally fearless types.

FREE THINKER

This is a house where we develop expansive beliefs about society and about even the Cosmos itself. Here we formulate our concept of God and of Universal Law. Once we are exposed to a thought system that inspires us, we often pump it up with the power of our faith and trust. This helps us to construct a sense of meaning about life, whether on spiritual (religion), intellectual (philosophy), or societal terms (politics). However, a potential problem with this approach is that some of us then want to spread such beliefs far and wide in order to reach the greatest amount of people, those whom we hope will reaffirm our beliefs. They, in turn, enthusiastically distribute these thought systems and, before you know it, a simple belief turns into a social movement. Some social movements become solidified and well-assimilated by the collective mind. What if the original belief is based on a completely false premise? What if those who redistribute the message embellish the content or add their own biases or misassumptions? These are some human possibilities that fire up transiting Uranus. We find ourselves eager to search for what is real versus what is societally endorsed propaganda.

Transiting Uranus gives us a lot of room to question big, social issues, because everything about the Ninth pushes for broader understanding—although sometimes widespread misunderstandings can also result. Neither facts nor rumors are contained within small, discreet circles in the Ninth—they disperse like wildfire, and become instant topics of global interest (especially in today's cyberworld of online data surfing). Uranus is fascinated with the media's potential to pollinate fresh ideas of futuristic promise. Convincing the Establishment that such ideas are worthy of respectful consideration becomes a problem for sometimes off-the-wall-sounding Uranus. Of course, this would be more of a problem for us if we were born with Uranus in our Ninth: our visions *throughout our lifetime* then become challenged or are subjected to ridicule for being "too far out." However, with the relatively short-term transit of Uranus, we are much too busy looking at life from excitingly different angles to worry about how others will cope with our large-scale views later. All that we know now is that conforming to standard ways of thinking and conceptualizing is unappealing to us.

OPEN WIDE

We begin to reconsider things on a broader basis in the hope of uncovering the real scoop behind all that we intellectually pursue (our vigorous Eighth House workout took us on a stimulating archaeological dig for hidden truths, and now we don't want to stop). Our quest is to reap meaningful rewards that help to open our eyes. Uranus enjoys truth-finding missions, so this transit gives us an opportunity to question everything that we currently believe to be true or right. This can, however, trigger disillusion-ment, as a few long-held, but unexamined, beliefs become shat-tered in light of new (shocking?) revelations.

This is nothing to worry about—it's just Uranus' way of open-ing up uncharted territory for us. We've probably been discon-tented with a few of our "cherished" beliefs and "unquestioned" assumptions for a while anyway, but couldn't fathom what would take their place if we ever let them go (and thus risk feeling an empty void inside). This is typically the case with religious and political beliefs. Uranus is now speeding up our mind's evolution. Refreshing concepts of a progressive nature will be available to excite us by arousing our sense of wonder. Uranus is a vision-awakener. The world can look bigger and brighter to us when we switch to a wide-angled, perceptual lens. Even our concept of God and the Cosmos can expand and become more exhilarating and unlimited.

MINDSWEEPING

What we probably won't be able to hang on to after this transit is over are rigidity, narrow-mindedness, programmed social preju-dices, and subjective intellectual biases. None of these fit into Uranus' future-oriented vision for us; they all must go. Let's hope that we experience a few Uranian trines and sextiles to our natal planets during this period that could help make this planet's house assignment more pleasurable. These transits heighten our excitement about learning better ways to view the puzzle of exis-tence without the tensional edginess more common of Uranus in conflict patterns. It becomes a thrill to dump the old and welcome the new, at least intellectually. We can play with concepts that amaze and delight us. However, the key is to initially make this an experimental phase, where we tentatively put our faith into such ideas and ideals. Let's have fun growing mentally without feeling obligated to faithfully adhere to any system of thought. We'll also need to be careful not to get entangled with any beliefs that demand a heavy dose of reverential treatment. Uranus is not much of a devout worshipper.

It's also best not to suddenly embrace any ideology too fervently or fanatically (as a square or an opposition from transiting Uranus might tempt us to do), at least not until we have tried these new concepts for a while to see if they truly fit us. Remember, Ninth-House energy makes us want to get on our soap-box and indiscriminately preach to the unenlightened whatever our revelations have "revealed" to us. However, this is not a reliable time for firm convictions, because we are still in a state of intellectual and spiritual flux. Nothing is to be made concrete or written on stone tablets until after our minds have returned home from their fantastic journeys far and wide—*if* we ever do come back home to familiar territory. It's probably safe to say that our world-view will be quite altered by the time Uranus is ready to transit our Tenth House. By then we will need to actualize some of the broader social and spiritual visions that have inspired us in our Ninth.

ONE STEP BEYOND

This is a house where the expansion of consciousness is encouraged. Here we begin to realize that reality is more mind-based than is commonly recognized by materialistic science. Already highly intuitive and eager to do something to prove it, Uranus will bring us experiences that show what it's like to go beyond normal brain-wave activity and to activate the mind's less frequently tapped levels of awareness. Such awareness is not mystical in tone (that's Neptune), nor does it arouse us deeply on an intensely emotional level (unlike Pluto). We simply are more apt to have sudden insights, quick overviews of situations, and even a dash of prophecy to help us light our future's path. All of it proves to be very exciting and stimulating stuff for us when it's happening. We are also highly alert and conscious when we are so tuned in, rather than in any kind of dreamy trance state.

I associate Uranus with telepathy (which is really a form of amazingly rapid communication—the psyche's souped-up version of e-mail). We might find ourselves picking up on the thoughts of others far away more often than we realize (that's when little impromptu dialogues with such folks suddenly pop into our heads, usually interrupting whatever else we were doing). Then they call us the next day or send us a letter—surprise!—saying that we were recently in *their* thoughts, too. Pay closer attention to this (and realize that transiting Uranus contacting our natal Mercury, Moon, or Jupiter may provide similar phenomena).

In general, taking classes in ESP and similar phenomena might be stimulating and educational, but it makes no sense to try to rush our inner development, or attempt to force the emergence of paranormal talents we may not even really have (wait for a strong, hard-angled Neptune transit to indulge in that kind of self-delusion—at least by then it will *feel* more real). Uranus is a planet that likes things to be spontaneous—it'll only repeat experiences sporadically, if at all. Thus, sometimes we'll feel so plugged in to the Cosmos we almost can't stand it, while at other times it may appear that even our pet iguana is more gifted with psychic awareness! Why does our hot intuitive streak come and go like this? Maybe it's because we also need to be appropriately grounded to get through the mundane business of functioning in the everyday world; therefore, life every so often demands that we snap out of it and stop cruising the "higher realms." However, in short spurts, Uranus comes in handy when we need a little extra intuitive *oomph* to help speed along our daily tasks. It's best to let this energy come and go naturally, without trying to manipulate the process.

226 • CHAPTER FOURTEEN

JET STREAM

In the Ninth, we can circumnavigate the globe and stop at many exotic ports of call, if we had the money and plenty of leisure time to travel abroad. This is not realistic for most of us, who can't even fit in a free weekend to drive to the mountains. Still, thoughts of international touring may come and go more than usual during this transit. The Ninth preaches "mind over matter"; perhaps we should first indulge in visualizing those special locations where we'd love to spend some quality time. Those born with Uranus in the Ninth probably already do this a lot—and probably end up getting out of town every so often, just before their pent-up Uranian energies become problematic (and before their mini-van's battery blows).

Anyway, we may find that vivid images of a faraway place suddenly coming to us in a flash, or perhaps we unexpectedly run into fascinating people who have lived in or have recently visited the destination of our dreams. We may not know why we have such an insistent desire to explore this area, but the urge is strong. If the current travel trend is to summer in balmy and beautiful Acapulco, we'd rather explore the volcanoes of Iceland! Our tendency is to pick those oddball places less glamorized by travel agencies and TV ads. And so, no, we probably *won't* be coming back to Jamaica this year. It's the Galapagos Islands for us and the thrill of kissing a giant tortoise (besides, there are a few active volcanoes there, too, and an unexpectedly varied climate).

If we do get to fly away to our ideal spot, we may have little interest in playing the well-behaved tourist who gets off and on the chartered bus with a structured group. We probably don't even want to be around familiar crowds, but instead are itching to go off the beaten path. However, we'll need to examine our transiting aspects before we globe-trot with the idea that "silly little rules" apply to others, not us—or that certain social customs only pertain to the natives, while we're gonna do what we want on our once-in-a-lifetime vacation, dammit! Uranus will quickly show us that too much willfulness and defiance can backfire on us when we flaunt our free style and disturb the locals or those who enforce the law. It's best to avoid mingling with cultures known for their rigid, archaic social values or dress code. We could, of course, simply play by the rules and make no waves. The bottom line is: don't expect everyone else to become instantly Uranian and liberal-minded just because we've breezed into town and want to have a good time.

Actually, there is a great likelihood that we will meet very progressive people, living at the place we visit, who don't seem to fit the social norm. They're considered unusual and even eccentric in some way. Our rapport with these folks can be uncanny and electric, a case of instant friendship. Uranus exposes us to what's different, but it also makes sure that exotic birds of rare plumage and song find each other. We probably will never look at "foreigners" the same way again because of this enlightening transit. We'll be less threatened by what at first appears culturally alien to us and others. Exposure to ethnic differences becomes a cool thing to experience, and very educational on a soul level.

UNIVERSAL CITIZEN

Before this transit is over, assuming we've handled it well, we could feel that we have much more tolerance for humankind's diversity than ever before. We intuit the higher purpose of collective variety as expressed through a wide range of social customs, religions, or general lifestyles. We are truly citizens of the world at this point, comfortable wherever we go and eager to learn about life from our travels. Global progress becomes exciting to us, especially along lines of constructive technology or of humane enterprise. We can adopt a "live and let live" policy, knowing that there's plenty of room for colorful individualists and conservatives alike. Uranus can play the role of freedom fighter for the oppressed; therefore, this transit can stir up strong political passions, and a determination to improve the social status of underdogs in society. However, it's best not to resort to lawless ways to shatter old regimes and to enforce overnight reform by using violence—which, unfortunately, is something that detached Uranus can ideologically justify, at times, as essential to fulfilling the greater cause at hand. It's best to first find educational ways to uplift the social consciousness of others.

URANUS/JUPITER TRANSITS
SHOOT FOR THE STARS

Did you know that mythological Uranus was Jupiter's grandpa? It's not too clear if they ever had any real family contact. Astrologically, however, Uranus and Jupiter share a lot in common. Uranus is a fiery kind of air planet, while Jupiter's an airy sort of fire planet. Both can get charged up with excitement, both think that being small-minded is a big waste of time, and both need lots of elbow room to explore life's greater potential. Also, each is key to how we face an unknown but hopefully promising future. Will any risk-taking or "going out on a limb" be involved? Almost certainly. Will we have to reevaluate elements of our past in light of new revelations? You bet.

When transiting Uranus aspects our natal Jupiter, the doors of perception may fling wide open for us. The future beckons and our aspirations may run high. We may feel unstoppable when it comes to reaching our goals, but not because we're human bulldozers plowing through all obstacles in our path. Instead, we may feel as though the angels are on our side, and the Cosmos is giving us a thumbs up! What we actually get out of this transit depends on what we are willing to freely give to the world. Jupiter connects us to our inner goodness, while Uranus makes truths about our ideal self known. Selfishness and greed won't help us to tap into the best this combo has to offer. However, feelings of altruism are well-supported. Although this is not a highly empathetic mixture (that's Neptune/Jupiter), it does trigger a generosity of spirit (typical of Jupiter), which often manifests as social activism. Uranus and Jupiter want to better the human condition. Both planets are advocates of societal enlightenment. Their understanding of human potential can be vast.

DO WE DARE?

Each planet is able to have far-sighted visions about tomorrow. We may feel a heightened degree of anticipation, sensing that something big may just be around the corner for us, some lucky break or something that might at least relieve the boring monotony of our daily grind. We hope that it's not just a temporary kind of relief, but one that ushers in a whole new world of exciting opportunity. Is the transit in question a square, a quincunx, or opposition? If so, be warned that self-discipline and common sense are two traits that we currently may lack as we attempt to apply this buoyant and high-spirited energy. However, sometimes the daring that we show is necessary to improve our life or the lives of others dramatically. While Saturnian attitudes could get us to carefully weigh the pros and cons of adventuresome action, we might end up taking no chances or making no great leaps of faith. This Uranus transit is not about playing it safe.

SPECIAL TREATMENT

Successful Uranus activity often demands bright but gutsy solutions, including a high degree of alertness and a cool-headed cerebral approach—nothing too zany or impractical, but nothing that's going to drag its heels, either. That's because Uranus lacks endurance, and simply won't want to wait around for results, while Jupiter could also lose steam when put on hold for too long (say goodbye to enthusiasm). With Jupiter we must fully believe in ourselves and those we draw into our personal circle in order to end up on the winning side of any project. We must have faith in a benevolent Universe that is willing to give us special privileges and to treat us very well at this time. Any hidden feelings of unworthiness must be routed out before we can ride the full wave of this transit.

The transiting conjunction, sextile, or trine may lack the extra push for sweeping change seen with more tensional aspects. We may be less driven to explore our future potential; we're satisfied with those more immediate comforts suddenly presented to us.

Nevertheless, we may find ourselves feeling in sync with the Cosmos at this time and thus willing to make a few personal dreams come true. Our hopefulness may be as strong as our positive thinking. Our optimistic faith can turn parts of our life around, very quickly, for the better. If we're not setting our sights unrealistically high (which is often a problem for the hard-angled aspects), our vision can be made workable. Success is more likely, and arriving at that level of success proves less disruptive to our overall lifestyle. The only problem—considering the know-it-all quality of a mismanaged Uranus and the arrogance of a mishandled Jupiter—is that we could convince ourselves that we are God's "chosen one" who's here to usher in some big social change without anyone else daring to question the rightness of our assertive actions or the practicality of our vision itself.

Million Dollar Winner?

Sudden windfalls are often associated with Uranus/Jupiter transits. However, before we run out and wildly buy fistfuls of lottery tickets or bingo cards, realize that our windfall doesn't have to be of the monetary or material kind. Sorry, Leo and Sagittarius, but this transit is no sure-fire ticket to living the sweet life on Easy Street! What *can* pour down on us in abundance is an inner enrichment provided by gaining the inspiration needed to live a fuller and more spiritually uplifting life, as well as arriving at a broader understanding of this complex world. Look at your chart and then look at yourself in the mirror and ask if that sounds like an even greater reward to have bestowed on you. Some would empathetically say "yes!" while others might just reluctantly nod (while thinking, "Ah, but vast sums of money would be really nice right about now.")

It takes all types to make up a interesting world, so we shouldn't feel shallow or guilt-ridden if we do indeed win a major sweepstake. Uranus/Jupiter teaches us the wisdom of tolerance. Maybe we can do big things for our community after we win that forty million dollars—who says we'd be greedy with our big bucks? This could also be an "easy come, easy go" transit for those who win great sums of money but who haven't been blessed with common sense. Uranus doesn't tolerate idiots very well, so the cash flow can suddenly dry up after a whirlwind of insane over-spending! My point is that not everyone during this transit feels compelled to make this a transpersonal experience; maybe that's not even our greatest need at the moment. So be it. It's okay for us to keep going to the race track and betting on our favorite horses! After all, the ability to magnetize great, financial abundance out of thin air is still an awesome act of spiritual power to behold!

BIG ITCH

What if some of us have no great dreams or goals to shoot for? What could be the downside of these aspects? Well, for one thing, there's an agitating restlessness that is not going to be easily satisfied with the routines of everyday living. How boring such routines are to both planets. With no broader horizons to capture our enthusiasm, aimlessness may result. We feel an inner discontent with the status quo of our lives, but may opt for quick relief in any activity that offers short-term excitement or novelty. This is nothing to get too concerned about if a trine or sextile is involved. We could take a weekend trip to Las Vegas, and try to get some of it out of our system. Yet with the square or the opposition, we could go overboard playing slot machines or roulette wheels and end up blowing a lot of our hard-earned money, because we usually are not on that hot, winning streak that we prayed for the previous night.

If we indulge ourselves in grand Jupiter fashion, we tend to do so erratically, either because of unexpected circumstantial detours or because we find it less fulfilling (there's too much focus on the level of earthly desires to satisfy Uranus). Uranus/Jupiter would rather that we aim our sights higher and wider. It is an especially important time to be careful not to add to our credit card debts. Again, self-discipline and common sense are not our strong points at the moment. We hunger for more out of life. Yet if we make this simply a quest to fulfill worldly appetites, we'll become manic about purchasing "toys" that we really don't need, and may end up with a whopping bill that we really can't afford to pay, at least not without disrupting our material stability. The thrills and spills of the inner freedom we seek can get expensive. A reckless attitude only compounds our problems.

TRAVELMANIA

It is typical for us to flirt with travel plans as we seek to get away from our mundane environment. The urge to get out of town can hit us quickly. We have an appetite for more exotic locales (such wanderlust is further emphasized with strong mutable or cardinal placements, but less so with fixed signs, except for Aquarius). It's going to be a judgment call whether we should buy that round-trip ticket to Tahiti or simply subscribe to *The Travel Channel* on cable and to *National Geographic*. What does our budget realistically allow? Of course, Jupiter seldom cares to think about budgeting, and particularly so at this highly impulsive and self-rewarding time. The temptation is to pay more for the opportunity to take a quick, spontaneous trip.

There are travel clubs where the members fly off on weekend trips to unknown destinations. The surprise factor of these mystery sojourns would be very appealing to Uranus, with the added attraction of having to be ready for anything at a moment's notice. This might be worth looking into if Uranus is trine or sextile our Jupiter, and most so if either one of them rules the Ninth (or the Third) House. However, if the rest of our chart is loaded with Taurus and Cancer placements, we'd likely want to prepare for any voyage we undertake. We'll need to also know that the place we're going to has most of the comforts of home, which leaves out that surprise, free trip to Mongolia!

This is, at the very least, a good transit for getting out of the house more often. The great outdoors is often where Jupiter feels most at home. Transiting Uranus is now hinting that our extended surroundings have a few surprising features that could excite and delight us, if we'd only enthuse ourselves to wander around a bit and take in the scenery. The key here is learning to follow our spontaneous impulse to head out the door. With Jupiter aspects, walking, hiking, and generally stretching one's body muscles are to be encouraged. This is a planet that enjoys living a robust life. While Uranus isn't into vigorous physicality as such, it can still pump up our body system with electrical energy that keeps us pleasantly on the go and eagerly active.

If nature calls us to come out and play for a spell, let's allow ourselves to have a few, fun get-away weekends. Just keep the dang beeper and darn cell phone at home! Much of what can make these trips exciting is not being fully prepared for them (although this could prove to be dangerous when we're exploring high-risk environments). If they can help it, Uranus and Jupiter really don't want to invite Saturn along for the ride, knowing that he'll bring his long list of "shoulds" and "should nots" with him— and he won't prefer a worn-out sleeping bag over a reasonably priced, safe-looking motel! So leave Saturn at home defrosting the refrigerator, and let's be as carefree as we can be having a smashingly good time.

THE THIRD DEGREE?

If we could find the time to squeeze it in, these aspects would be great for going back to school in some manner, even if we already have a master's in sociology or whatever. Our mind is eager to be fed new information. Maybe weekend seminars and workshops will do the trick, or even adult educational classes at night. Can some of us reasonably juggle the details of our life and still work toward yet another academic degree (our mid-life crisis)? If so, we'd better also check our Saturn aspects to see if we have support for the stick-to-it-ness required. What we don't need to do right now is to go into a high state of activity overload. However, mental boredom is the kiss of death for some of us during this period. Whatever we choose, opportunities for learning something of value abound. We just need to pick our topics of study wisely, especially if our education is going to be costly. Uranus and Jupiter typically don't consider the practical elements of such stimulating endeavors before taking the big plunge.

In fact, Jupiter always thinks it can handle more than it really can; it promises what it often can't deliver on time or even at all. Uranus is not the planet to tell Jupiter to be cautious and to slow down its pace, because Uranus itself believes in acceleration. Our expectations regarding higher education can be unrealistic. We feel a sense of urgency and assume we to have to take as many courses as we can before we talk ourselves out of the whole thing. This is why Saturn has to be a part of this picture if we are to reach our academic goals, rather than just waste time with costly false starts. Why put all this energy into our goal, just to suddenly drop out later?

A FICKLE FATE

Much of the time we are not even in the control seat when it comes to channeling this transit. Although we may try to take over and willfully direct matters, these quickening energies seems to have a life of their own. Maybe that's a blessing, because the Cosmos is thinking in bigger terms regarding our future than we probably are at the moment. The protective Hand of God is on us, helping us to make quick detours to avoid certain unforeseen pitfalls (Neptune/Jupiter can feel like this, too). That's why the road we travel suddenly takes a sharp turn, but in a favorable direction that we'd normally never consider. Any Uranus/Jupiter transit saves us from stagnation, from getting stuck in ancient mud pits of our own chronic inertia. This time, Jupiter doesn't get much of a chance to procrastinate. We will take inspired action, even if the results are erratic, or are much too unstructured to last. When fate's fickle finger points the way, we can follow a new path both somewhat innocently and somewhat knowingly. It's an odd state to be in, but one that helps us optimistically move forward toward life's next exciting chapter.

Meanwhile, we only benefit in growth by "keeping the faith," and by realizing that life doesn't always have to be a case of uphill climbing. If we truly awaken to new levels of Jupiter consciousness, we'll know it by how much warmer and friendlier we feel toward the world. We are less uptight around strangers. Children with this transit, however, will need to be careful. Our disposition can seem more cheery, especially when nobody's trying to steer us away from our current interests. We are enthusiastically alive and well—but maybe getting a little fatter. Even such body fat can be quickly burned off with the right amount of involvement in healthy social activity, plus some vigorous leg-kicking, hip-swinging exercises.

CHAPTER FIFTEEN

URANUS TRANSITING THE TENTH HOUSE

REINVENTING THE ROLE

Transiting Uranus entering the Tenth House marks a time of identity crisis, but one that's different than what we are likely to experience when this planet crosses our First House. In the Tenth, our role as a responsible participant in society is highlighted; therefore, the professional image we create usually becomes our main focus. Assuming that we are law-abiding citizens, who take a constructive approach to career endeavors and know the limits of our ambition, our status in the eyes of the world is ready to undergo a change for the better. In some cases, even a complete overhaul is possible if we have not already found reasonable fulfillment in our line of work.

This major reorientation doesn't happen "overnight," except in a few, rare cases. Signs of unrest are evident long before things shake up. Yet Uranus doesn't allow for too much foot-dragging if it's clear that we are miserable with our current professional role in the world. Our circumstances during this transit dare us to try something new, and this often requires we take big chances that won't offer firm guarantees of success.

Our reflections on our inner beliefs and the possibility of expanding ourselves in the bigger world around us when Uranus was transiting the Ninth can now help generate the inspiration and idealism we will need to make new waves in our Tenth House. We need to go against the grain of conventional wisdom, which would soberly warn us to stay put rather than risk losing all that we've worked so hard to achieve. Not rocking the boat is fine if we are already thrilled about our career and its future promise. However, should we be discontented, feeling trapped in repetitive work that offers no creative challenges, Uranus is ready to come up with exciting alternative plans. We could be at a dead-end road in our current job situation. Therefore, recognizing a viable way out of this frustrating condition is vitally important. Whatever the case, we are less willing to obediently function day-by-day in the nine-to-five world the way society expects us to.

Driver's Seat

Much depends on how gutsy and willing we are to take certain career risks. The new changes in store are not always going to be smooth for us or psychologically painless. In fact, they may put us in contact with deeper insecurities—as Uranus opposes our vulnerable Nadir—regarding the possible lack of reliable safety nets, economically speaking. This is especially so if we feel a compelling urge to completely strike out on our own in some fashion. Uranus is probably the patron god of the career free-lancer. We begin to toy with the idea that we need a working situation where we can operate independently from authoritative rules and regulations. We would prefer to create our own policies and procedures, rather than cater to any boss or impersonal corporate power structure. We seek to work when *we* want to, at hours or on days that *we* choose. It all sounds like a nice set-up, if we can get it.

If self-employment is truly in our future, our current attitudes and our sense of timing will determine whether we get to bow out from a long-held professional position gracefully (often surprising others when we make our exit), or whether we instead get booted out of a stable job due to our insubordination, rebelliousness, rule-breaking behavior, or spotty work performance. Maybe we give off a different vibe to those in key positions that they find threatening or alienating. We'll need to be conscious of and honest about the career we are in and where it's going. What facets of this career do we have a hard time tolerating? Where do we feel straitjacketed or stifled in expression? Does this career go against our inner ideals? Do we sense we could do better elsewhere if given the chance? Well, guess what, we *are* going to be given the chance! Our periodic thoughts of escaping from our present predicament hold more power to recreate our professional future than we realize. Action follows thought; therefore, we need to see where such musing leads us.

WALKING PAPERS

If we can no longer stand what we do for a living or how we are doing it, but nevertheless adopt a passive, self-avoidance approach, Uranus becomes especially geared up to give us its famous zap (without the benefit of any "flash" of insight first). Things can suddenly fall apart—perhaps we walk into our place of employment one morning to find out we've been terminated because our company unexpectedly is in a major restructuring cycle. It's nothing personal, they tell us, because a lot of others are getting dumped, too! It's just a board or an upper management decision made after much research into the matter. We've been downsized out of our position, and suddenly our dreams for the future have been shattered. Oddly enough, we may feel stunned and relieved at the same time. We're stunned because of the shocking quickness of it all, yet we're also relieved because we realize we weren't happy anyway, but didn't know if or when we should start looking elsewhere for work. The Universe therefore draws that Uranian wild card out of fate's deck, and suddenly one morning we find our desk is to be cleared by noon. Our dismissal by the company can feel cold and impersonal.

Sometimes our job is still there for us, but it's our employer who has either unexpectedly quit or has been replaced (usually due to personality clashes or "creative" differences). He or she has become a manifest symbol of the Uranian unrest building inside us. This in turn can make us feel shaky and disoriented—unless we hate our boss. Then we feel elated and excited at the prospect of finally having someone new in charge of operations. Yet I've had clients with this transit tell me that their new employer is just as disruptive or as temperamentally difficult, only in a different manner. Still, they're a pain-and-a-half to deal with. It's at this point that we start to feel restless and at odds with ourselves. Knowing our chart and zeroing in on this particular transit, we can suspect that, in due time, we will also be zapped hard by Uranus if we stall and do nothing about our situation. The calm, "pretend nothing's happening and it may go away" approach doesn't work when this planet's ready to roll.

BREAK-OUT TIME

The part of us that has usually listened to our Saturnian practicality in the past (sometimes out of guilt and fear) now finds itself pitted against that less-heard inner voice of self-willed defiance that urges us to take this job and shove it! If it's a bona fide career at stake—not just a transitional stage but something that once felt like a true calling to us until now—we may need to let go of the honor and prestige that came along with our title (as well as the reputation and name we made for ourselves). Uranus is giving us a special opportunity to be free of the trappings of our status. We can break away from the pack and do something that is highly individualistic. However, with such Uranian-triggered individualism can come a bit of controversy (although perhaps this is more of an issue for public figures who are highly scrutinized by the media).

The transiting trine or sextile helps to make this transition more adventurous than nail-biting, because we are unwittingly led to many unfamiliar doors that unexpectedly open for us (as long as we already are on a conscious path of unfoldment). People enter the picture, maybe briefly, just to get us revved up about whatever the next step of our career's path is to be. As they quickly leave the scene, others may take their place. Let's hope that they offer us astute advice. Things can feel very lively and electric for us during this time, although not necessarily stable. The transiting opposition sometimes works this way, too.

However, with the tensional patterns (for example, Uranus in the Tenth square the Sun or Moon), we still may be driving down the right road in life, although the road now seems a bit rockier and the street lights are screwed up. We could feel unsteady or inwardly uncertain about taking any direct course of action (with no reliable road map to guide us). One day we flash on the overview of our life and the future looks bright. The next day we feel as if nobody's behind the wheel of this speeding car and that a big bad crash is imminent. Actually, this is probably our typical response to Uranian energy in the beginning. Unless we ignore too many red lights and stop signs along the way, we just need to carry on with our vision while we endure the momentary ups and downs. Uranus makes fate's workings seem fascinating but forever unpredictable.

YEARS OF WONDER

During this transit, we don't have to quit our job to discover our inner potential, nor do we have to unconsciously behave in ways that get us fired (thus liberating us from a stifling, going-nowhere career). However, we'll need exciting new roles to play, if our career allows for it. We cannot afford to grow stale, bitter, or become "bored to extinction!" (to quote an old, Uranian college music professor of mine). Life is telling us to experiment with different avenues of opportunity to see what develops. We could connect with dormant or underused talents and skills that may result in remarkable achievements we never thought possible. Uranus is not called the "Great Awakener" for nothing. The cosmic alarm clock is ringing, so it's time to stop sleeping on our potential. If we are on the right track (and good luck in the beginning figuring that out!), amazing things can happen that rocket us to some kind of stardom, even if we're not in the entertainment business. Instant celebrity can be ours, usually because a few of us present the public with something novel or earth-shatteringly important. To get this far, we probably had to pay a few Saturn dues while struggling behind-the-scenes in relative obscurity. Now it can be pay-off time for our efforts.

It's apparent that not each of us will explode into the limelight, nor would some of us even want to. However, if we've been sitting on our genius potential all of these years, wondering if the world will ever notice us and the special things that we could contribute, now's the time to let our talents shine. Stress aspects during this transit simply suggest that we should slow down and be less obstinate about how individualistically we're trying to make our mark. It's important to meaningfully awaken people to greater realities, not shock them senseless. If we're flamboyant and reckless enough, we could stir fiery controversy, outraging conservative social watchdogs who wield enough power to get us in trouble. In olden days, they would have gathered kindling for dangerous "social misfits" like us—unrepentant Uranians were always easy targets for persecution and execution. In today's world, we won't be burned at the stake for being weird, but our credibility can nevertheless go up in flames if we become too defiant and alienating—and this *can* happen overnight!

UPROOTED

This is another two-for-one transit, because Uranus crossing our Midheaven also opposes our Nadir and any natal planets in our Fourth House. What we subconsciously identify as our true, subjective source of personal security (symbolized by our Nadir) can seemingly be under attack at this time, as Uranus teaches us how to let go of our dependencies or at least become more objective in our dealings with them. "Who still needs security blankets that should have been outgrown by now?" asks Uranus. At the beginning of this transit, we certainly can feel as if our self-protective instincts don't want to go where our flashes of intuition are taking us. The lure of the unknown clashes with the comfort of the familiar. Often, whatever happens in our career zone impacts our family life on some level (a typical Fourth-Tenth House dynamic even if Uranus is not involved). Some of the more interesting Uranian opportunities for career advancement at this time may require relocation to a completely different environment. If we're the type who devotedly cling to our roots (born and raised in Brooklyn perhaps?) this becomes a huge dilemma.

Transplanting ourselves only works if we're sincerely ready to try out different domestic lifestyles. Our Nadir-consciousness is given a special chance to expand its ability to be at home with new experiences, ones that can turn out refreshingly nurturing. It helps if we realize that we'll never have the future of our dreams if we stubbornly live in the past. It can't be done. Besides, Uranus hates carrying old emotional baggage needlessly. Sometimes an impulsive career move can turn out to be disastrous, if we have fixed expectations of what's to come. Stubborn types usually suffer under pivotal Uranus transits, while those who can bend and flex make the most out of their situations. Still, it's good to know how we feel about security before quickly uprooting ourselves. In addition, Uranian career opportunities are not known for longevity. They can, however, become springboards to even greater success in different professional areas. It takes a spirit of adventure and a lot of guts to willingly explore where this unusual path leads us.

LETTING DAD GO

Both the Tenth and the Fourth Houses involve parental themes. How we perceive our father is usually a Tenth House issue (at least for American astrologers). Uranus is pushing for the breakdown of power over behavior that all authority symbols have or have had, including those in our family. Maybe there is now needed a healthy distance between us and our father, if we have felt restricted by him in the past. Perhaps a surprising breakthrough awaits us regarding what may have already been a shaky or strained parent-child relationship. We could be unexpectedly challenged to let go of any former negative images of and feelings about our dad in favor of turning over a new leaf. Perhaps all that we want and need right now is to be validated by him as the individualistic person we know ourselves to be.

For a few of us, Uranus might mean, depending on our age and that of our dad, that our time with this parent is limited. Uranus is a "here today, gone tomorrow" planet; therefore, we don't have forever to work out our differences with our father. It behooves us to resolve any long-standing feuds or resentments while he's still around. He may not actually die anytime soon, but he could become inaccessible to us in other ways. We need to have our long-overdue dialogues in an attempt to work out our differences (knowing that any Uranian insight into his personality patterns helps us to see and understand this relationship in a different light). In some cases, Uranus in the Tenth could suggest that it's our father who's ready to adopt a better outlook on old, troublesome matters that have created tension for both of us. He could now be the one eager to address former misunderstandings. It's good to take advantage of this galvanizing Uranian period and finally clear the air of all stored-up grievances, which allows a healing process to begin. Tomorrow could be a brand new day for both of us.

URANUS/SATURN TRANSITS

SMELL THE FREEDOM

Natal Saturn suggests where great resistance to change can be found. This planet is a key to our long-standing inhibitions, doubts, and fears. Uranus transiting our Saturn will provoke us to release parts of ourselves that have been kept in check for whatever reason. At this time, Uranus will not support further suppression of our higher potential, and it doesn't even care if it has to meet a hard-nosed pragmatist like Saturn to make its point. We are now to liberate our capacity to be confidently individualistic. However, until Saturn permits itself to wake up and smell the freedom in the air, Uranus' energy creates an inner tension that can be explosive (especially when its drives are thwarted by continual Saturnian rigidity). However, that's usually only in those instances where we are too set in our ways to allow for real growth. Uranus has no choice then but to blast away.

It's not Uranus' intent to attack and destroy Saturn (although mythologically, Saturn did castrate and thus devitalize his daddy Uranus). Uranus needs Saturn in order to manifest itself in less abstract terms. Saturn translates Uranus' energy, converting it into more socially understandable forms and thereby making this outer planet less bizarre and alienating. Idealistic concepts, therefore, become shaped in ways the world can work with. Saturn also helps step down Uranus' high-voltage expression so that it becomes more accommodating and less overwhelming to others. Uranus will better materialize its intent by using Saturn's talent for structuring form—except that Uranus is not interested in putting its electrifying power into the same, old structures (especially those that have outlived their purpose or are much too short-sighted to keep pace with future growth). Uranus seeks to reform whatever Saturn has previous constructed. We'll probably want to keep those elements of structure that are not blocking our growth, but the rest we are prompted to eliminate swiftly.

BENDING THE RULES

What have we built for ourselves in life using our natal Saturn thus far? We need to look at this planet's house position to determine if our structures are sound but flexible enough to invite improvement. Can we bend here and yield to progress when life ushers in new and stimulating alternatives? Or do we distrust ever deviating from the norms determined by our society or even by our inner parameters? Often, with Saturn, we have shaped our life according to the dictates of outer authorities (even in minor issues of dress and appearance). Our obedience to standard rules of conduct and behavior may make us look like exemplary citizens, but are we too robotic in our pursuit of safer avenues of expression? Have we conformed in ways that have silenced the free-spirited voice of our idealistic youth?

If so, Uranus will have none of this. As the trigger planet in this temporary but potent relationship, it is the one to call the shots and determine the agenda. After all, Uranus has already determined this to be the time in our life to make its presence known. It's ready to stir up a little of that unstructured energy that always makes natal Saturn initially insecure, even though Saturn needs the kind of attitude adjustment that only Uranus can stimulate. We don't necessarily have to make a radical shift in our lives (the triggered natal planet does have a say in the final outcomes of any transit). Saturn certainly doesn't favor anything too disruptive, but there need to be a few areas where we risk becoming less habitual in our response.

A SUDDEN AMBITION

If we are undergoing a transiting sextile or a trine, Uranus appears to be less of a threat to Saturn's status quo, while Saturn seems less inflexible and vision-deprived to Uranus. A workable alliance can be forged with a little effort and a willingness to experience contrasting drives. This usually manifests as fresh options entering our life that energize whatever we have already effectively stabilized, especially in the area of professional matters or in other major responsibilities undertaken. We are able to retain the form we have worked so hard to create, but now we can add a few ingenious enhancements that can really help things take off fast! Sudden spurts of ambition are also likely, which drive us to accomplish worthwhile things in relatively short periods of time (especially if our Saturn is in a cardinal sign or already forms a natal aspect to our Uranus or Mars). One key to our success is knowing just how far to go with these Uranian energies, which means recognizing the limits of unconventional and somewhat chancy routes in a world that has already set up established procedures. Equally important will be to know how and when to move beyond those limits that fear and self-doubt have created. Uranus demands that we take new risks, and to do so means mustering up the courage to move into the unknown.

What about those other, more frictional aspects? The square, quincunx, opposition, and even the powerful conjunction all suggest a problem with balanced Uranus/Saturn expression. Uranus, when denied its proper outlets, can arouse elements in the environment that operate with greater urgency and forcefulness—abruptly so, giving us unwelcome surprises. Intensity builds to uncomfortable levels, and we feel as though life is unfairly throwing us too many curves. Maybe some of us are too driven in our ambition to make a clean break with the past and defiantly forge a new future. Saturnian prudence may go out the window if Uranus takes over completely. That includes the seasoned wisdom that warns us never to burn our bridges behind us professionally. We need to stay in good standing with career people from our past, because they may play a unforeseen role in our advancement in the future (this is advice that even shrewd Capricorn would offer). In other words, we mustn't use this Uranian energy to unduly aggravate those in important positions while we're in an overly confident state of rebellion against tradition and authority. Some of us may think we don't need our current professional connections; that is, until some future date when we're stunned to discover that we need them very much indeed! At all times, Saturn must be included in Uranus' game plan if we are to have brighter tomorrows that we can count on to last.

No Guarantees

How are *we* contributing to any unsatisfying scenarios we're undergoing? We only add to the discomfort that change sometimes brings by feeling too scared to take action, as if almost paralyzed by Saturnian fears of failure. No one can be given firm guarantees of secure outcomes with Uranus involved. Besides, that would make things too easy—where's the rapid, inner growth based on *taking risks* that Uranus wants to promote, if we know beforehand that no risks will be involved? How experimental can things be, if we already know the outcome ahead of time? No siree, all of this goes against the principles of Uranus. We'll therefore have to struggle with our nagging, inner uncertainties as we quickly react to out-of-the-ordinary challenges that suddenly pop up at this time.

This is a good transit to find out just how much of a fuddy-duddy we really are concerning mundane matters of little consequence. Saturnian perfectionism can drive us crazy if we insist on everything being organized just so! Mischievous Uranus will make sure that minor issues that we try to control too much in life either break down or rebel against us. If we have become an efficiency-freak and are too impatient with life's ability to be messy (because Saturn hates disorder), Uranus will introduce further states of disorganization that suddenly disrupt our busy schedules and may even make some of us come unglued.

If we're too hung up on time, then mischief-making Uranus makes sure that our precious time is wasted in the weirdest of ways. Uranus is not trying to torture Saturn, but it's letting that planet know that unpredictability in life is invaluable, even if Uranian timetables don't seem too trustworthy. Saturn begrudgingly learns to accept that far-sighted Uranus knows what it's talking about. We'll need to get a grip on our anxieties and built-in doubts, especially when minor events are not going according to our set plans and expectations. Virgo needs to hear this, too, because it also can be thrown for a loop when life takes detours, and when "perfect" plans change without warning.

A CLASH OF WILLS

Tensional Uranus/Saturn transits are not great times to pick a fight with the boss, become defiant on the job, challenge traditional policies, or rebel against authority in its various forms, even though that is exactly where these transits are eventually leading us—to the overthrow of existing control-structures that limit us. Certain, long-time inner blockages will have to be removed so that Uranus can breathe new life into our soul. However, with the frictional square or opposition, how we attempt to liberate ourselves can be uneven, choppy, agitating, and offensive to the powers that be (or maybe that's how we view others operating in the control seat). A confrontational battle of wills is possible. With Saturn involved, stalemates sometimes result, aggravating inner pressures already felt. When the energy finally breaks loose—and it will—we find that civil compromises may no longer be feasible. Uranus ultimately wins out, even when the results are destructive. Change will occur in swift and direct terms.

If we have not cooperated or remained adaptable throughout this process of change, unambiguous alterations are imposed on us. It's such a paradox to think of freedom-fighting Uranus imposing *anything* on us (implying no conscious choice in the matter), yet it does so at times. Evolutionary development is Uranus' ultimate goal, rather than upholding the values of our self-contained ego. (Saturn is the dutiful, stiff-lipped guardian of our ego-structure.) This is a time to transform ourselves in inventive ways and along different paths of clearer self-definition. Why fight this necessary pattern of unfoldment just so we can continue to preserve non-productive "support" systems that no longer offer real support? Both planets want us to see things as they really are, free of illusion. Uranus adds to that the need to become immune to the intimidation tactics of others.

TOO ALERT?

Uranus' energy, while enlivening, can be hard on the body if too much raw voltage is coming through. Its electrical surges are unpredictable, upsetting sensitive body rhythms (regulated by both the Moon and Saturn). If Saturn is in any way natally connected to the First or Sixth Houses, or even the Moon, transiting Uranus can interfere with physical stamina. Saturn conserves energy so that it can be used properly when most needed later, with none of it wasted. However, Uranus, in its manic state, can drain energy resources due to overactivity or to our inability to rest properly. Uranus is a very jumpy planet and is no friend of deep sleep. We can feel energized and ready to get out of bed hours before Saturn (and sunrise) says it's time to wake up. It seems Uranus' "great awakenings" can sometimes become a big nuisance.

With Uranus always wanting to be up and about, Saturn's energy reserves can be quickly depleted and its timing mechanism can be thrown off. One result is a strangely wired-up state that masks fatigue—we seem vibrant to others but we could actually be suffering nervous exhaustion. We are continuously pushing beyond sensible limits. Especially with the conjunction or the square, we have a hard time unplugging from our mental activity ("off" buttons are ruled by Saturn). Constant Uranian input does not allow for proper Saturnian synthesis, which requires stillness and concentration.

Saturn demands some degree of self-discipline, so we'll need to learn when to stop over-exciting ourselves with physical or intellectual stimulation. Thrill-seeking has its limits, and Saturn will make sure this fact is known to us. If we don't pay attention to the warning signs of tension building up, the results could be explosive as reckless, speeding Uranus crashes into Saturn's brick wall. Accidents can be a big, dramatic way to release tensions. However, why fall down a flight of stairs and break a leg, just to find out the hard way that we simply need to be still for a while? Why willfully push for more action and suffer nervous overload as a result? Why even flirt with triggering temperamental outbursts? We need to find ways to calm down and chill out.

FORCING ISSUES

Any transiting Uranus/Saturn aspect suggests that we may attempt to ignore the the reality that certain situations are immune to our reformative efforts, yet putting our planned changes on hold is unappealing. Refusing to wait for a better time to act on our ideals, some of us instead try to urgently to move forward with surprising force. Transiting Uranus trine Saturn poses few problems for us here, because the environment also is ready for a little restructuring. This aspect can place people in our path who show us how easily such remodeling can be done. Even the sextile means that we can alter circumstances once we've thought things out and have come up with several options. In both instances, our circumstances facilitate making smart changes. We're given the green light, with only some degree of caution to observe.

However, with the conjunction, square, opposition, or quincunx, Uranus wants to smash its way through Saturn's barriers. Saturn will staunchly defend its status quo. Both planetary forces thus run the risk of a head-on collision due to mutual inflexibility. If we meet prolonged resistance to our new goals, and weird stumbling blocks suddenly appear to dog our every move, this could be a sign to back off and review our objectives. Again, our timing is thrown off even if our goal is very much on target. The answer is not to get pushy. Delays happen for sound reasons, and we need to accept this as part of the broader, cosmic plan of our ultimate success. Otherwise, rebellion is carried out without the benefit of the intuitive sense of knowing that more balanced Uranian expression provides. In addition, it's not a good idea to needlessly annoy Saturn, a planet that can penalize us for any unwise use of freedom. Pushing too hard to inflame authorities can result in certain matters blowing up in our face. We are seen as a major source of uncalled-for disruption, and people in power will not wish to help us achieve anything. But they will be glad to call in a security guard or two who then forcefully shows us out the door!

AQUARIAN BUDDIES?

You'll recall that, in the Greek myth, Uranus and Saturn had an adversarial relationship, filled with father-son, high-voltage tensions that culminated in violence. Despite this, the Zodiac has evolved to allow these planets to co-rule truth-finding Aquarius. What makes this fixed, airy sign work well is the teamwork spirit of both these planets—something that's easier said than done. Whenever Uranus and Saturn form an aspect, we are challenged to pull their diverse energies together to build new and improved life structures for ourselves and for the world. One planet is not to dominate the other. They work best when we consciously and cooperatively use the energies of each. It's good to keep this in mind. Saturn is willing to work hard to make sure that innovative Uranus inspires us to use its ideas in progressive but useful ways. Uranus is always ready to help Saturn take reality just one step further. Let's use these transits to make these planets the best of friends within us!

CHAPTER SIXTEEN

URANUS TRANSITING THE ELEVENTH HOUSE

HOME FREE

The Eleventh is already an odd-ball house, in that it is associated with two planets that seem to be very dissimilar: Uranus and Aquarius' old-time ruler, Saturn. A wide range of contradictory behaviors and activities can be found here. This is the house of the non-conformist (Uranus) whose social activism and burning idealism pose a threat to an established community's conventional standards. With strong Eleventh-House emphasis, a few of us can feel like rebellious crusaders on a mission to free society from stodgy cultural traditions. If we're more politically-minded, we may idealistically work to liberate the populace from oppressive authoritarian rule. This is the house where dissenters conspire to overthrow aging institutions that no longer best serve the needs of the people.

Another facet of the Eleventh House supports the concept of group power, where people come together to advocate relevant social issues central to mass organization and control (that sounds like airy Saturn's influence in the collective sense). Even very outer-fringe Uranian groups can structure themselves

according to strict rules of behavior, showing unquestioned loyalty to the group's cause and obedience to that group's central
leader (again, a Saturn response).

One Uranus/Saturn adage worth repeating is, "Freedom without responsibility is no freedom at all!" Not just this planetary
combo, but even Aquarian energy needs to keep this in mind.
Actually, the airy side of Saturn does share common denominators
with Uranus: both have a sense of social consciousness, an interest
in collective development, and a feeling of being separate from the
herdlike masses (which appear unappealingly Neptunian or perhaps lunar to both these somewhat aloof planets). Each is eager to
contribute something special and timely to society. They also share
an interest in truth-seeking through scientific channels. But that's
where most of the similarities end. Their style of achieving such
goals is quite different. Their sense of clock-time is also clearly different. At least for Uranus, attachment to anything—including the
intoxicating fruits of success—is to be avoided at all cost. Uranus
likes the non-stick approach to living, where we don't possessively
latch onto anything, and where nobody owns us.

GROUP VISION

In our Eleventh House, we constantly do a juggling act to cooperatively fit into large social frameworks without losing our unique
identity. Leo, in the so-called Natural Wheel, ruling the Eleventh's
opposite house—the Fifth—also underscores this polarity. When
Uranus transits our Eleventh, we are more inclined to feel the
influence of the "free-soul" qualities of this house, but not necessarily because of impulses emerging from our internal self. It is more
likely that we start to connect with colorful individualists, or
maybe we begin to network with open-minded, far-sighted groups
that have an exciting vision about society's future. They provide
new roles for us to try on, according to our willingness to re-invent
ourselves.

Anything progressive in thought and action is more appealing to us during this transit, even if we are not quite ready to
join a group. We attract unusual input from the world that may

fascinate us, yet we'll still need some time (Saturn-style) to digest this new influence—especially if we have a lot of fixity in our chart. Taurus, Leo, and Scorpio don't like too many new experiences quickly pushed at them. Still, it is likely that we are ready for progressive, social models to explore, considering the insights we've gained about our place in the world when Uranus was passing through our Tenth. Uranus wants us to assimilate these social experiences quickly and then bravely seek out even more challenging ones, especially those that allow us to further experiment within a group dynamic.

A WIDER SPECTRUM

Uranus is an impersonal planet now moving through an impersonal house. This suggests that life will not unfold itself in the next several years according to any personal, self-centered expectations. What we get to do here will not always directly or immediately benefit us (although independent, self-contained Uranus typically wants us to innovate for ourselves before getting others involved). Instead, the Eleventh House represents a complex of egos that need their share of attention and the right to be heard—this is "the voice of the people" house. Maybe this transit will get us in touch with any personal separatist attitudes that may have previously prevented us from sharing in the spirit of camaraderie. We are now developing greater individuality so that we can broadly and freely share ourselves with others. This will require an ego adjustment as Uranus picks only strong-willed, free souls to cross our path and help to alter our mindset.

Our consciousness may be raised by dealing with people from unfamiliar walks of life. Maybe we were even prejudiced about certain classes and types until now, when we have to rub shoulders with them in real life situations, perhaps in group settings, and find out they're not so strange after all. Or perhaps we discover they're strange in "good" ways. They're even more like us than we were led to believe by those who held influence over our mind in our early years. This transit, therefore, can be a time when negative, social images we've never questioned before are shattered. We begin to realize the these biases are unfair and suppressive to others. Shattering such false images is an unexpected eye-opener for us, and we tend to quickly reform our thinking once we've had a direct encounter with real people who have been hurt by prejudicial stereotypes. Some of us may feel shocked by our intolerance over the years. It was easier and safer to go along with the opinions of the majority than it was to think for ourselves and stand apart from the crowd by boldly attacking social injustice. Now we learn more about how those unfairly targeted have always felt, and what we can do to prevent further social intolerances.

Don't Sign Me Up

Uranus, in my opinion, is not a joiner. It doesn't want to be just another member of the gang. It feels too different from others to allow itself to identity or merge with the group-mind. Much of the required, hard work that people do to support Eleventh House team projects and social causes comes from dedicated Saturn. Uranus initially sparks others to wake up and realize what has to be changed in the name of progress, but it doesn't wait around long enough or plod on as these social ideals slowly take shape and become accepted by the masses. That's because the masses assimilate change at a snail's pace.

Some of us undergoing this transit may become involved with associations and societies that spur our interest in the world's unexplored potential (even our own). However, we may not be completely ready to become fully committed participants. We'll show up at a group's monthly meeting when we feel the urge, not because of any sense of duty. Maybe we haven't always been like this, but it's probably how we choose to operate now (because Uranus in the Eleventh can signify a double-dose of the freedom theme). We may also hop from group to group, sampling all social options supporting alternative lifestyles and futuristic vision. (Gee, maybe we'll finally get to attend that local UFO club meeting we've been thinking about off and on, or even visit that holistic nudist camp during our next summer vacation!)

GOT TO HAVE FRIENDS

Although friendly alliances in less-personal group settings are part of the Eleventh House experience—such as with colleagues you network with at professional conferences—friendships of a more casual nature are also to be explored here. Unlike the Seventh, where our "best" friends, who know us intimately, are to be found, the pals we connect with in our Eleventh are better termed "acquaintances." With Uranus, we can be drawn to one-of-a-kind individualists precisely because they are so different from us and our social background. They find us equally unusual, with something stimulating to offer them, which could come as a surprise to us. Sometimes, colorful Uranian types we barely know throw big parties or social get-togethers where all types of people mingle. By a simple twist of fate, we get invited and thereby become exposed to an amazing assortment of free souls; a few of them find us interesting, and thus new and unusual friendships develop. Little do they know that some of us are in a transitional stage and are doing a lot of Uranian things for the very first time!

We'll have to be ready to circulate, however, for any of this to happen. No matter what our social history has been, we have an ability to attract others who can open our eyes to ways of thinking and living, usually in self-revelatory ways. This is especially so regarding our unlimited capacity to alter our lifestyle. With stress aspects involved, however, we'll need to be careful not to link up with people whose vision of an ideal tomorrow is too extreme; or with people whose underlying anger regarding society drives them to come up with radical solutions that suggest mental imbalance on their part. Anyone who comes off as too pushy, with a universal message that's too urgent in tone, is probably someone we should avoid (or, at least, curiously observe at arms-length). Basically, we only need a few scintillating friends to hang out with anyway, not political revolutionaries looking for other zealous comrades to help to overthrow systems they deem obsolete or hopelessly corrupt! However, a few of us who have been malcontent for a long time and who are disgusted with mainstream society will want to politicize everything related to this Uranus transit. It's a different type of party, then, that we want to join.

HOPES AND WISHES?

The Eleventh House has traditionally been associated with our "hopes and wishes," a key phrase I've always found to be too vague. If I "hope" for a brand new car, it that an Eleventh House issue? If I "wish" for a pay raise at work, is this an Eleventh House yearning? Apparently not. Very little about the Eleventh supports exclusive self-interest. Here we learn that we are a dynamic part of a social matrix that has a collective mind of its own. Individual ego-desires are to fit into the greater needs of a group's will, which then enables community standards to be formulated and enforced (overriding individual self-will). This collective mind also unfolds unconsciously, as demonstrated by the emergence of new generational outlooks that disturb the status quo while they further reshape society. In the Eleventh, we foresee enlightened societal harmony as the goal to be reached, where betterment applies to all people and not just to ourselves.

From this perspective, hopes and wishes pertain to a futuristic blossoming of human potential. Uranus transiting this house helps us see far ahead into an ideal world of almost utopian vision (maybe with a lot of snazzy, high-tech innovation thrown into the picture, because Uranus is quite interested in amazing inventions—sometimes even more so than in uplifting the human condition). Whether or not our current reality is ready for such a radical metamorphosis, at issue is our ability to be open to vast and progressive possibilities. Of course, the farther ahead we see, the more discontented we can be trying to live in the now, especially when Uranus makes transiting squares to our natal planets from this house.

Uranus enlightens very quickly, albeit abruptly. That famous Uranian sense of being zapped means everything in life is going to be perceived differently now that we're wide awake! We could run on more intuitional energy than ever before. The trouble is that few are also tuning in to our new wavelength. Patience with living in this seemingly thick-headed world and tolerance for its painfully timid approach to progress will be important during this transit. Otherwise, we may feel that we really do belong to a different galaxy of higher beings. Why aren't those UFO guys beaming us up by now, some of us may wonder, since we feel more ready than ever to leave this idiotic earthbound existence behind? Such feelings are less due to escapism than to our intellectual disgust with humankind's stupidity and stubborn resistance to knowing the Truth. Unfortunately, Uranus cannot stand dealing with chronic states of imperfection, human or otherwise. This explains why the socially discontented can get even more worked up during Uranus' transit of their Eleventh House.

NOBEL PRIZE WINNER?

We are going through a phase where our humanitarian urges are coming alive, maybe even reawakening. Our natal Uranus' house position might tell us what aspects of social reform interest us the most. This is now a time to put certain ideals into action, a time to be an activist on some level. Society may have convinced us that our youthful hopes of changing the world were the result of idealistic naiveté. After all, we weren't full-fledged adults—just starry-eyed dreamers with a passion for peace and a simplistic vision of global tolerance. As we aged, the idea of getting on with the business of mundane living by producing something really useful to the material framework of society was drummed into us, while we sadly realized that the world was not yet ready to embrace a new age of love and light.

Now, however, our old rebel spirit reemerges, even though we may be more grounded and pragmatic than in our earlier, free-spirited days. We've incorporated more Saturn-awareness due to the pressures of maturing (especially if we are over forty when this transit begins). The interpretation will be quite different if we are only fifteen, when Saturn also opposes Saturn, a time that sounds very rebellious and peer-focused. Anyway, maybe we won't win the

Nobel Peace Prize this time around, but in some way we can and should find outlets for constructive social participation. Perhaps we can volunteer our time and energy toward civic-minded causes or anything else that requires a crusading spirit. What community services that we believe in need our involvement, as long as they are not too bogged down by bureaucratic structures? We may find ourselves easily taking a controversial stand on current issues that divide the public. Uranus in the Eleventh means we can go out on a limb to promote or defend our principles. We may advocate things that frighten others.

Personal aspirations in general will change during this transit, as our long-term goals become less worldly. Our interest in the broader picture of a planet of great, human diversity can increase, leading us to conclude that universal cooperation between nations and their people needs to be high on the list of global priorities. We may support whatever organizations take a progressive stand on human rights, nature's rights, or whatever else helps ensure that humanity will evolve along more enlightened areas of thought and action. All in all, this is one Uranus transit that can enable us to look at life's universal issues very differently in the future.

URANUS/URANUS TRANSITS
DIFFERENT DRUMBEATS

When Uranus transits itself, we are dealing with *life cycles* that occur in everyone's chart at relatively the same chronological ages. I'm not going to interpret each and every possible Uranus/Uranus cycle, but there are four that stand out in my mind: the first square, the opposition, the second square, and the Uranus Return. The main thrust of these cycles revolves around confronting all things Uranian, especially on internal levels. These can be periods when we receive wake-up calls telling us to follow our path of freedom. Maybe we find ourselves rebelling against self-identifying labels that we've bought into in order to conform and be accepted by society. We've usually taken the safe route rather than risk social rejection. During any one of these cycles, we may suddenly feel that those images are false and misleading. Our true self wants to break out in some fashion and make its individuality known. The environment is ready to provide a few shake-ups to match our altered, inner rhythms. Something's gotta give!

Technically, this is a transpersonal transit (Outer Planet to Outer Planet); therefore, a lot of it won't be under our conscious direction. This enables Uranus to introduce a wider range of unfamiliar experiences that we normally would not attempt to magnetize in our day-to-day situations. The Cosmos, working through our environment, will usually orchestrate how the message of liberation will be heard. People entering our lives shake us up in ways that help to bring greater self-expression. Situations force us to show our true colors. We may marvel at how alive and well we can feel, if we freely experiment with this rushing energy and don't waste time trying to resist the educational nature of such mind-altering change.

First Uranus Square Uranus (ages 18–21)

The first Uranus/Uranus square is notable because it coincides with a time when most young people feel they have reached true adult status—finally!—not just in the eyes of their parents, but also in the eyes of society. Of course, nobody really gets to claim adulthood until after the first Saturn Return (ages 28–30); that is, *if* they pass the life exams given at that time. However, anyone who's almost twenty-one certainly doesn't want to hear that. This planet takes a little less than seven years to pass through a sign; therefore its first major tensional pattern happens after high school. (My first Uranus/Uranus square was, to the minute, just a month before my twentieth birthday. Although I was already studying the subject for a while, I took my first official class in astrology around that time and learned to calculate actual charts. As stated before, my natal Uranus is in my Third House of educational outlets.)

During this square phase, we envision that we'll have autonomy and be able to break free of parental rule, especially during our twenties. Any system that seems authoritarian is rejected. We assume that we now can do whatever we please, determine our own life approach as we see fit, experiment with jobs without feeling firmly committed to them (it's a restless period anyway), and basically keep all options open-ended. We also can and do formulate a few aspirations as we explore the structure of adult society.

We may even embrace the unknown with a sense of excitement. All future prospects become very attractive and compelling (or maybe nerve-racking, if we started off on the wrong foot with this transit). To get the most out of this thrusting phase of Uranus, we will need to be as educated as we can be, because this planet thrives on mental stimulation. We need to realize that our society is more willing to open the doors of opportunity for those possessing valuable knowledge. People who have had inadequate schooling may not feel this to be a liberating, freedom-oriented period, but instead one where stability and economic survival become paramount, making circumstances feel more Saturnian. Indeed, Saturn squares our Saturn around the time of our first Uranus square. Our direction feels uncertain during this Uranus/Uranus phase, and we are not too sure how many "adult" values we want to adopt (such a double Uranus influence could suggest emphasized rebellion and defiance).

This is a cycle that can start off as difficult for those who already had a tough time with authority during the mid-teen angst and resentment generated by Saturn opposing Saturn (a rebel period also fed by transiting Uranus sextile itself). However, with both Uranus and Saturn simultaneously squaring their natal positions, clashes with family or community values can erupt. Our urge to break away from our past then becomes too undisciplined to be handled in a productive manner. Some of us learn the hard way that there are many parental substitutes in the world who do not love us, but who are willing and able to stop us in our tracks when we run wild. In actuality, the number of those who get in trouble with the law during this cycle is relatively low. Still, for those criminally inclined, this becomes a vulnerable period.

Breaking society's rules of conduct can have serious legal conse-quences (Saturn square Saturn makes sure of that).

As we reach our mid-twenties, transiting Uranus trines Uranus—a time when we may have the clarity we need to foster a workable life vision. At least the transiting Uranus square gives us the guts to leave behind our childhood and the nest our par-ents have created, and launch into a bigger and sometimes anxi-ety-inducing world, where we can learn more about who we really are as individuals.

URANUS OPPOSING URANUS (DURING OUR MID-LIFE CRISIS)

This cycle occurs as early as our late thirties. It is one of the hall-marks of our infamous "mid-life crisis" phase and one that most astrologers show a keen interest in exploring. This transit describes those facets of our mid-life years that are marked by an urgent restlessness to bring something or someone fresh into our stable but monotonous life patterns. Up until then, we may have played out an orderly social role that now feels too tightly defined and wrong for us. We need space to breathe again while recover-ing lost parts of our individuality. We may even try to revive the youthful gusto for adventurous living we may have felt during our Uranus/Uranus square—which is not always a smart idea. The daily grind with all of its mounting responsibilities has taken over how we have shaped our identity, and now we sense that it's time for some sort of internal revolution. A reordering of very per-sonal priorities is due.

We first need to define where we seem to be the most trapped. This transit could be a timely turning-point of self-awakening, during which we can feel a resurgence of the spirit of experimen-tation. It can symbolize a second adolescence phase, where our freedom-craving, rule-breaking ways turn some of us into "teenagers" who thumb our nose at the conventions of "proper" adult conduct. Some of us may be tempted to pull out and wear our bell-bottomed pants again, once we lose about twenty pounds! We feel a strong urge to let adventure have its way with us, an adventure that defies logic at times and that allows us to be a bit wild and crazy (less so if our natal Saturn has a dominant presence in our chart, or if we are loaded with earth placements and/or much natal fixity).

Usually, other people help to set this spark-flying cycle in motion. Their changing situations become catalysts for how we repattern our lives. Maybe a spouse wants to leave us, something we dread. Perhaps, after a destabilizing divorce, we realize we have gladly outgrown much of what our marriage was based upon, and yet it took our partner's unrest and pressing urge to break loose to get the ball of freedom rolling. Within time, we realize that we are becoming our true selves without that union. This is not something easily embraced, but that's why this transit's called an opposition: we oppose it until we've had time to adjust to the new life forces entering our world.

On the other hand, sometimes those of us committed to the single life have an unforeseen change of heart and suddenly leap into a relationship (even marriage) with a very special individual whom we never dreamed of meeting. During this unpredictable period, wallflowers may blossom, sexual hot-shots fizzle out, and, in general, reversals of status occur that surprise everyone. If not due to another person, then the environment at large forces change upon us in ways that accelerate our growth. While this period is usually called a "crisis," it's really a turning point that allows us a greater range of self-expression as we move into the second half of our lives. The double dose of Uranus symbolism means that the courage to push forward is needed, even when mixed feelings split us apart in ways that make us crazy and unhinged for the moment.

SECOND URANUS SQUARE URANUS (AGES 59–65)

The theme of breaking away from the yoke of authority is here repeated, except we're not restless teens anymore, innocently eager to explore the adult world of freedom. We now know that the adult world has built-in limitations as part of its fundamental reality, and most of us have accepted that. Who or what is to take on the role of the authority we oppose at this point is less clearly defined. Age sixty-five rings a bell, because this is legally (at least in America) when workers can begin their retirement years and collect their Social Security checks. Retirement could be acknowledged as a time of shock for many of us, even though we may have claimed to look forward to it. At best, it's a time to manifest idealistic plans by which we expect to live out the remainder of our lives. However, such ideals often do not match up with reality at this time, something which generates the full tension of the square.

This period should be freedom oriented, giving well-deserved breaks from routine, as well as allowing us to pursue new areas and cover fresh territory with a greater sense of leisure. Unfortunately, few in this culture are prepared for such a dose of "freedom." Instead, some of us may feel cut-off from the accustomed rhythms of society due to the abrupt change in our daily pattern. Obviously, those of us who have long ago made alterations by which we became "free agents" on a professional basis are less apt to feel that the rug's been pulled out from under us because, technically, we do not have to prepare to retire at this time. No company is formally terminating our services.

However, for the majority who have followed the standard straight-and-narrow path toward success within established parameters, this period can sometimes mark a time of inner chaos and even alienation. It all depends on the extent of our inner discoveries during our Uranus opposing Uranus phase. The more self-knowing we have become, the more hopeful this square phase. Otherwise, disorientation can take place along with anger and bitterness, as we feel we're being "put out to pasture" against our will. Many can feel impotent under these conditions (like the fate that awaited mythological Uranus). Feeling non-effectual on the social level, we can withdraw altogether and resign ourselves to an unsatisfying final chapter of life.

Astrologers are not accustomed to thinking of this dynamic Uranus phase in such bleak terms. However, this can be how a sudden change of social status impacts some of us at this time.

Still, even for those of us depressed or distressed, much understanding of the meaning of our total life process can flash before our eyes. This can be an insightful time for all undergoing this cycle. Much depends on what we are becoming awakened to. Our environment also offers outlets to us that can get us out of our doldrums if we are willing to do a little exploring.

URANUS RETURN (AGES 82–84)

Until recently, the Uranus Return was an astrological oddity, in that only a small percentage of people ever lived long enough to experience it. Most were probably recluses by then, so astrologers never got to know directly what was really going on during this life cycle. In the past several decades, however, it's become obvious that many folks are living longer and with greater vitality even at advanced ages; they are not necessarily confined to their homes as once was the case. A few can even be found having a blast with the slot machines in Atlantic City. (Heck, even aging Leos just wanna have fun!)

This is good news for Uranus, which always has an easier time working with live-wires who are ready for anything than those dead-beats who aren't. The biggest obstacle at this juncture, however, is our darn old physical body (as Saturn now claims complete control, almost with a vengeance, over the body's continued decline and eventual demise). Assuming we will live this long, what can we expect? Believe it or not, there is still room for both excitement (within reason) and further self-discovery (even if we never do make it to to the gambling casinos).

The Uranus Return seems to mark a symbolic climax for our soul's development. We have officially completed our assignment on Earth regarding our mundane "contract" with society and the roles we felt somewhat roped into playing (for whatever circumstantial reasons). A lot of people become very bitter about the roles they've had to play long before they reach age eighty-four (perhaps the bad feelings kick in around the retirement years). By our mid-eighties, there's not much more we can extract and use from our social environment that will help us to grow further. Now our challenge lies in how we embark on our inner journey in consciousness, something that will require a greater detachment from direct involvement with worldly matters. Our physical body also shows signs of wanting to withdraw from the frenetic stimuli of our outer surroundings.

However, even if we are to be confined to our home, thank goodness we still have the world at our fingertips via our TV's remote control or the Internet. We certainly won't have to suffer the isolation that some of us did as elderly shut-ins in dreary past lives. Electronic and digital media stimulation will be our constant companion if we wish. Uranus will make sure our brain remains active and still eager for new mental excitement. It does boggle the mind, however, to think of those Pluto-in-Leo oldsters who'll still be rocking and bopping during their Uranus Return to their scratchy Rolling Stones and Jefferson Airplane albums, assuming their hearing isn't totally shot by then!

Ideally, we'd get the most out of this life cycle if we didn't have to concern ourselves with mundane responsibilities, some of which have now become either burdensome or uninteresting to us. We wish to be free of these tedious tasks and instead give our mind permission to explore whatever interests it. For some of us, this can mean a time of mental disorientation (we've also had Neptune opposing Neptune just before our Uranus Return). I suspect that a way to prevent this is to remain constantly flexible, adaptable, and unshockable. After all, we've witnessed a lot about society during almost a century of living. With the right amount of detachment (not to be confused with withdrawal), we can learn to de-emphasize our emotional reactions to all that's constantly happening beyond our control around us. Curiosity about life will keep us sharp and alert. It is clear by now that we are only *in* the world, not of it!

CHAPTER SEVENTEEN

URANUS TRANSITING THE TWELFTH HOUSE

BETTER TO GIVE

Traditionally, the Twelfth House has been associated with dungeons, prisons, hospitals, nursing homes, asylums, and other places of physical confinement for disfunctional people. At first, it would seem that Uranus transiting the Twelfth would have a miserable time with this constant threat of being imprisoned on some level. In addition, it's a planet that possesses "dangerous" qualities that conventional society would find objectionable, and therefore punishable by solitary confinement or exile. Luckily, Uranus isn't much into either self-pity or scapegoat-identification, two things that could immobilize us in this ego-deflating house. Uranus, not easily undone by social condemnation, would rather go down in flames as a clear-eyed heretic than to turn its back on its principles and be absorbed by a corrupt System it deems a *real* source of danger.

Generally, the Twelfth seems to be more troublesome regarding issues totally rooted in selfish, worldly experience. It's simple: you rob a bank in defiance of society's set rules of conduct, and you land in the slammer for a long stretch—stealing other people's money is a good example of a selfish worldly desire. Those who only want to take from the world and never give back to society will suffer soul-eroding predicaments in this house. Although Uranus typically is less involved in the trappings of desire and greed, its rebellious streak can nonetheless jeopardize its cherished freedom by outraging authority. Uranus doesn't want to give in to the will of society and, therefore, can get itself booted out of general membership—with all privileges suddenly revoked. Yet this is also a house where we can, in good faith, surrender our "all" to a higher guiding power—whether it be the Divine or a great social cause. However, surrendering is not something Uranus (or Pluto) does easily. The potential for losing one's individuality altogether in the process also bothers Uranus—a planet that's not going to jump into the Twelfth House's melting pot willingly. However, in this house, if so desired, Uranus can turn invisible and de-emphasize what makes it stand out from a crowd and become the target of ridicule (or hate mail).

BEYOND THIS WORLD

The Twelfth is associated with watery Pisces and Neptune (examples of Principle Twelve). Therefore, selfless activity, working for a common universal cause, realizing the oneness of all people, and being empathetic to the struggle of the human condition on Earth are all to be encouraged. This is a very transcendental house (with its underlying theme of soul-liberation). Here we eventually learn to safely contact and merge with "higher" invisible realms of existence. All that's required is the right inner preparation and motive. The Twelfth House doesn't focus on the tangible results of materialistic ambitions. In addition, this is where the Unconscious (both its personal and collective facets) likes to hang out, and to blow our mind every so often with its otherworldly imagery and its constant supply of potent archetypal energies.

Uranus is not particularly preoccupied with physical structure or with dense material expression, so we may find that now we can almost boundlessly soar to greater heights of mental exploration. Imagination is given free reign to unfold its spirit-empowering visions, helping our consciousness to expand quickly. However, we'll need to be careful not to encourage such unfold-ment from occurring too rapidly and without sufficient grounding on our part. In general, the Twelfth is much less confining when we creatively express ourselves on its abstract levels. What we can paint here on our canvas is unlimited. However, things can get tricky when we try to pour incarnate energy into physical forms and stable structures. Think of the Twelfth as the won-drous Land of Virtual Reality. Once computer programmers enable us to affordably enter total digital or holographic realms, a whole new universe of Twelfth-House activity will open up for us. Expect Uranus to play a hugely innovative role here, especially during its passage through Aquarius and Pisces.

GHOST BUSTING

The Twelfth is also seen as a house where fantasies can warp our perceptions. With Uranus mismanaged, we could find ourselves willing to reinvent reality to suit our *un*realistic yearnings, espe-cially when Uranus conjuncts a natal Twelfth House planet that may already have a history of tricking itself into believing its illu-sions. Things could get bizarre when Uranus produces chaos in this surrealistic realm known for its ever-shifting landscapes. Clinging to illusions eventually backfires, because Uranus doesn't want to be part of any self-induced fraudulent situation. It will, instead, insist on arming itself with the harsh and unrelenting light of Truth. Still, it's tempting for some of us to think our ticket to unconditional freedom may involve ignoring life's limitations by abandoning common sense in favor of rebellious escapism—usually in the form of quick thrills and risky highs. Our head-strong impulses here can lead to our self-unraveling, usually through out-of-the-ordinary circumstantial means.

The inner ghosts of our past that psychologically haunt us can be found wandering through the dark chambers of our Twelfth. If these trapped internal elements still try to slowly drain life out of us in order to survive, Uranus can play the role of an exorcist—not in the ritualistic religious sense, of course, but in terms of stimulating us to make a clean break from such disturbing elements of our past. The problem is that living in the outer world and following its rules of behavior do not prepare us to face a troubled inner world that now may have to be opened up and reviewed.

During the many years this transit is in effect, the outer world knows nothing about the upheavals we may undergo. Our psycho-explosions in consciousness and the resultant breakdowns are internal and private for the most part. If anything, outer situations can provoke odd and unexpected responses that set off chain reactions concerning buried memories that need to be dislodged. As discomforting as this can be, Uranus tries to convince us that it's for the best. The past is not to hold us back any longer. We are going to experience some big breakthroughs regarding our identity when Uranus transits our First, but this Twelfth House transit prepares us by first sweeping away unhealthy, spirit-clogging emotional debris. The health of our soul is at stake.

MISSION NOT SO IMPOSSIBLE

Uranus behaves a bit like Pluto in this house, in that its mission is to shatter illusions, break up long-term hang-ups, dispel phobias, and, in general, help us to more clearly evaluate the most subjectively damaging weak-links in our psychology. Negative attitudes that have remained hidden, insidiously working behind-the-scenes, may now become identified and eliminated. Any crisis that occurs serves as an effective catalyst, helping us to expose even deeply entrenched complexes. Uranus would usually be transiting a planet during such times (particularly a Twelfth-House planet), and that planet would play into the dynamics of a Uranian release from the inner suffering our psychological bondage has created. In this respect, Uranus assists in healing old inner wounds (although Uranus itself is no comforting healer or redeemer—that's Neptune).

Typically, when any planet goes through our Twelfth, we feel we are all alone in experiencing the challenges of that planet, for better or worse. We assume no one will really understand what we are going through regarding that transiting planet and the natal planets it aspects. Some of this is because, with our inclination to internalize conditions, we are not giving others clear signals telling them that we need their help and support. We're not reaching out to people. How can we expect anyone or everyone to read our minds and sensitively tap into our unspoken needs? Unconsciously, we create conditions that force us to fall back on our hidden resources (our Twelfth House assets). However, this can be a very good thing for us to do in the long run.

Uranus can at least awaken us to those hidden resources. Sometimes, the sign on this house's cusp gives additional clues. Even in the Twelfth, Uranus is an activist at work trying to spark us to energize those back-up strengths normally not used in our daily waking state. Sometimes this means turning on our psychic skills and allowing intuitive energy to flow freely without too much logical analysis getting in the way. Remember, we're in Neptune's realm now, where boundaries in consciousness are non-existent—if only we would believe this to be so. This constructive inner activity prepares us for our next astounding Uranian adventure—Uranus crossing over our Ascendant. How successful we are in altering our identity or in revamping our self-image depends a lot on what we do with our Uranus in the Twelfth opportunities for self-illumination. For those of us willing to let go of all that weighs our soul down, there is much inner light here to be channeled into Uranian ways of being.

ALL WE NEED IS LOVE

Uranus probably wonders "what's love got to do with it?"—especially if it's a mushy, sentimental, hearts-a-fluttering kind of love. Erotic love is not what Uranus understands very well, but it can work with more inclusive *agape* love—the altruistic love for humankind. Remember, with this planet, the more that a concept can be applied on a wide and impersonal basis, the more Uranus' interest in piqued (as long as Uranus is itself not bound by such a concept)! Uranus has little trouble seeing itself as a humane light-bearer, but it will empathetically reject being seen as a savior to be exalted and worshiped. Yet a common pitfall in our Twelfth is to look for such saviors to indiscriminately follow and adore, at the expense of losing our identity. Uranus wants us to avoid such glorification, whether we are attempting to do the savior act or the devotee role. Uranian spirituality likes to give others the tools needed to be more godlike themselves (as shown by the myth of the Uranian enlightener Prometheus, who rebelliously stole fire—jealously guarded by the head god Zeus—to then give to humanity). Uranus' is a more clear-headed, democratic approach that doesn't lend itself to devotionalism.

Perhaps one potential Uranian breakthrough in consciousness is to realize how our inflexibility creates separative conditions in our lives. While we may enjoy being unlike anyone else, we could also suffer unnecessary isolation due to our reluctance to blend into the broad social mix symbolized in this house (here's where all of humanity comes together, devoid of social class distinctions). We may discover that we harbor unconscious, elitist attitudes that have kept us from feeling at one with others. Maybe we were born with Uranus square our Sun or our Mars—two astrological signatures that may suggest we don't feel we have a lot in common with the masses but are glad we're independent, one-of-a-kind types.

In this instance, Uranus can either intensify our self-perception, so that we experience a greater sense of remoteness and alienation, *or* Uranus says we're due for a major melt-down of such aloof individualism—one that allows us to move from cold, abstract intellectualism to warmer and deeper emotional human engagement. This becomes a breakthrough for the heart and a triumph over artificial social barriers produced by mind and will. By learning the meaning of Twelfth-House love, our fundamental loneliness may thus be dissolved, which then becomes our gift to ourselves for loving others on a soul level.

CRAZY CLOSET

On a lighter note, the Twelfth has often been likened to our psyche's closet, where we store away all sorts of eccentric and even anti-social things about ourselves that we don't care to have lying about on our coffee table in plain view. Others are not permitted, especially in public situations, to easily observe these less-obvious qualities we own (we're afraid the world won't accept our closeted traits and urges). Maybe we like to vacuum the house while in the nude in the middle of the night with Rachmaninoff playing in the background. It's an odd pleasure, we realize, but it's also none of anyone else's business, now is it? Our 3:00 A.M. cleaning ritual becomes a secret Twelfth-House issue (only if we close our drapes, that is, to avoid encouraging someone else's little nighttime Twelfth-House ritual involving binoculars). Maybe some of us like to cross-dress and slip out after sundown to window-shop at the mall just for the thrill of it. That could end up being someone else's business if we eventually get caught using the wrong public restroom, but our festive masquerade feels very Twelfth House-ish, and maybe becomes almost an out-of-our-body experience for some of us. Therefore, this becomes a pure Twelfth House escapade.

When Uranus enters this house, our secret self-expression enjoyed behind the scenes can be very unusual indeed, what some love to call kinky. We unleash a wilder part of ourselves that the public would either probably frown on or be utterly amazed by. At least that's how we see it and why we hide it. However, with Uranus taking a walk on the wild side, perhaps we are learning to associate inner liberation with the urge to expose some of our guarded, private interests. Of course, with the wrong transit in full swing (like Uranus square our natal Neptune), such exposure can lead to an overnight scandal. Uranus can shock people better and faster than any other planet! We'll need to analyze our situation and determine if it's worth the risk to "come out of the closet" just to feel freer doing what we've been happily indulging in our underground world for so long without a hitch. Sometimes what we've been hiding is nothing more racy than a yen for karaoke performances in swank retro lounges, or maybe we want to master conscious astral travel and visit the actual planet Pluto. The world may not understand our uncommon needs, but the Twelfth House is surely not going to stop us from exploring them as long as we're discreet. The problem here is that Uranus is seldom discreet, and there will always be plenty of grim Saturnians waiting to catch us in the "shocking" act of finally being ourselves.

ROMANCING THE SOURCE

In general, Uranus is closing a cycle of our experiences with the freedom urge before starting up on its exploratory mission again in the First. In the Twelfth, we may tap into the loving source of our spiritual origin, a dimension where limitations and confinements become unreal. We can't stay connected to this source of ultimate being forever because we have more work to do (that is, as long as we're in this earthly body and haven't paid our end-of-the-month bills yet). Our soul's vast and unlimited potential can be sensed in brief but deep flashes of Uranian insight or in electrifying visions of startling clarity—but only when we learn to stop putting unnecessary hurdles in our path to higher consciousness. Self-love and empathy for humanity can carry us far and wide on our journey of self-discovery. Let's not be afraid to mentally go where Uranus and the inner spiritual fires of Prometheus take us!

URANUS/NEPTUNE TRANSITS
SNAP OUT OF IT

Here is a case, as with Uranus/Pluto, of two Outer Planets connecting. This becomes generational, in that many will experience these same transits at the same time. However, even if others are simultaneously having transiting Uranus aspect their Neptune, that doesn't make this pattern any less personal or relevant for us. This transit is not just some collective challenge in the impersonal world. It is also a force of inner awakening that can be profound in nature, although often too subtle to observe on the everyday surface of a busy, mundane life.

Actually, Uranus would love to tackle the job of straightening out poor "lost-in-space" Neptune (ah, but just wait until it's Neptune's turn to do a magical make-over on cocksure Uranus)! Uranus is now in a position to eye-ball how well we have processed our Neptunian potential up to this point. It's reviewing this matter with cool objectivity and emotional detachment, two things Neptune sorely lacks. Think of Neptune as a beautiful soap bubble, with its swirling colors and its ability to float effortlessly in the air. Then think of Uranus as a huge, sharp, gleaming straight pin suddenly coming out of nowhere to prick that iridescent bubble—pop! This is Uranus' basic role at this time: to burst our own self-created bubbles of illusion or those set up by our environment that lull us into a false sense of serenity, security, or complacency.

DUMBSTRUCK

When any bubble of long-held idealism bursts, a confusing period of disillusionment and sadness can occur, perhaps including a slump of depression. If transiting Saturn was aspecting Neptune, this feeling of being down in the dumps could become prolonged, with signs of sinking disappointment more evident. With Uranus, however, nothing is allowed to drag on hopelessly, not even the shock of reality's cold water splashed in our face. Therefore, with a Uranus/Neptune transit, we can feel stunned and truly bewildered whenever we find our faith and trust in somebody or something has been misplaced. A belief system, especially one that has been a blind spot for us, can break down and dissolve as new and contradictory information is revealed. Unrealistic dreams we have hung on to suddenly go up in smoke as outer events beyond our control insist that we deal with uncompromising realities. Usually, a third personal planet is involved in the action when situations start turning out emotionally weird like this—is it our Moon or maybe our Venus? The more our personal planets are tied into this transit, the greater the impact it has on us. This goes for all Outer Planet to Outer Planet transits.

278 • Chapter Seventeen

A Fate Attack

Sometimes our high-minded objectives are inexplicably forced to come to an abrupt end. (An aspiring ballerina badly breaks both legs in a freak car accident and ends a promising career; little does she know that years later, at a time when she's devoted to writing epic poetry to help assuage her wounded spirit, she'll win a Pulitzer Prize for her inspiring work.) Many disturbing Uranus/Neptune incidents can seem tragic and grossly unfair at the time. We seem to be innocently caught up in something set into turbulent motion by people we don't even know, or by some force of nature we cannot control. These events also make little sense to those of us who try to figure out such Neptunian predicaments logically and rationally. It won't work, because this is a transiting planet that can baffle us (Uranus) aspecting a planet that often confuses us (Neptune). Together, both generate states of impermanence, the kind that get earth signs panicky and water signs hysterical. Nothing remains stable and secure, as a strange fate enters our lives and forces on us an entirely new perspective.

Of course, it can turn out to be a wonderful kind of fate that shatters our less worthy dreams and instead opens doors of awesome promise. However, both planets represent a consciousness that is little concerned with satisfying our ego demands (the more self-centered our desires, the less workable the end-results). Deeper soul-gratification can be our outcome when launching humanitarian projects designed for the betterment of all—that's quite a tall order for most people who don't have a social destiny similar to that of Mother Teresa, Martin Luther King, Jr., or Gandhi.

That Vision Thing

So, how else can we experience transiting Uranus aspecting Neptune, short of being stuck in a sea side vacation spot where El Nino's wreaking havoc? Neptune's natal house might help us to figure out where we can refocus ourselves, or maybe even effectively establish a clear focus for the very first time. Neptune can natally show where our life situations and our attitudes remain chronically fuzzy and ill-defined. Nonetheless, this is where a new and better-realized dream is to emerge. The house where Uranus transits is also important, giving us needed here-and-now clues. Uranus is the trigger; therefore, something from our current life experience is to stimulate our inner development in such a way that new Neptunian awarenesses are able to surface more freely.

This is a process similar to all transiting planets: the transiting planet provides the outer conditions needed to stimulate the growth of the natal planet contacted. In our case, Uranus is bringing new people and conditions into our life to either enhance our Neptunian visions (if our ideals are on the right track and are in harmony with practical realities), or unexpectedly turn our world upside down abruptly, if we've been losing touch with ourselves for too long, stumbling down the wrong Neptunian paths.

Visions we have at this time are filled with clarity and a sense of urgency, but they may lack practical current application. Still, both planets are eager to open the vault of heaven long enough to capture numinous ideas of great universal import. Perhaps what we envision is not yet supposed to manifest in concrete terms. We'll have to simply enjoy our insights, with the knowledge that these possibilities may someday have their chance in a future world ready for such transformation. Nevertheless, when these two planets come together, divine discontent results as we see things way ahead of their destined time to unfold. Such is the fate of visionaries plagued by the impatience that their out-of-the-ordinary passions create.

COMING CLEAN

Do we have deep, dark, self-destructive secrets? That's as much a Neptunian issue as it is a Plutonian theme. Uranus doesn't want to read our laundry list of excuses (abuses that were done to us or by us in past lives or in this one). Uranus just wants this nightmare to end now, before we get further pulled down into our darkness and start taking hostages with us. (Neptune in trouble typically entangles others in its sticky web.) I sound a bit dramatic here but, even in less serious predicaments, Neptune draws people into situations that sometimes add to the confusion. It is this complication that Uranus tries to sidestep. Because both planets relate to the workings of the unconscious, we certainly do not feel that we are able to manipulate matters at the moment or manage events. They just happen. We simply have to deal with them squarely, and perhaps struggle to keep ourselves conscious about the odd dynamics at play.

Uranus will not allow our escapism to succeed should we get further caught up in delusion. What can happen, and probably should happen, is that we finally make major emotional breakthroughs with ourselves, and thereby put a halt to any further inner torment. Uranus short-circuits this self-destructive energy. Life gives us a good shaking, snapping us out of our trance. We need to redeem ourselves on some level, and Uranus offers us a key to unlock the dungeon cell and walk out into the bright light. Spiritual awakenings, perhaps meaningful to us and no one else, can occur during these transits. This can be an exceptional time to feel alive and well, and in tune with our spiritual essence. After all, both Uranus and Neptune are not very worldly planets. They have more transcendent goals in mind. First, though, we'll have to come clean regarding our energy-sucking, self-deceptive ways of dealing with others.

UFO ALERT?

Some of us might wait for some stunningly otherworldly event to happen in our lives, like being abducted by aliens and taken aboard the mothership. Perhaps we're ready for something equally dramatic to come into our world that can remove us from everything that has become tediously familiar and banal. This transit could provide an escape hatch from the all-too-ordinary patterns of practical living. Although UFO aliens may not be levitating us out of our bed each night at 2:30 A.M., there's probably a part of ourselves that wishes to be teleported to fantastical realms of sensory wonder. We suddenly might have a hankering for whatever we deem to be of rare and sometimes unappreciated beauty.

Why not channel this desire into art or music in ways most accessible to us? Our spirit longs for aesthetically captivating stimulation at this time. Maybe we can't play piano, but we can buy a few CDs of Vladimir Horowitz doing his finger-magic on the keyboard. We can also visit local museums and galleries more often to whet our appetite for uncommon artistic visions. Even taking up photography or film-making, and learning about creating special effects, could prove exciting during this transit. The key thing is to break away from our safe perceptions about the world in ways that enrich our soul. Most of our daily routines just can't take us to Uranus/Neptune levels (unless you count the frenetic chaos of Los Angeles rush-hour traffic, but that's certainly not an inspiring manifestation). A Uranus sextile or trine can stir our imagination in fruitful ways, even though we find ourselves leaning toward the unusual in our taste. With more frictional aspects, both planets become exaggerated in expression and our visions can turn truly bizarre, à la Salvador Dali.

THE DEEP END

If a few of us are already spiraling downward in the grip of drug addiction or alcoholism, these transits unravel our structure even further. We could be flirting with madness as well, where the little green men we see are strictly those that our stressed-out imagination and paranoia have created. We may feel we have strong reasons to suddenly distrust the world. Our inner radar may mistakenly tell us we are being betrayed by others (most so with an agitating Uranus square or opposition). Darker fears can seize us, and our instinct is to break away from whatever we feel has trapped us. Neptune can respond by inviting further psychological terrors, which lead to oblivion. Barely alive and definitely not well, we'll need emergency measures to keep us from extinguishing our life force altogether. Admittedly, this is an extreme scenario that few will undergo.

Even in less drastic predicaments, we can suffer a rapid deterioration of psychological boundaries that have held us together in the past. Becoming unglued opens doors to strange and unsettling perceptions, keeping us in an unstable, unworkable state of mind. We'll need to study what our natal and transiting Saturn (our urge to reglue what's broken) are doing at this time. We'll need to draw extra ego-strength from Saturn and other grounding chart factors also in operation. The scary features of the deep end do not await those who have already plumbed their depths carefully without fear and denial.

Fog Lifting

If we keep in mind that Uranus demands clarity, not fogginess, plus seeks to blast away at deception, we may better attempt to fulfill our Neptunian needs with a dose of self-honesty. We'll need to carefully evaluate those who suddenly enter our world and present us with their techniques of enlightenment. We especially need to beware of spiritual hype, or overblown promises of instant cures. No one can provide us with the sudden miracles that we may seek at this time, especially if we haven't done the inner work that real transformation requires.

However, by turning within and listening to our inner voice, we can awaken our own intuitive gifts and healing powers, which then help to restore our faith in our spiritual identity. Uranus wants us to be enlightened, but with our ego-integration intact. We are not to passively and totally surrender our selfhood to anyone else at this time (a typical Neptune trap) no matter what is being promised as our ultimate reward for obedience. To do so would risk inner upheavals and a sense of being completely undermined down the road, when our eyes finally open and we see the damage we've allowed to be done to us. To stay focused during this transit is to stay awake and clear while we enable our special dreams to unfold.

CHAPTER EIGHTEEN

THE LONG WAY HOME

Well, it's been quite a while since we first rocketed away to the far-flung reaches of vast inner space. Not surprisingly, we failed to keep from coming down with attacks of those rare states of mind the Saturnians at Control Central warned us about—and now they want to put us through a battery of reality tests to find out why! It was tricky navigating through alien force fields that awaited us in the depths of our unconscious—especially when having to adjust to the effects of Uranus' dramatic tilting of our mind's axis. We made the best of it and even sometimes the worst of it—but we finally emerged as individualists with much more courage and self-reliance than ever.

However, one bittersweet realization we have after undergoing a few successful Uranian transits is that we really can't go home again, at least not back home to our original state of self-ignorance where we once felt safe in a sheltered world of minimal challenge and risk. Feeling safe was often more important to us than was feeling alive and well. This previous state of mind and heart may now no longer exist, and it would be spiritual suicide to attempt to return to any such psychological place of self-limitation.

Unfortunately, there may be people we still love who desperately want us to come back to our former levels of awareness so that we may permanently resettle ourselves there. They still can't figure out why we ever wanted to leave behind comforting familiarity to venture off into an unknown future of uncertainty. Some people didn't realize how soul-suffocating our old patterns were for us. These people were baffled and skeptical when we made a few radical but positive lifestyle changes (at least *we* thought they were positive). Our bold actions scared them, and their adverse reactions to our plans for greater personal freedom proved equally unsettling to us. Still, we felt driven to do what we had to do even when we weren't sure where it would all lead. Uranus guided our unfamiliar path more protectively than anyone realized or appreciated at the time. This we will know to be true after many years have passed and we more fully appreciate the breakthroughs we have made with ourselves.

Once Uranus has altered our consciousness—and usually it's a lasting alteration—we can no longer accept the confinement of old frameworks of societal and familial conformity. We now wield greater power to create more of the self we are meant to be, our soul's unique unfoldment of who we are. However, this process often invites a certain loneliness, because now we may observe our prior background—and all the conditioning factors entailed— with the detachment of an "outsider." Some of us can't comfortably return to our roots and blindly fit in with our family's natural rhythms as perhaps we once pretended we could. A few of us may now be labeled "black sheep" as a result, or else we're just called the "eccentrics" of the community.

If we are truly becoming enlightened in a Uranian way, we'll find ourselves more able to accept the various levels of consciousness of those we know and love (especially those close to us who are trapped in negative Saturnian and Martian states of awareness; maybe we, too, were operating on those levels once). Such people may not have a clue about us these days, but we seem to understand them and their motivations quite well. Our need for kinship now becomes better met by different individuals from broader worlds of possibility. As we begin to recognize universal levels of human connection, we understand the human race to be one big cosmic family. Still, it can sadden us to realize that people we grew up with, or others in our past with whom we've shared intimacy, are potentially those who least understand who we really are now.

The long way home to our real Self is a path that can lead us to great inner rewards. Our earned enlightenment cannot be taken away by those skeptics and pessimists who doubt the depth and sincerity of our inner changes. It is not easy to obtain these new awarenesses. Any commitment to authentic living requires bravery and determination. Ideally, Uranus will no longer seem like an unearthly, alien force from some galaxy many light years away. Let's all learn to develop Uranian energy in highly personal ways that will feel quite natural to us.

I hope this book has allowed you to consider the benefits that can be gained by openly embracing and identifying with at least some of the principles of Uranus. We can't afford to sit back and view this planet as strictly an impersonal, generational force that only indirectly has an impact on our lives. Twenty-first Century Astrology will further support the concept that all the Outer Planets are becoming even more personalized expressions of our being because, individually, we are already clearly learning to assimilate their energies in more conscious, self-directed ways.

It seems that a strong, well-formed, resilient ego best supports our safe passage through the realms of awareness symbolized by Uranus. Rigidity and selfishness mixed with Uranian energy create tremendous problems, the kind that invite distorted self-concepts and malcontent behavior. This is why ego-adaptability is the finest tool we can use to employ Uranian principles in our lives. Also, let's not forget to approach life with an offbeat Uranian sense of humor, because this helps make much of the duller, day-to-day routines of practical reality more tolerable. Saturn will simply have to adjust!

Well, I hope you have enjoyed reading Volume One of my *Alive and Well* trilogy (after all, you made it all the way to the end of this book)! The next stop on this cosmic safari takes us to Volume Two—*Alive and Well with Neptune*—where we get to deep-sea dive into the quiet depths of our psyche so that we can explore the more fluid and enchanted worlds of our unconscious landscape. Then it's on to Volume Three—*Alive and Well with Pluto*—where we plunge into the psyche's volcanic underworld to find hidden treasures that can renew our soul and revolutionize our life. Believe me, the journey only gets more fascinating!

BIBLIOGRAPHY

I am listing the titles of some books that either cover information about Uranus transits or that describe the nature of Uranus at length. This is certainly not a complete list of what's available; I tried to select mostly books that are not out-of-print and thus hard to find. Happy reading!

URANUS

Negus, Joan. *The Book of Uranus*. San Diego, CA: ACS Publications, 1996.

Paul, Hadyn. *Revolutionary Spirit: Exploring the Astrological Uranus*. Dorset, England: Element Books Limited, 1989.

Tarnas, Richard. *Prometheus, the Great Awakener: An Essay on the Archetypal Meaning of the Planet Uranus*. Vol. 21. Woodstock, CT: Spring Publications, 1995.

URANUS, NEPTUNE, AND PLUTO

Ashman, Bernie. *Roadmap to Your Future*. San Diego, CA: ACS Publications, 1994.

Arroyo, Stephen. *Astrology, Karma, and Transformation*. 2nd Rev./Expanded Ed. Sebastopol, CA: CRCS Publications, 1993.

Forrest, Steven. *The Changing Sky*. 2nd Ed. San Diego, CA: ACS Publications, 1999.

Greene, Liz. *The Outer Planets and Their Cycles*. 2nd Ed. Sebastopol, CA: CRCS Publications, 1996.

Hand, Rob. *Planets in Transit*. Atglen, PA: Schiffer Publishing, Ltd., 1980.

Marks, Tracy. *The Astrology of Self-Discovery*. Sebastopol, CA: CRCS Publications, 1985.

Rodden, Lois. *Modern Transits*. Tempe, AZ: AFA, 1978.

Rudhyar, Dane. *The Sun Is also a Star—The Galactic Dimension of Astrology*. New York, NY: E.P. Dutton & Co., 1975. (Hard to find.)

Sasportas, Howard. *The Gods of Change: Pain, Crisis and the Transits of Uranus, Neptune, and Pluto*. New York, NY: Arkana—Viking Penguin, Inc., 1989. (This is the only book I'm aware of that gives an in-depth coverage of all three planets in a single volume; I definitely recommend that you have this book on your library shelf.)

Thorton, Penny. *Divine Encounters*. London, England: The Aquarian Press, 1991.

Tompkins, Sue. *Aspects in Astrology*. Dorset, England: Element Books Limited, 1989.

Tyl, Noel, ed. *How to Personalize the Outer Planets: The Astrology of Uranus, Neptune, and Pluto*. St. Paul, MN: Llewellyn Publications, 1992. (This is an anthology presenting the works of seven astrologers.)

MYTHOLOGY

Aldington, Richard and Delano Ames. *New Larousse Encyclopedia of Mythology*. New York, NY: The Hamlyn Publishing Group Limited, 1978. (Hard to find.)

Bolen, Jean Shinoda. *Goddesses in Everywoman*. New York, NY: HarperCollins Publishers, 1989.

———. *Gods in Everyman*. HarperCollins Publishers, 1989.

Gayley, Charles Mills. *The Classic Myths in English Literature and in Art*. Atlanta, GA: Ginn & Company, 1939; Cheshire, CT: Biblo-Moser, 1991 (paperback edition).

Morford, Mark P. O. and Robert J. Lenardon. *Classical Mythology*. New York, NY: Longman, Inc., 1977.

Richardson, Donald. *Greek Mythology for Everyone: Legends of the Gods and Heroes*. New York, NY: Avenel Books, 1989.

TWELVE FACES OF SATURN
Your Guardian Angel Planet
Bil Tierney

Astrological Saturn. It's usually associated with personal limitations, material obstacles, psychological roadblocks and restriction. We observe Saturn's symbolism in our natal chart with uneasiness and anxiety, while intellectually proclaiming its higher purpose as our "wise teacher."

But now it's time to throw out the portrait of the creepy looking, scythe-wielding Saturn of centuries ago. Bil Tierney offers a refreshing new picture of a this planet as friend, not foe. Saturn is actually key to liberating us from a life handicapped by lack of clear self definition. It is indispensable to psychological maturity and material stability—it is your guardian angel planet.

Explore Saturn from the perspective of your natal sign and house. Uncover another layer of Saturnian themes at work in Saturn's aspects. Look at Saturn through each element and modality, as well as through astronomy, mythology and metaphysics.

1–56718–711–0, 6 x 9, 360 pp.   **$16.95**

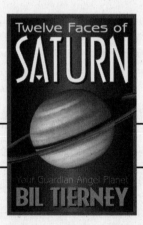

Synthesis & Counseling in Astrology
The Professional Manual
Noel Tyl

One of the keys to a vital, comprehensive astrology is the art of synthesis, the capacity to take the parts of our knowledge and combine them into a coherent whole. Many times, the parts may be contradictory (the relationship between Mars and Saturn, for example), but the art of synthesis manages the unification of opposites. Now Noel Tyl presents ways astrological measurements—through creative synthesis—can be used to effectively counsel individuals. Discussion of these complex topics is grounded in concrete examples and in-depth analyses of the 122 horoscopes of celebrities, politicians, and private clients.

Tyl's objective in providing this vitally important material was to present everything he has learned and practiced over his distinguished career to provide a useful source to astrologers. He has succeeded in creating a landmark text destined to become a classic reference for professional astrologers.

1–56718–734–X, 7 x 10, 924 pp., 115 charts   **$29.95**

To order call 1–800–THE MOON
Prices subject to change without notice.

THE HOUSE BOOK
The Influence of the Planets in the Houses
Stephanie Camilleri

What gave Marilyn Monroe, John Lennon, John F. Kennedy and Joan of Arc their compelling charisma—could it be that they all had planets in the Eighth House? Find out why someone with Venus in the Fifth may be a good marriage partner, and why you may want to stay away from a suitor with Uranus in the Second.

Now you can probe the inner meaning of the planets in your chart through their placement in the houses. *The House Book* provides a solid base for students of astrology, and gives advanced astrologers new ways of looking at planet placement.

The author culled the similarities of house qualities from 1,500 different charts in as intensive and as scientific a method as possible. The most important feature of this book is that each description was written from the perspective of real charts with that location, without referencing preconceived ideas from other books. In some places, the common wisdom is confirmed, but in others the results can be very surprising.

1-56718-108-2, 5 ³/₁₆ x 8, 288 pp., softcover   **$12.95**

To order call 1–800–THE MOON
Prices subject to change without notice.

ASTROLOGY: WOMAN TO WOMAN
Gloria Star

Women are the primary users and readers of astrology, yet most astrological books approach individual charts from an androgynous point of view. *Astrology: Woman to Woman* is written specifically for women, by a woman, and shows that there *is* a difference in the way men and women express and use their energy. It covers every facet of a woman's life: home, family, lovers, career, and personal power.

Whether the reader is new to astrology or an old pro, there are new insights throughout. Those of you who don't know what sign their Moon is in or which planets are in their seventh house can order a free natal chart from Llewellyn that will tell you everything you need to know to use this book.

Discover what's at the heart of your need to find a meaningful career, understand your inner feminine power, own up to your masculine self, uncover your hidden agendas, and much more.

1–56718–686–6, 7 x 10, 464 pp. $19.95

ASTROLOGY: *Woman to Woman*

GLORIA STAR

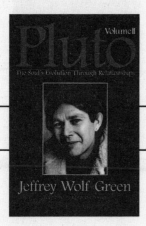

WHEN PLANETS PROMISE LOVE
Your Romantic Destiny Through Astrology
Rose Murray

(Formerly *When Will You Marry?* Now revised and expanded.)
Never before has an astrology book so thoroughly focused on
timing as a definitive factor in successful romantic partnerships. Written in language the beginner can easily follow,
When Planets Promise Love will engage even the most advanced student in search of love. Identify what you need in a
partner and the most favorable times to meet him or her based
on transits to your natal chart. Then learn how to compare
your chart with that of a potential mate. This premier matchmaking method is laid out step by step, starting with the ba-
sics through fine tuning with Sun-Moon midpoints, chart
linkups, and Arabian parts. Confirm with exactness whether
or not someone is "the one!"

Includes a 50% off coupon for a marriage year chart.

1–56718–477–4, 6 x 9, 256 pp.   **$12.95**

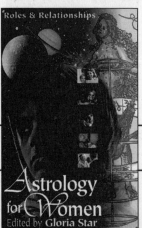

ASTROLOGY FOR THE MILLIONS
Grant Lewi

First published in 1940, this practical, do-it-yourself textbook has become a classic guide to computing accurate horoscopes quickly. Throughout the years, it has been improved upon since Grant Lewi's death by his astrological proteges and Llewellyn's expert editors. Grant Lewi is astrology's forerunner to the computer, a man who literally brought astrology to everyone. This, the first new edition since 1979, presents updated transits and new, user-friendly tables to the year 2050, including a new sun ephemeris of revolutionary simplicity. It's actually easier to use than a computer! Also added is new information on Pluto and rising signs, a new foreword by Carl Llewellyn Weschcke, and introduction by J. Gordon Melton.

Of course, the original material is still here in Lewi's captivating writing style all of his insights on transits as a tool for planning the future and making the right decisions. His historical analysis of U.S. presidents has been brought up to date to include George Bush. This new edition also features a special "In Memoriam" to Lewi that presents his birthchart.

One of the most remarkable astrology books available, *Astrology for the Millions* allows the reader to cast a personal horoscope in 15 minutes, interpret from the readings and project the horoscope into the future to forecast coming planetary influences and develop "a grand strategy for living."

0–87542–438–4, 6 x 9, 464 pp., tables, charts $14.95

OVER 250,000 COPIES SOLD!

ASTROLOGY for the MILLIONS GRANT LEWI

UNDERSTAND THE PAST! PREDICT THE FUTURE!

To order call 1–800–THE MOON
Prices subject to change without notice.

ONE OF THOSE
MALIBU NIGHTS

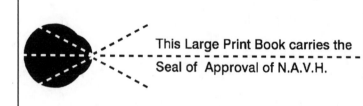

This Large Print Book carries the
Seal of Approval of N.A.V.H.

ONE OF THOSE MALIBU NIGHTS

ELIZABETH ADLER

LARGE PRINT PRESS
A part of Gale, Cengage Learning

GALE
CENGAGE Learning

Detroit • New York • San Francisco • New Haven, Conn • Waterville, Maine • London

GALE
CENGAGE Learning

LIBRARY OF CONGRESS CATALOGING-IN-PUBLICATION DATA

Adler, Elizabeth (Elizabeth A.)
  One of those Malibu nights / by Elizabeth Adler.
    p. cm. — (Thorndike Press large print core)
  ISBN-13: 978-1-4104-0466-4 (hardcover : alk. paper)
  ISBN-10: 1-4104-0466-8 (hardcover : alk. paper)
  1. Private investigators — Fiction. 2. Actresses — Fiction.
3. Americans — Foreign countries — Fiction. 4. California —
Fiction. 5. Mexico — Fiction. 6. Rome (Italy) — Fiction.
7. France — Fiction. 8. Large type books. I. Title.
PR6051.D56O54 2008
823'.914—dc22   2008008164

ISBN 13: 978-1-59413-323-7 (pbk. : alk. paper)
ISBN 10: 1-59413-323-9 (pbk. : alk. paper)
Published in 2009 in arrangement with St. Martin's Press, LLC.

Printed in the United States of America
1 2 3 4 5 6 7 13 12 11 10 09

For lovely Aunt Bebe Sell,
still reading and still enjoying
at the age of ninety-two

CHAPTER 1

It was not the kind of night, nor the kind of place, where you'd expect to hear a woman scream. It was just one of those Malibu nights, dark as a velvet shroud, creamy waves crashing onto the shore, breeze soft as a kitten's breath.

Mac Reilly, Private Investigator, was walking the beach alone but for his dog. His lover, Sunny Alvarez, had taken off for Rome after a slight "disagreement" concerning their future. But that was an ongoing story.

Mac lived in the famous Malibu Colony, habitat of movie stars and showbiz moguls and megabucks persons of every sort, each one richer than the next, give or take a couple of million, or in some cases billion. Their fancy beachside mansions didn't look so fancy from Mac's angle, but then the beach was also not an angle from which most people ever got to see them. In fact the public rarely got to see them. The Colony was gated and guarded, one gate in or out, and though the beach had free access it was only along the water's edge with no loitering. Any unknown caught prowling along it at midnight would be in for some tough questioning.

The Colony's mansions were mostly the simple second or even third homes of rich people, understated in their beach chic and with the narrowest bits of oceanfront deck known to man, at a cost per square foot that boggled the accounting.

Mac's own place was a more modest dwelling, a forties bungalow he had bought cheap years ago in the big real estate slump and which had once been owned, or so he'd heard, by the old-time movie star Norma Shearer. Or was it Norma Jean? Norma or Marilyn, it made no difference. A shack was a shack whichever way you looked at it.

The house's saving grace, apart from its ritzy location and the view, was a small wooden deck with steps that led directly to the beach. It wasn't unknown in a winter storm for the ocean to come thudding at the wooden pilings under that deck, slapping over the rails until Mac felt as though he were on a boat, but he liked the excitement and even the possible danger. He was happy in Malibu, he wouldn't live anywhere else if you paid him. Except maybe Rome for a week or two, in the company of Sunny.

Mac kind of looked the PI role, six foot two, longish dark hair still thick on the head, thank God, even though he was forty. Dark blue eyes, kinda crinkled from too many days on the beach and too many nights spent propping up bars in his youth. No facial hair — Sunny didn't like it. A lean athletic build, which since he was a lazy guy gym-wise, was mostly earned from jogging along the beach with his rescued three-legged, one-eyed mutt of a dog, Pirate, who was pretty fast when he had the wind behind him.

Pirate was Mac's best buddy, and you've never seen a more perky little tyke. With his long spindly legs and ragged gray-brown fur, plus a severe underbite that left his bottom teeth exposed in a perpetual grin, he'd win Malibu's ugliest dog contest easy.

Of course Sunny adored Pirate, even though she wouldn't let him near her Chihuahua, Tesoro. Strong on the claws, quick with a bite and weighing all of three pounds, Tesoro outsmarted Pirate at every turn.

Sunny believed it was the animosity between their dogs that was preventing their marriage, but Mac was not quite certain on that score. Why spoil a good thing? Sunny and he were *good* together just the way they were, i.e., unmarried.

Sometimes Mac thought maybe it was his alter ego that appeared on your TV screens Thursday nights, in real-life documentary style reinvestigating old Hollywood crimes, of which there were more than you might imagine. His show was titled *Mac Reilly's Malibu Mysteries,* with yours-truly looking extra-cool in the Dolce & Gabbana black leather jacket Sunny had bought him.

When she'd told him it was a Dolce, Mac had had no idea what she meant. It sounded like Italian ice cream to him. Later, he'd discovered it was an Italian designer and the jacket was without doubt the coolest garment he owned. Soft and pliable as wet putty it had become part of his on-screen image, though God knows he was more usually to be found in sweats slouching up Malibu Road to Ralphs supermarket in search of beer and dog food, or breakfasting in Coogies coffee shop in a T-shirt and shorts rather than decked out in black leather.

Anyhow, the show, which took old murders and reckoned to solve them, had given him some kind of fame. It was all relative of course, because as everybody knew in Hollywood, once your show went off the air you were as forgotten as last week's dinner. And now it looked as though Mac's time had come and gone and the show was likely not to return for another season. Too bad, because the income had come in handy and he'd gotten to keep his day job, investigating for all those nice rich folk. And surprisingly many of them were genuinely nice. Plus they had the same troubles as everybody else. Sex and money. In that order.

He gave Pirate the low whistle that meant get the heck back over here, and the dog came running from whatever exciting secrets he'd found on Malibu's most expensive bit of shore. Together, they turned and headed for home. They were strolling along, minding their own business, listening to the crash of the waves, breathing in the salty ocean air and keeping an eye out for shooting stars, all that romantic stuff. And then they heard the scream.

High pitched. Quivering. Terrified.

It didn't take a PI to figure out that the screamer was female. And that she was in trouble.

CHAPTER 2

Mac quickly scanned the houses. All were in darkness save for a glimmer of light on a deck a couple of houses back. He stumbled through the soft sand toward it, followed by Pirate.

He paused at the foot of the wooden steps leading to the house, listening, but there were no more screams. What he did think he heard though was a sob. Muffled, but nevertheless a definite sob.

Telling Pirate to stay put, he inched his way up the steps onto the deck, which was only about ten feet deep, a usual size for the acreage-tight Colony. The house loomed in front of him, a glass-and-limestone cliff that was more modern than the millennium and more stark than the architecture of Richard Meier, famous for the design of the L.A. Getty museum among other things. It was also as dark as the night outside.

Suddenly a lamp was turned on. Through

the window he glimpsed a woman. A redhead wearing a sheer black negligee and, if he was not mistaken and even though it was at fifty paces, very little else. Now contrary to the popular belief, this was not your normal bedtime attire in Malibu and nor was midnight a usual hour to retire. Most everyone in the movie biz had an early call and were in their flannel PJs, curled up in a bed, learning the next day's lines by nine.

Mac knocked on the window but the woman didn't seem to hear. She just stared down at her feet as though there was something fascinating there. Like maybe a body, Mac thought.

She was young, maybe twenty-three, and beautiful, with everything in the right place as revealed by the sheer bit of black chiffon and lace she wore. Plus she had the face of a naughty angel. Mac felt glad to be of help. He checked, saw that the glass doors were unlocked, and in his knight-in-shining-armor role he slid it open.

Her head shot up and he flung her a reassuring smile. "Hi," he said, "I'm Mac Reilly, your neighbor. I thought I heard a scream. Are you in some kind of trouble?"

The woman tossed her long red curling hair out of her tearful green eyes, lifted herself to her full statuesque height and pointed

13

a gun at him.

"Get out," she said in a throaty whisper.

Mac eyed the gun. It was a Smith & Wesson Sigma .40, and definitely not to be messed with. He paused long enough to wonder why she was not pressing the button to summon Security from the gate instead of threatening him. And then the gun went off.

The bullet ricocheted from the polished concrete floor near his foot, shattered a crystal vase then buried itself in the back of a nearby sofa.

Mac didn't wait around for a second shot. He took off down the steps, sprinting back along the beach a couple of paces behind the cowardly dog.

"Oops, sorry, my mistake," she called after him, her voice floating eerily on the breeze.

CHAPTER 3

Sunny Alvarez was lying on the bed in her room in the Hotel d'Inghilterra in Rome, dialing L.A. every ten minutes and wondering where on God's earth Mac Reilly was. It was nine a.m. in Rome, which meant it was midnight in Malibu. Could Mac be out on the town the minute her back was turned? When the truth was she'd only come here to stir him up a little. She'd figured a little jealousy wouldn't hurt. They'd always said absence made the heart grow fonder. Now she wasn't so sure.

Restless, she got up and began to pace, running her hands distractedly through her long hair that swung around her shoulders like liquid black satin, with just enough wave in it to give it bounce. Sunny's eyes were amber brown and fringed by lashes so thick they were like miniature shades on the windows of the soul. Her skin was golden, her legs long and her skirts usually short. She was, as Mac often told her, in between kisses, a knockout.

"Even though you're ditzy enough to drive any man mad," he'd once said to her, causing her to swat him with a handy cushion, which in turn sent Pirate into a barking frenzy because nobody — not even Sunny, who he loved — was going to harm his "father."

The only positive news Sunny had were the postcards Mac sent her daily. At least she believed they were from him but since there was no signature she couldn't be absolutely sure. Except she knew nobody else who would be FedExing pictures of Surfrider Beach, and Zuma, and Paradise Cove with the anonymous message "From Malibu, with Love." Sunny was saving those postcards. She planned to stick them in her Memories book to look at when she was old and gray. And also, unless she could get Mac to the altar soon, still single.

Her room looked as though a typhoon had hit it. Her method of unpacking was to take everything out of the suitcase and toss it over chairs and the bed, then sort out whatever she needed from the various piles. Her apartment was kept mostly in the same state of chaos. It was a leftover habit from her college days when it had seemed the easiest —

and quickest — method of getting dressed, and it drove Mac crazy. To compensate she would point out her kitchen to him, immaculate as an operating room, and where she would cook him delicious meals — plus she always did the dishes afterward. Food was her first passion. The second was clothes, as evidenced by the shopping bags from Rome's boutiques scattered around the room. The third was her Harley chopper, but that unfortunately was back in L.A. Rome was a city full of Vespas, and they were definitely not the same thing.

She picked up the phone, got Mac's voice mail one more time, slammed it back down again and lay back on the bed contemplating her coral-pink toenails and her life.

Of course her name was not really Sunny. That's just what Mac called her. Her real name was Sonora Sky Coto de Alvarez. Quite a mouthful, as she was only too painfully aware. In fact she was truly grateful to be designated as Sunny. At least it let her off the hook of constantly explaining those names, which were the direct result of having a hippie-style mother who'd communed with nature as well as with the spirits, in the desert around their adobe-style ranch outside of Santa Fe, New Mexico.

Sunny's mom was still dreamy and offbeat

beautiful and still prone to wearing floating shapeless garments with long strands of crystal beads, and often flowers in her smooth blond hair. Yet, oddly enough she'd always been a terrific mom, even if her daughters did have to spend nights with her out in the desert, communing with nature while keeping a nervous eye out for rattlers. Mom didn't even think about things like snakes. Her mind was on a higher plane, one that sadly Sunny and her sister never reached.

Their feet were more firmly planted on this earth. As kids they loved riding horses, chasing boys and raising hell. Later, they'd graduated to riding motorcycles, chasing boys and raising hell. That is until their father took them in hand, straightened them out and packed them off to college, where he hoped real life would not deal them a killer blow between the eyes after the gentle ministrations of their otherworldly mother.

Sunny's papa was something else again. Handsome? You don't know what that word means if you haven't seen her dad. He's Mexican, with that polished-tan skin, thick silver-gray hair, soft brown eyes and a trim mustache. Kind of like Howard Keel used to look in *Dallas*. Astride his black thoroughbred he was the epitome of the Mexi-

18

can ranchero.

He'd thought Brown was the perfect college to tame a Harley-riding, boy-mad eighteen-year-old, and true, it had opened up Sunny's world to a kind of life she had never seen. But she'd missed her family, and she'd cried thinking of her beloved *abuelita,* her Mexican grandmother, and of the tamales, cooked the way only Abuelita knew how to cook them. The tamales were a Christmas Eve staple at the ranch and everyone from the workers and the cowboys and the local families gathered to enjoy them, along with a large amount of tequila and Corona beer and Mexican music and dancing.

Of course Mom also cooked the traditional turkey, albeit in her usual haphazard way. Sometimes it wasn't quite done and had to go back in the oven for an hour or two; and sometimes it was too well done and Papa said you needed horse teeth to get through it. Either way it was fun.

At college it hadn't taken long for the golden-limbed raven-haired Latina in black biker leathers zooming around on her Harley to get noticed. Soon, she was cooking tamales and handing out the Corona at her own parties. By the time she graduated, magna cum laude, with her proud parents and her sister beaming in the audience,

Sunny felt almost ready to tackle the world. But before that came the Wharton School and a master's in business.

Later, she'd found herself a job in Paris, working for a fragrance house. After a year there she moved on to Bologna and a job with the Fiat corporation. Then back home and on to California, where she'd opened her PR business, which was doing very nicely, thank you.

She'd met Mac at a press party for his TV show. He told her he'd noticed her across the room. "How could I miss you, in that outfit," was what he'd actually said.

It was winter and she had on a tiny white miniskirt, her tough-girl motorcycle boots because she'd driven there on her Harley, and a black turtleneck. She was all long golden legs, sexy curves and tumbling black hair. She was always careful about drinking and driving and was sipping lemonade when he'd come up behind her and tapped her on the shoulder. Swinging round, she found herself looking at this rugged guy in jeans and a T-shirt, whose deep blue eyes were taking her in like she was the best thing he'd seen all night.

It's him, she'd thought, thrilled. *The man I've been waiting for all my life.* Of course she was smart enough not to tell him that and it was true, they were total opposites: Mac, dragged up by his bootstraps from the streets of Boston and the Miami crime scene to the PI and TV personality he was now. And she, the wild child brought up on the ranch, beautiful and brainy and ditzy, but with the determination to be her own woman.

In fact life seemed set fair, romance-wise, until she'd invited him for dinner at her smart high-rise apartment in Marina del Rey, a few miles from his home in Malibu. Even her home-cooked tamales were no match for that first disastrous encounter between Tesoro and Pirate.

Let's face it, Sunny thought, sighing, Pirate was willing to be friends. Tesoro was not. And rather than have his dog harassed, Mac had departed, leaving the tamales uneaten. "Next time I won't bring the dog," he'd said, shielding Pirate from the marauding Chihuahua.

And that was how things now stood. She went to his Malibu house without Tesoro. He came to her Marina apartment without Pirate. "And never the twain shall meet" was Mac's motto. Which of course left them in their current uncertain limbo.

Sunny checked the time. It was after midnight in Malibu and Mac still wasn't home.

21

She should get off this bed right now. Get out there in the vibrant bustling streets of Rome, pick up a charming handsome Italian and let him sweep her off her stilettos.

Heaving a sigh that this time came from her gut, she decided that she would call Mac no more. The hell with the diet. She could practically smell sugar and cinnamon as she ran her hands hastily through her long dark hair, pushed her feet into black patent sandals and headed for the door.

The phone rang. She swung round, staring at it.

It rang again. Of course it wouldn't be *him*. How could it? Hadn't she been calling him for the past hour, damn it?

She picked up the phone. *"Pronto?"* she said sulkily.

CHAPTER 4

"Sunny?" Mac said.

"This is Sonora Sky Coto de Alvarez."

Mac felt a sudden frost in Malibu. She was giving him the full-name treatment. He was in real trouble. Only thing was, he didn't know why.

"You sound so Italian," he said. "Maybe I should be calling you Signorina."

"How would you know I'm not already a Signora now? Neglected as I am by you."

He grinned. "Okay, *Signora,* who popped the question?"

"Certainly not you. Where were you, Romeo? I've been calling for the past hour."

"Would you believe the beach? Just me and Pirate. Gazing at the stars and wondering whether you were looking at the same stars, all those miles away in Rome."

"Huh. Some story. Anyhow, it's breakfast time here. There are no stars, except the ones out at Cinecittà Studios, where I expect

23

to be spending the day surrounded by the cream of Roman manhood."

Mac's grin widened. "Don't let it go to your head, honey. Hang out with me and I'll introduce you to the cream of Malibu manhood."

"Like yourself, you mean? Thanks, but I can do without it."

"Listen, Sunny, something odd just happened. . . ."

"Don't even bother to tell me." She sank back onto the bed, legs crossed, dangling one black patent sandal on the end of her toe, contemplating it as though it were the only thing of interest in her life.

Her indifference permeated the telephone miles, registering like the knell of doom in Mac's brain. "Aw, come on, Sunny, honey. . . ."

"And don't call me by that ridiculous rhyming name."

It was Mac's turn to heave a sigh. "Okay, so you don't even want to know that somebody took a shot at me."

Unbelieving, she swung the sandal back onto her foot, uncrossed her legs and stood up. She was heading out that door right this minute. Espresso and sugar buns awaited.

"I'll bet it was a woman," she said.

Mac was genuinely astonished. "How did

you know that?"

"Just call it feminine intuition. And I have no doubt you deserved it."

"Well thanks a lot for that vote of confidence. Really, Sunny, I expected more from you. Y'know, like a little concern for my well-being, a touch of compassion, or at least an inquiry as to whether I might be bleeding to death from my wounds."

"She *wounded* you?" Sunny's knees were suddenly shaky. She sank back onto the bed. "Oh, Mac, darling, are you all right?"

Mac was laughing as he said, "Well, actually, no, she didn't get me. But she had a darned good try, I can tell you. And it was a Sigma .40 handgun I was facing."

Sunny gritted her teeth. "You rat," she said shakily. "Setting me up like that."

"How else was I to get your attention? Look, Sunny, it's the truth." He told her quickly what had taken place just fifteen minutes earlier, taking care to eliminate the red hair, the sexy black negligee, the spectacular body and the face like a naughty angel.

"What I don't get," he said finally, "is why she didn't summon help. I mean, why was she shooting at me? Her would-be savior?"

"Perhaps she's the murderer."

"What murder? I didn't see a body. But I'll bet she was the one who screamed. Plus she

25

was sobbing. I saw the tears on her face. Look, babe, I'm in a dilemma here. I already called the guard at the gate. He called the house, got word from the owner, Ron Perrin — as in Ronald Perrin, billionaire investment mogul — that nothing was wrong. Said it was probably just that the TV was too loud. Now that's b.s. I *know* what I saw. So do I call the cops? Or do I let her get on with whatever it is she's up to and keep my nose out of their business, because it's probably only the usual domestic quarrel and she was just making her point?"

"With a Sigma .40? Some point! If I were you, Reilly, I'd keep my nose clean and stay away where you're not wanted. Unless of course she wants to hire you for some fabulous fee that you can't refuse, especially now the TV show might be canceled. I mean, why work for free?"

Mac thought worriedly about it. "What if she's really in trouble?"

"It seems to me she knew exactly how to take care of herself. And so, I guess, did Mr. Perrin. Do me a favor, Mac, you're talking Malibu Colony. Nothing bad ever happens there, everything is sweetness and light. Just don't be the one to make waves."

Sunny was sitting on the bed again. The phone was clamped between her shoulder and her ear and she was wishing she had a cup of coffee and that Mac would talk about something other than business. Like them for instance.

As if in answer to her wish, he said, "I miss you like crazy. I couldn't take it tonight, just me, mooching along the beach with Pirate. And you not beside me, not there waiting for me, not in my bed . . . in my heart."

Sunny's own heart shifted pace to an incredulous little jiggle. "What did you say?"

"I miss you, Sunny. How about I catch the next flight to Rome?"

"Oh, Mackenzie Reilly," she said, tremulously, "that would be heaven. I know this café with the best espresso. . . ."

"Forget espresso. Make a reservation at your favorite restaurant. I'm taking you out on the town tomorrow night. It's nothing but the best for my woman."

Sunny sighed happily. All was right in their world again. And for her, the billionaire Ron Perrin and the woman with the Sigma .40 were temporarily forgotten.

CHAPTER 5

Early the next morning Mac strolled up the street to Ron Perrin's house. Of course he knew all about Perrin. Who didn't? He was a big shot who'd made his first money by successfully investing for the insurance business, and had then parlayed his investment firm into one of the Wall Street majors. And though he now seemed all power and success, the man had a past. He'd divorced his first wife amid a great scandal because of his relationship with another woman, who happened also to be married to a prominent man. Plus he had once been accused of mishandling funds, though he had emerged, at least on the court records, as clean as a whistle.

Now Perrin was CEO of a string of high-profile companies, and even more powerful. And much, much richer. He was also married to a famous movie star, the blond, petite and beautiful Allie Ray.

As well as the Malibu house, Perrin owned a mansion in Bel Air and a desert compound in Palm Springs, a couple of hours' drive from L.A. It was nothing but the best for Ron Perrin. He lived like the king some folks claimed he believed he was.

From the street angle, Perrin's house in the Colony was a simple blank sheet of windowless limestone. The door was a lofty slab of unpolished steel that looked like a pewter coffin lid and without a knob or a handle of any sort. A discreet button set into the wall invited the visitor to Press.

Mac did so but heard no distant electronic chime. Even the doorbell was silent.

Over the top of the steel slab of the gate he could see the ruffled fronds of a couple of tall palms and some branches of bamboo. Like most of the houses in the Colony he guessed the gate led into an entrance courtyard, beyond which would be the front door proper.

He pressed the bell again and glanced round, waiting. No cars were parked on the yellow lines in front of the house that marked the owner's parking spaces, and the blank steel door to the garage was shut. He wondered what kind of car Ron Perrin drove. A silver Porsche? A Bentley? A red Ferrari, perhaps? It was sure to be expensive and flashy because that's the way the man was.

At last a male voice answered. "Who is it?" He sounded out of breath.

"Mac Reilly," he said into the speaker. "Your neighbor."

A pause, then, "Come in," the voice said.

The steel slab slid to one side and disappeared into a recess in the limestone wall. No crazy paving for RP, only a straight dark blue concrete walkway leading past a midnight blue reflecting pool, through a tropical courtyard where the jungle foliage reached out to grab Mac as he walked by.

Perrin was waiting at the glass entry. He was a short man with the wide shoulders and hard stare of an aggressive primate. He also had the slight forward stoop of a man who lifts weights, as though permanently about to bend and pick up a two-hundred-pound barbell.

Perrin's brow was wide, his hair was dark with a slight wave; his eyes were a light molten brown and his thick eyebrows were what a writer like Dickens might have termed "beetling." That is to say, they joined over his nose in a distinguished frown. His nose had a sharp look to it but his mouth was full-lipped and sensual. He was in good shape and even now, in a sweat-stained tee

30

and gym shorts, Mac could tell he was definitely a man who knew his Dolce from his Italian ice cream. He was also attractive in an offbeat way and Mac could see why beautiful Allie Ray would have been drawn to him. Power combined with money made a formidable combination.

Perrin said, "I know you. I've seen you on TV. Come in."

Mac stepped inside and took a quick look around. The entry soared thirty feet to a beveled glass dome. The house itself was open plan and sleek. An all-steel kitchen to the rear; a jutting staircase of free-floating steps with no visible means of support to the left; and in front a wall of glass through which Mac could see, though not hear, the crashing ocean waves. All the windows were closed and the air-conditioning was blasting, as was a recording of Roxy Music's *Avalon.*

Perrin's expensive Malibu walls held a collection of even more expensive art, whose value was apparent even to a nonconnoisseur like Mac. And the furnishings were unbeachy, with serious antiques, soft leather chairs and fine silk rugs on the dark, lacquered-concrete floors.

An odd feature was the model railroad that ran around all four walls, vaulting over the glass doors, undulating through the open-

31

slab stairs, sneaking along the baseboards and climbing the limestone in layers of splendid, and pricey, miniature rolling stock. It was a child's, or in this case a grown man's, dream. Mac was immediately intrigued. But Mr. Perrin had matters other than model railroads on his mind.

"Take a seat, Mr. Reilly," he said.

Mac perched on the edge of a slippery green leather chair. He glanced at the place where he'd been standing when the redhead took a shot at him. There was a large chip in the polished concrete floor. The remains of the crystal vase had been removed but he guessed the ricocheting bullet was still buried in the back of the bronze velvet sofa where Perrin now slumped opposite him. He looked distinctly pale as well as haggard and not at all like the rich, successful party animal everyone was used to seeing in the glossies.

Mac noticed a shredder, the cheap kind you see in drugstores, standing next to its up-tipped cardboard container. Perrin had obviously bought it recently and Mac had interrupted him in the process of shredding documents, a pile of which still waited on the floor.

"I wonder if you know why I'm here," Mac said.

Perrin slumped forward, hands clenched

between his spread knees. He nodded, still staring somberly at Mac. Then he said, "Mr. Reilly, you are looking at a frightened man."

Now of all things this was not what Mac had expected Perrin to say. He'd thought he would bluster it out about the girl, tell him he had been mistaken, that he had been seeing things. He'd thought Perrin would offer him a drink, slap him on the back, invite him to a party and advise him to forget about it. For once he was lost for words.

Perrin was staring at him with that intense molten brown gaze. It occurred to Mac that perhaps Perrin didn't want it broadcast around, especially to his wife, that there had been a beautiful half-naked woman in his house last night, let alone one toting a gun.

"Mr. Reilly," Perrin said finally, "someone is trying to kill me. He's been following me for the last few weeks. He's on my tail wherever I go."

Again he surprised Mac. "How d'you know he wants to kill you?"

Beads of sweat trickled slowly down Perrin's neck and Mac wondered whether it was from a workout he might have interrupted or from genuine fear.

"I just know it," Perrin said.

"So why haven't you reported it to the police?" Mac knew this would have been the move of an innocent man. Or at least a man with nothing to hide.

Perrin lifted his shoulders in a bewildered shrug, spreading his arms wide. "You've no doubt heard my wife is divorcing me? What if it's her? Maybe she wants me killed? How can I tell that to the cops? Her attorneys would have me locked up in a second. They'd say I was trying to screw her out of the money they claim is rightfully hers."

"I hear it's half your fortune." Mac kept his tone light but he was still wondering what Perrin was hiding.

"Mr. Reilly, do you know my wife?"

"I've seen her movies."

"Hah. Of course you have. The famous Allie Ray. One of the world's most beautiful women. But behind that elegant blond façade she has turned into the most grasping avaricious soul on this earth. And maybe on a couple of other planets too."

Mac stared at him, surprised. This was not the image Allie Ray, America's girl next door, projected.

Perrin fell silent, obviously still smoldering. Then he added bitterly, "She married me for my money and I was dumb enough to fall for it. She'd married two other rich guys before she got to me. Sure, I wanted the trophy wife, the one everybody else wanted."

He glared at Mac. "I scratched my way up from the gutter, Reilly, y'know that?"

He got up and began to pace, twisting his hands together, as though he were hurting inside. "I thought she loved me," he said, sounding almost piteous, if a man that powerful was capable of such a thing.

Mac sat silently, waiting for him to spill the truth, which he knew was what usually happened when you acted like a fly on the wall and just let them get on with it. Perrin was a sad man, no doubt about it, but he still hadn't mentioned the Naughty Angel with the gun.

Perrin was pacing again, actually wringing his hands now, agonized, Mac guessed, at the thought either of parting with his wife or his money.

Perrin glanced up. "Y'know how much I've offered her, Reilly? Eighty million bucks. *Eighty mil,* buddy. Plus the Bel Air house that I shelled out twenty-five mil for and on which she lavished another fuckin' fortune. But is that enough for Mary Allison Raycheck? The girl from Texas with a no-good alcoholic father who beat her with his belt every Saturday night when he got back from the bar? And the hard-drinking depressed mother who neglected her?"

Perrin shook his head vehemently. "For the

35

media's sake I helped invent and perpetuate the story that she was raised a lady, all sugar and spice. And of course now she is that lady. She heads up a couple of worthy charities, though I'd bet she doesn't part with a cent unless it's tax deductible."

He slumped onto the sofa again and put his head in his hands. "She wants it all, Reilly. And to get it, I believe she wants me dead. And that's why you have to find out if she's after me."

Mac thought about the blond movie star he had seen in magazine photographs, though there were none of her here in the beach house. Always smiling, often pictured holding the hands of sick children in hospital beds, or hosting parties at her Bel Air mansion in aid of the latest political candidate, and patronizing the smartest restaurants in town while arranging for them to donate their leftover food to homeless shelters. And always with a photographer handy. That was one way of looking at Allie Ray.

The other way was of a girl from a poor and abusive background who knew as a child what it was to have no money for doctors, never to have enough food and no pretty clothes. And no love. Maybe Allie was only giving back some of what she had been fortunate enough to acquire. Maybe Allie really cared and Perrin was maligning her so he could hang on to his fortune.

One of them had to be a liar. Mac thought he would certainly like to hear Allie's side of the story.

Meanwhile, RP had not called on Mac. Mac had called on him. So why was the guy suddenly spilling his guts to a perfect stranger? Plus there was still the matter of the Naughty Angel to be explained.

"I would like to hire you to find out who is following me," Perrin said, fixing Mac with that sad-puppy gaze again. He went to the desk, took a business card and handed it to Mac. "I don't want to end up dead. And I don't want my wife to be known as a killer."

He sat there, waiting for Mac to speak and maybe pass judgment.

He twisted those hands anxiously again. "I'll pay double whatever your usual fee is. Triple, even." His eyes clouded as he thought about what he'd just said. "Make that double," he amended hastily. RP was a businessman first and foremost.

Mac got up. He walked over to the up-turned boxes of files. "You trying to hide financial records from your wife?" he asked. "Is that what this is?"

Perrin hurried after him. "Yup, yes, that's all it is," he said quickly.

37

Looking down at him, Mac realized that Perrin was considerably shorter than he was. "Mr. Perrin," he said finally. "Who was the tall red-haired woman who was here last night? The one with a Sigma .40 handgun."

Perrin's face was suddenly suffused with color, a vein throbbed in his neck, then he got himself under control and said, "I was out until one o'clock last night, Reilly. There was no one in my house. There was no redhead with a gun. I already told that to the guard at the gate."

Perrin turned his eyes away. He knew that Mac knew he was lying. He walked back to the sofa and slumped down again.

"Would you care for a drink?" he asked with a sigh.

Mac shook his head. "I don't drink in the mornings." He thought quickly about the job offer. With the TV show likely not to be picked up he could surely use the money, but there was something he didn't like about this scene. Perrin was lying to him about the redhead, and probably also about his wife, though Mac would bet, under all his bluster, he was still in love with her.

"Thanks for offering me the job," he said, "but I can't do it. I'm off to Rome in a couple of hours." He walked to the door. "I'll be away for about a week."

38

Perrin hurried anxiously behind him, sneakers squeaking on the expensive modernist lacquered floor.

"You're going to *Rome?*" His voice was as squeaky as his sneakers. "But you *can't.*" He was shouting now. "I've offered you the job. I need you to find out what the hell's going on."

"Thanks a lot for the offer, Mr. Perrin." Mac turned and they looked at each other. Mac handed him his own business card. "Call me when I get back and we'll talk about it some more. Meanwhile my advice to you is to go to the police, tell them you suspect you're being followed. They'll take care of things for you. They'll come up with the truth."

"Don't they always," Perrin said bitterly. Then added, positively, "No police."

Mac felt Perrin's eyes following him as he strode down the deep blue cement path. The steel gate remained shut and he waited, without looking back, for Perrin to open it for him. Soundlessly, it slid back.

He stepped out into the fresh clean air and the sunshine. The gate slid shut behind him, locking in a frightened man.

CHAPTER 6

Allie Ray Perrin was on her way to Malibu. The morning was gray with the kind of fog that left droplets on your hair and was known euphemistically in California as a "marine layer." Driving slowly over Malibu Canyon, Allie knew from experience that it was unlikely to shift before three p.m. even though behind her in the San Fernando Valley the sun was shining just as always. It was one of the penalties — or delights, take your choice — of living at the beach.

Still, she drove with the top down on her Mercedes 600 convertible, snuggling into her cashmere hoodie, uncaring about the mist. The large dark glasses were L.A. de rigueur, sun or no sun, though she doubted anyone would recognize the unmade-up woman they glimpsed quickly in passing as the glamorous movie star they knew from the screen. Except for the paparazzi, of course, who hounded her like bats out of hell. But today she had escaped them, taking the back route out of her Bel Air home, cutting across Mulholland to the Valley, then over the canyon to Malibu.

It wasn't her husband she was going to see though. She had been watching Mac Reilly's program Thursday nights and she thought he seemed a man of integrity. A man you could trust. And if anyone needed that sort of man right now, it was Allie Ray.

Turning left onto Pacific Coast Highway, commonly known as PCH, she idled through the fog, denser now she was at the shore. The guard at the Colony gate recognized her car immediately and the bar swung up. She waved to him as she drove in and he waved back. There were no grand iron gates here, no high stone walls. Low-key was the watchword at the Colony. Everyone liked it that way. Life at the beach, Allie thought wistfully, was nice.

At the T junction leading to the Colony's only street, she made a right rather than the left she normally took to her own place, heading instead for Mac Reilly's house. Checking the number, she found it at the very end, stuck like a worn green barnacle to the Colony's glossy façade. A shiny red Prius was parked outside. She parked behind it, glancing doubtfully at the exterior of the house. It appeared to have been painted recently, yet somehow it looked shabby.

Still, Mac Reilly's housekeeping was not what she was here for, and she tugged on the bellpull, which was in the form of a cracked brass captain's bell, long since oxidized to an almost matching green. She waited. And waited some more. She jangled the bell again, anxious now. He *had* to be here.

Stepping from the shower, Mac heard it ring. He glanced at the clock. Because of Perrin he was running late. As it was he would just about make the flight. He knew it couldn't be his assistant, who was due to arrive any minute to drive him to the airport, because Roddy had his own key.

Cursing, he wrapped a towel around his loins then ran to open the door. And found himself looking into the eyes of one of the most famous and most beautiful women in the world.

She was wearing a gray cashmere hoodie, matching pants and Reeboks that were so pristinely white, Mac figured she must have bought them specially for the occasion. He corrected himself quickly. A woman like Allie Ray probably had a dozen pairs, all new, sitting in her closet, and she probably never wore them twice.

Allie stared back at him, waiting for him to

take it in that it was really *her* standing on his doorstep. Then she gave him the smile that had charmed moviegoers for over a decade.

"I'm so sorry," she said, "I must have caught you in the shower."

Brought back to reality, Mac hitched up the towel. He apologized for his appearance and invited her into the tiny square that constituted his front hall.

Allie stared at the dog as it bounded lopsidedly toward her. Pirate gave her the usual investigative sniff, then sat back on his haunch, allowing her a full view of his one eye and goofy smile.

Catching her shocked look, Mac said quickly, "His name is Pirate. After Long John Silver. In *Treasure Island.* Y'know: the wooden leg, the eye patch." She frowned and he added, "Hey listen, it was better than the alternative. He was almost dead when I found him."

Allie bent to pat Pirate's head. Sliding the cashmere hood from her pale blond hair, which was pinned in a loose ponytail, she walked into the living room.

With her hair like that, Mac thought she looked like the college version of Barbie. Except for those eyes of course, which when she fixed them on him, had a haunting quality, like diving into a turquoise tropical sea

43

where troubling undercurrents tugged at you.

Excusing himself, he hurried to put on shorts and a T-shirt.

When he came back he found her looking round his comfortable, if shabby, domain. At the squashy old sofas covered in a variety of plaid rugs, most of which Pirate called home, when he wasn't sleeping on Mac's bed that is. At the beat-up black leather La-Z-Boy with a cup holder where Mac stashed his beer and pretzels, with the flap on the side for the remote, and that little leg-lifting device that, if the Lakers game wasn't so hot, relaxed a guy so much it could put him right to sleep.

Allie's gaze moved to the old surfboard that in a fit of artistic triumph Mac had painted gold and converted into a coffee table. Then on to the mélange of wicker chairs surrounding the squat, solid-looking oak table, bought by Sunny at a flea market and which she swore was a valuable antique. Mac had told her he was hanging on to it so when that rainy day came he could make his fortune.

Allie had moved on to his eclectic art collection, if so proper a term could be used for the colorful canvases on Mac's walls, most of which he'd picked up on visits to new young

44

artists in their Venice Beach studios, at prices that had left him worried, and wondering if he'd left them starving in their garrets.

Allie took in the sea grass rugs on the wooden floors, the shutters flapping at the window, the faux-zebra rug in front of the fifties white-brick fireplace, the unmatched lamps, and the collection of candles and votives, courtesy of Sunny.

She gave Mac that haunted turquoise look again. "I envy you," she said, surprising him. "You have exactly what you want. You're a lucky man."

"There's no need for envy, particularly coming from a woman like you."

She perched on the edge of the dog-hairy sofa, looking up at him. "Tell me, Mr. Reilly, what exactly do you know about women like me?"

She had put him on the spot. Did he tell her the truth about what he'd heard she was? Or did he go for the comfort factor? Tactfully, he took the middle path.

"I know you came from a poor background, that you married well. Several times. I know that you're a famous actress."

Allie ran a hand through her pale blond hair, lifting her ponytail and shaking it free from the folds of the cashmere hood. "Do you know what despair is, Mr. Reilly?" she asked quietly. "Do you know what it is to arrive at the realization that there is only one way you are going to get out of the stifling small town you lived in, away from the alcoholic father, and from the worn-out, depressed mother you dread becoming?"

A shudder shook her slender frame and she frowned. "Away from the brutality and the grinding poverty, and the stifling grayness of a life with no prospect of a silver lining. Away from the small-town high school football heroes who lied about their conquests with the shy pretty blond teenager simply to add to their own macho luster. And away from the preacher who from the pulpit kept on spelling out a life of damnation for the fallen, and then afterward would try to grope you?"

She stopped and gave Mac that haunted look again. "Do you know what it feels like, Mr. Reilly, to wake up to the fact that in order to get out of there, there is only one thing you can sell. And that is your beauty. Because that's all you have."

Her sigh dredged up like an ill wind from her past. "At least," she said, looking squarely at him, "if I had to do it, I decided it would be to the highest bidder. There was just one rule. He had to marry me."

"And you stuck to your plan?"

"I married rich men. And I kept my part of the bargain. I was a good wife for a while. But eventually they got bored with looking at me, I guess. Anyway, I had always wanted to be an actress — a movie actress. And now I have all the money in the world — and possibly even more when Ron Perrin comes through with the divorce settlement. Not that I need it. I'm successful in my own right, more successful than some men. And y'know what, Mr. Reilly? I'm still not a happy woman."

Her eyes met his. "You're judging me, because I told you the truth." She lifted a shoulder in a delicate shrug. "Obviously you don't know what it is to have no choice."

Mac said nothing because it was true. Yet he understood.

She got to her feet and went and stood close to him. "Look into my face," she said. "What do you see written here?"

Actually, Mac could see nothing. Although she must have been coming up to forty, there were no time wrinkles, no laughter lines, no marks of sorrow. Just a very beautiful face that photographed really well.

"Discontent, Mr. Reilly," she said quietly. "That's what you see. I'm the archetypical woman who has everything. Oh, believe me, there are dozens of us in this town, maybe even hundreds. And we all have the same expression. As though life passed us by. Real life, that is."

She walked away, staring through the picture window at the ocean throwing itself lustily onto the shore in a flurry of spray. "But one day," she said softly, "one day I'm going to find that 'real' life, y'know that?" She swung round. "I'll be me again. Mary Allison Raycheck."

Mac said, "Back to that."

"I see you know everything about me, even my real name. I guess I should have expected that from a private investigator."

"Actually, your husband told me."

She gave a short bark of a laugh. "Of course. But then, he would, wouldn't he?"

She slumped onto the La-Z-Boy and flipped the lever, stretching out as it lifted into position under her long, slender limbs.

"God, I always loved these things," she said, half to herself. "Once, I thought the epitome of being rich was to have a leather La-Z-Boy and a coffee table from Sears with a glass top and gold legs. My, my, how times have changed."

A tear trickled down her lovely face and dropped with a splash off the cliff of her cheekbone onto the gray cashmere. "Now I have all this furniture an expensive decorator

48

chose for me because it's in perfect taste. I wear designer clothes because that's what I'm supposed to wear. I eat the right diet at the right restaurants, attend the right parties." She glanced despairingly Mac's way. "You see, Mr. Reilly, what my trouble is, don't you? I just don't know who I am anymore. It's still the way it's always been. What you see is *all* you get."

And then Allie Ray, movie star supreme, curled up in Mac's tattered black leather La-Z-Boy and, to his horror, burst into floods of tears. She pounded the chair with her fists, howling and sobbing.

Pirate hauled himself to his feet and ran to her. That dog hated a scene, he'd probably had had too much of it in his previous life, before Mac became his father. He sniffed Allie anxiously, whining and pawing at her. And to Mac's astonishment the movie star leaned over and scooped him onto her knee. "Sweet doggie," she whispered as Pirate began to lick away her tears.

"So now you see why I envy you, Mr. Reilly," she said between hiccups. "I don't even have a Pirate. I just have the *right* dog, the one we're all supposed to have this year."

"Not a Chihuahua!"

She shook her head, scattering tears all over Pirate, who shook his head too, to get

49

rid of them. "A miniature Maltese. By the name of Fussy. And believe me, she is."

Mac offered her a box of Kleenex. He thought regretfully that he would miss his flight to Rome, but knew he had no choice. Allie Ray needed his help.

"Tell you what," he said, "why don't we go for that walk along the beach? Now I know you better you can tell me why you need me. And why you are so desperate."

CHAPTER 7

Pirate loped along the shoreline while Mac and Allie followed at a more leisurely pace. After all, they were not there for the exercise.

Allie took off her sneakers and brushed the drops of mist from her hair. Digging her toes into the wet sand she said, "I'm sorry. I didn't come here to cry on your shoulder. I came to ask for your professional help. I'm a rich woman, Mr. Reilly. I'm willing to pay lavishly for your exclusive time."

Mac raised his brows, surprised by the offer. It was surely coming at a handy moment, when his TV income might be suddenly cut off.

"Anything I can do," he said.

She stopped, then turned to face him. "For the past week, I've been followed. It's a black Sebring convertible with dark windows so I can't see the driver. There's no license plate at the front, and since he's always in the back of me I never get to see it. But lately it's always there, on my tail. I don't know whether it's my husband having me watched to see if he can get any dirt on me. Or if it's the same crazy stalker who's been after me for the last few months. He sends me letters — love letters he calls them, though it's all just filth. Of course I don't look at them anymore, I just burn them without opening them. Anyhow, it's scary."

Mac didn't like the sound of those letters, nor the black Sebring. He thought it strange that both Perrin and his wife believed they were being followed. He wondered if they were tailing each other, but when he asked her Allie denied it.

"Then why not go to the police?"

"Because I'm Allie Ray," she said simply. "You can only imagine what would happen. I'm terrified though. I feel eyes on me, as though I'm being watched wherever I go. I don't know what to do."

Mac made a quick decision. It wasn't only the offer of good money that attracted him. Allie Ray was vulnerable, and she was hurting, and it was more than just a scary stalker and a husband who no longer appreciated her. He got the feeling Allie was a desperately lonely woman who needed not only his help but also his support.

"So why don't I find out who it is, and if

it's your husband or not."

She threw him a grateful smile then turning away she walked along the edge of the shore where the waves hit the sand, uncaring that she was getting wet. Picking up a stick, she threw it for Pirate who galloped joyfully after it. Wagging his butt, he dropped it at her feet making her laugh and she picked it up and threw it again.

"That's the first time I've heard you laugh," Mac said.

She shot him a mischievous glance. "Except in the movies, you mean. Then I laugh all the time."

"Unless you're kissing someone."

She laughed again. "You're right, I am always kissing someone. It used to drive Ron mad. 'You're my woman,' he'd say, whenever we had a fight, 'and all I see is you in bed, half-naked with some poncey actor.'"

Mac could just imagine Ron Perrin making that remark. Ego had no boundaries even when it came to the fact that acting was his wife's job.

"I didn't tell you everything," she said. "It's not only that I'm afraid of the crazy stalker. You see, Mr. Reilly, my husband is having an affair. And this time I think it's serious."

"It's not the first time, I assume?"

She shook her head, sending the ponytail and the drops of water flying. "It's not. But the truth is I still care about Ron. He's the only man who ever bothered to look behind the façade. The only man who wanted to know the real me. Without him, I don't know who I am." She sighed as she tossed the stick again, then turned to meet Mac's eyes. "Look at me," she said sadly. "What you see is all you get. I'm a public success. And a private failure."

Remembering the redhead with the gun, Mac thought Allie had a right to be worried. "Let me see what I can do," he said. "I'm off to Rome for a week, but I'll get my assistant right onto it.

"Anything else on your mind?" he asked as they made their way back to his house.

"Plenty," she said, smiling ruefully. "Unfortunately, it's nothing you can do anything about. My new movie is to premiere in Cannes in a couple of weeks and I know it's a mess." She shrugged sadly. "Of course, they'll say it's all my fault, that I'm difficult, that I made changes, that I'm getting older."

She stopped at the top of the steps, watching as Pirate hobbled back up. "So now you understand why I envy you, Mr. Reilly."

"I think it's time you called me Mac."

She smiled. "The simple life has so much

more to offer, don't you think, Mac?"

A car honked outside. Mac knew it was Roddy, waiting to take him to the airport, though even if they got lucky with the traffic he guessed he'd miss his flight and would have to reschedule.

At the door, Allie turned to hug him good-bye. "Please help me," she said.

And of course Mac promised he would, even if he had to do it long-distance from Rome.

CHAPTER 8

Allie drove slowly back along PCH lost in her thoughts. The congested highway offered glimpses of the ocean, glinting between low buildings whose doors opened directly onto the road, and which surprisingly were mostly extremely expensive homes.

Revisiting the past was not something Allie did frequently, in fact never, if she could help it, and then only in her dreams when she had no control over the memories that drifted into her mind.

She felt exhausted from the explosion of tears, again something she never did. Nobody ever saw anything but the public Allie, the one smiling for the cameras. She had let her guard down to Mac Reilly and now she was wishing she had not.

The past was the past and that was where she wanted to keep it, locked away in a safe place where nobody could find it. Except Ron, of course, because Ron knew everything there was to know about her. It was as though he'd known right from the beginning, without ever having to ask.

She shook her head, pushing the thought of him away, as she had done physically the night she told him to leave. "Get out of my life," she'd yelled. "Or I'll get out of yours."

"Oh? And exactly how will the famous movie star, America's 'good girl,' manage to do that?" he'd asked.

"Same as all the other women," she'd snapped back, picking up a vase of roses, ready to hurl it at him if he so much as came near her.

He had laughed. "You don't have to go that far," he said. "I'm outta here."

As he'd marched to the door she'd shouted after him, "Back to her, I suppose."

He had turned to look at her. For a long moment their eyes had connected. Then he'd lifted a shoulder in that familiar dismissive gesture of his. "Have it your own way," he'd said. And he'd walked out, closing the door quietly behind him.

After he'd gone Allie had slumped into a chair in the massive front hall with the double curving staircase and the antique chandelier that had sparkled like Christmas stars on their ugly fight. The chair was a designer *fauteuil* with plaid silk upholstery in soft

57

shades of celadon green and walnut arm-rests. Expensive of course, as everything in this house was. Still clutching the vase of pink roses, Allie had stared blankly at the door where her life, her future, her very reason for being, had just walked out. Ron Perrin no longer loved her and she did not know what to do about it.

Now, heading home along the coast road, she still didn't know what to do, and what she had told Mac Reilly was the truth. Her very private truth. She liked Reilly. He had something that was rare in her circle. Integrity. She could tell that, even from their brief meeting.

Remembering suddenly why she had gone to see him, she glanced in the rearview mirror. Her heart jumped into her throat. It was there! The black Chrysler Sebring, the kind of inexpensive convertible tourists liked to rent so they could put the top down and enjoy California's rays, soaking up the sun and a lungful of gasoline fumes along with the sea breeze. Except tourist cars didn't have windows tinted so dark you couldn't see the driver, and *this* Sebring driver never put the top down, never stopped to admire the view, and never overtook no matter how slowly she drove. He just sat on her tail a couple of cars back and waited to see what she would do next.

A shiver of fear trickled down Allie's spine. She had experienced stalkers before. Usually they were more insistent, wanting to get next to her, to try to make conversation as she waited at Malibu's Country Mart Starbucks for her double latte, two shots, skinny. Of course she didn't go there anymore because the paparazzi lurked around the place, cameras at the ready, waiting to snag celebrities in their lenses, hopefully doing something naughty. But this stalker was sending letters filled with threats of violence.

She drove faster, to the fringes of Santa Monica. When she got to Topanga Canyon, at the last second she made a quick turn left across the traffic into the parking lot of the Reel Inn, a beach café that served fish in many varieties and almost any form.

She turned to look back. As she had hoped the Sebring driver had been taken by surprise and forced by the traffic to drive on. There were no U-turns on PCH, and he could not come back until he reached Sunset and made a legal turn. She had given him the slip.

She pressed the button to put up the top on her car, then made a left out of the parking lot and a right at the light. She was taking the same road as the Sebring knowing she would pass it coming back the other way in search of her. Then she would take Sunset Boulevard, the long road that led all the way from the beach to Bel Air and Beverly Hills, and to Hollywood and beyond. Meanwhile, she punched in Mac Reilly's number.

"Hi, it's me, Allie," she said when he answered, liking the sound of his strong voice. This man knew who he was. She only wished she could learn from him.

Mac was in his car on his way to the airport. Pirate rode shotgun, head sticking out the window, ragged ears flapping, while Mac's assistant, Roddy Kruger, was in the backseat, negotiating a new flight to Rome and complaining about lack of legroom. Roddy would drive the car and the dog back to Malibu after dropping Mac off.

"So what's up, Allie Ray?" Mac asked, noticing that Roddy fell suddenly silent at the mention of the famous name.

"He's on my tail again. I just lost him at the Reel Inn. I'm on my way home now, via Sunset."

"Okay, no need to panic." Mac's voice was calm. "Your new 'tail' — your personal one hired by me, will be with you by the time you reach home. He will be driving a souped-up black Mustang and he'll have a camera around his neck, looking like the rest of the

60

paparazzi. He's fortyish, bald as a coot, the usual aviator sunglasses, a Tommy Bahama flowered shirt, jeans, sneakers. Six one, in good shape — good enough to take on all comers. You can trust me on that. He's a triple black belt in karate and trained with the Israeli Special Forces. He's also organized round-the-clock surveillance. No need to be afraid, Allie, I promise you. He'll soon find out who the follower is, whether it's the stalker, or some PI hired by your husband to keep tabs on you and dig for any dirt. Not that I expect there is any," he added casually.

"No," Allie said shortly. "There is not."

"Glad to hear it." Mac was smiling. "It makes life easier all around, especially from a divorce lawyer's point of view."

Out of the corner of her eye Allie caught a glimpse of the Sebring with the darkened windows speeding in the opposite direction. Heaving a sigh of relief, she said, "What's his name?"

"Your bodyguard? His name is Lev Orenstein. You can't miss him, and trust me, he won't miss you. You're in good hands," he added gently.

"Okay," she said in a small voice. "See you when you get back from Rome then."

"One week," he said, "We'll get together right away."

"I leave for Cannes a few days after you return," she said. "Please don't forget to call." She was almost begging him, wishing he wasn't leaving, wanting him to stay near her, wanting his strength. It was not often you got a man who understood; a man who listened; a man who saw beyond the façade. A man like Ron had once been.

"Okay, don't worry, I'll call and we'll set a date. You're in my thoughts, Allie."

"And you are in mine," she whispered as he rang off.

Turning in to her street, Allie saw the black Mustang parked discreetly under a tree opposite the gates. She knew Reilly must have informed Security that Lev Orenstein was here for her protection, otherwise the patrol would have moved him on. A couple of other vehicles nosed slowly by, but there was no black Sebring convertible and she breathed a sigh of relief. Slowing opposite the Mustang, she rolled down her window and a tall guy, whippet thin with shoulders like a halfback and, she guessed, probably a full six-pack of abs, got out and slouched over to her.

"Ma'am," he said in a rich dark voice, "I'm Lev, here for your protection. Mr. Reilly probably told you about me."

He leaned an arm lazily on the roof of her Mercedes. "Yes, he did," she said, managing a smile. "I'm very glad to see you, Lev."

"Don't worry about a thing, ma'am. I'm here for you." He stepped back, lifting his hand in farewell. "You need to let me know when you're going out and where you're going. You ever need me, up at the house, wherever, you call this number." He passed a card through the window. "Put it on your cell," he said. "Tuck it in your bra, drill it into your brain. It's your lifesaver, ma'am. Your link to me."

"I will," she promised, shakily. Then she pressed the button that opened the electronic gates and sped down the straight-as-an-arrow driveway that led to the twenty-thousand-square-foot soulless mausoleum she called home.

CHAPTER 9

Of course Sunny was not waiting for Mac at Rome's Fiumicino airport, as he had hoped, but he took it in his stride. That's women for you, he thought, smiling. Perverse as all get-out. Plus of course, due to his unexpected chat with Allie Ray, he'd missed his original flight and had been forced to make a couple of connections via D.C. and New York to get to Rome. He was late, tired and jet-lagged, but happy.

He took a cab into town which after a lengthy drive deposited him at the venerable Hotel d'Inghilterra, with its charming tearoom, and the wood-paneled bar where Ernest Hemingway used to hang out, and where the driver now relieved him of which seemed to Mac to be a great many euros for his trouble. The bellman took his bag and the desk clerk informed him that the Signorina Coto de Alvarez was taking coffee at Tre Scalini in the Piazza Navona, a short walk away.

He pointed him in the right direction and Mac sauntered happily along the narrow crowded streets. The sun bounced off the ruins in a golden glow, the air smelled of fresh coffee and the chic Roman women, always intent on presenting a *bella figura,* looked like Armani models. Not a bad day in Italy, he thought, pleased. Still, when he stepped from the side street into the vast expanse of the Piazza Navona, it took his breath away.

He stopped to look at the ancient buildings in faded shades of ocher and rose surrounding the arena that, centuries ago, had started life as a stadium, and that now lay buried deep beneath the cobblestones. The many cafés with their striped awnings were crowded, and Bernini's glorious Fountain of the Four Rivers pulsed sparkling jets of water into the air. Tourists milled around taking photographs, while Borromini's Sant'Agnese church, domed and turreted, pedimented and columned, ruled over all.

Mac sauntered past several cafés until he came to Tre Scalini, near the big fountain with a grand view of the church and an even better view of his Sunny, wearing a pale green dress with a neckline that he immediately decided was too low-cut for Rome.

Now Mac knew he might not be your typical detective from the crime-novel genre, but Sunny was another matter. She was your Raymond Chandler woman all right. Long silky black curls brushed smooth from a heart-shaped face, smoky eyes, amber-brown under winging brows; straight-on perfect nose; and the pouty red mouth all Chandler's heroines and villainesses had. Add to that a delicious cleavage and golden legs that went on forever, and Sunny was some kind of woman.

She was sitting in the shade of the café awning, leafing through a copy of Italian *Vogue* and sipping a cappuccino.

"What, no sticky buns?" he asked, depositing a kiss on her sleek dark head.

"You're late," she complained.

Mac sighed as he sank into the chair opposite. "Nice greeting for a guy who's just spent sixteen hours on several planes in an effort to be with the love of his life."

But then she gave him the smile that lit up her face with a thousand candle-watt-power. "I'm glad you came," she said simply and she leaned forward and kissed him. Everything was right in their world again. Temporarily, of course, because that's the way their relationship went.

"I'm sorry," Sunny said, "but I have to go

to the studios. It's a longish cab drive out of the city, it's better if you get some rest and I'll catch up with you later."

"Don't worry," Mac said, pushing jet lag away as a memory. "I'm getting my second wind. I'll come with you."

In the car taking them to the studios he slid his arm around Sunny's shoulders. He dropped a kiss on her hair, reaching with his other hand to smooth it from her neck, snuffling her familiar scent.

"I missed you," he murmured. "Do we really have to go to the studios? Can't we just turn around and go back to the hotel, where I can get you alone?"

Sunny was staring nervously ahead at the tangle of traffic that their driver was negotiating with loud honks of the horn and swift sideways maneuvers that seemed typical of Roman driving.

"You seem to forget I'm here on business," she reminded him.

Mac studied her delicious profile. "You're right," he admitted. "Somehow I thought I was here with my lover, enjoying the beauty of Rome, tasting the good red wine and wonderful food, exploring the ancient ruins . . ."

She turned her head a fraction to look at him. A smile lifted the corners of her pretty mouth. "Is that *really* why you're here?"

He lifted his shoulders in a shrug. "Why else, baby?" he said, as she slid into his arms and began to kiss him. Properly this time. None of that cold shoulder stuff. Just real kissing, like two people in love.

Fifteen minutes later as they drove through the gates at Cinecittà, Sunny quickly reapplied her lipstick and combed her hair. "Do I look okay?" she asked.

Mac's eyes were warm with love. She looked flushed and sparkly, a woman ready to make love and not at all businesslike.

"You look beautiful," he said.

Cinecittà Studios were famous for the years it took to film *Cleopatra*, the Taylor-Burton epic in the early sixties, and for the even more famous love affair between the two stars. Now they were more often used for smaller films, though many of the old sets remained standing.

Sunny's client, a young actor by the name of Eddie Grimes, was making a sci-fi epic produced by the eminent Renato Manzini that, it seemed to Mac, could easily have been made anywhere on the planet. Still, he guessed Rome was as good a place to make a sci-fi as Hollywood or Mars, and the sets were certainly stupendous.

Still, jet lag was taking its toll. He sank a

couple of espressos for sustenance, lounging in a chair in the shade while Sunny chatted to Eddie, making a few notes in the yellow legal pad she always carried.

He fell asleep in the cab on the way back to the hotel, leaning dopily against her as the tiny elevator whisked them slowly upward. In their room, he didn't even bother to complain about Sunny's clothes strewn about. He took a quick shower, flopped onto the bed and dropped off the edge of the world into a black abyss of sleep.

So much for romance, Sunny thought tenderly, as she watched him.

She caught him on the rebound though, a couple of hours later. She stretched her long naked body next to his, his hand reached out for her and he turned his face to hers, breathing in the scent of her, his mouth searching for hers.

"This," he whispered, his arms gripping her close, "is what I came to Rome for. I can't do without you, Sunny."

CHAPTER 10

Mac's assistant, Roddy Kruger, age thirty-five, short bleached-blond hair, good-looking, gay and very popular, was staying at the Malibu house babysitting Pirate. He was sitting on the deck on an old metal lounger from Wal-Mart, which was about in keeping with the rest of the furnishings in Mac's home, a Diet Coke in one hand, the *L.A. Times* sports section in the other.

Every now and then he would glance up from the newspaper to check the Perrin house down the beach. Mac had filled him in on the events and put him in charge of the Allie Ray case, though "case" was hardly the correct term for finding out who was following her, and the anonymous letter writer was more of a problem. Allie had sent Mac a couple of the letters and they were not at all happy about them. Still, Roddy was a long-time fan and the thought that he was working for the star gave him a distinct buzz.

There was no activity at the Perrin house though, and he went back to his newspaper.

Half an hour later he glanced at the bright blue rubber-encased diver's watch, waterproof to a depth of three hundred feet, that Mac had given him the previous Christmas. It looked like a piece of junk but he knew it had cost a small fortune, and since he was an avid surfer, he loved it. It was time to polish the Prius. Those pesky seagulls were constantly flying overhead and their droppings could take the paint off a car in no time flat. Twice a day, had been Mac's instructions, and Roddy was conscientious about it because he knew how much Mac loved his customized car. Even more than the black Dodge Ram that had gotten the same treatment, and that had been his previous passion, but now, like most of Hollywood, Mac and Sunny were passionate about ecology.

Malibu Colony might be a beachy suburb grown rich but it still retained its old-world charm. Every house was in a different style, from traditional picket fences to concrete modern. Telephone and utility wires still looped shabbily along the only street, unchanged since the forties, and several cars were usually in the process of being washed, only now it was the detail guys giving the Mercs and the Porsches the tip-top treatment. Kids Rollerbladed and uniformed maids walked the dogs, stopping for a chat with the Mexican gardeners who kept the tiny expensive patches of lawn and the floribundas in immaculate shape. Joggers, looking as sweaty as any regular joggers, even though they were movie stars or just plain rich, trotted past, and vans delivering flowers and groceries lurched slowly over the speed bumps. It was like any other upmarket suburb in America.

Roddy carried a bucket of water and a chamois leather out into the street, sloshing off the latest seagull deposits, cursing the birds under his breath. Pirate sat next to him hoping for a ride but today he was out of luck. Roddy dried the car off with paper towels, gave it a quick polish, emptied his bucket down the drain, then checked the car's door. As he'd thought, it was open. He sighed. Mac never locked his car or his house. "Which of my neighbors is gonna steal my Prius?" he'd asked with a grin, and Roddy guessed he was right. Still, he checked the interior to make sure everything was okay.

Smoothing his palm approvingly across the custom black leather, he opened the glove compartment, then took a quick breath.

He was looking at a Sigma .40 handgun.

Now he knew Mac never carried a weapon unless he was heading into dangerous territory, and he would certainly never leave one in the car. Anyhow, as far as Roddy knew, Mac's only gun was a Glock semi. He had never seen him with a Sigma .40. Ever.

Roddy put his polishing cloth over the gun, slid it from the glove compartment, put it in the empty bucket and carried it back into the house.

In Rome, Sunny was lying on her back, gazing at the ceiling, a happy post-lovemaking smile on her face, her hand linked with Mac's, when the phone rang. Groaning, she reached for it.

"Pronto," she said, Italian-style. Then, "Oh, hi, Roddy, how are you? Good. Yes, great. It's wonderful. Yes, Mac's here, I'll put him on."

Handing Mac the phone she propped herself on one elbow, watching him.

"Hi, Rod," Mac said lazily.

Sunny saw him frown. She wondered what was going on.

"Okay," he said. "I know where the gun came from. Miss Naughty Angel. So wrap it in the chamois leather and leave it in the bucket under the sink. It's as safe a place as any I guess, until I can give it back to her."

"Crafty woman," he said to Sunny when

73

he'd said goodbye. "Dumping the weapon in my car. Now I wonder why she did that."

Sunny got up. She put on a hotel white waffle-weave robe, took a bottle of water from the minibar and climbed back onto the bed. Unmade-up and with her long dark hair all tumbled Mac thought she'd never looked more lovely.

"Why do I get the feeling I don't know *everything*?" she asked, giving him the keen amber long-lashed look he knew meant business.

"I was going to tell you all about her, but somehow I got diverted."

She grinned forgivingly at him, upended the bottle and took a slug of the water. "Better tell me now. And make sure you tell *all.*"

Mac got up off the bed. "Can't I even take a shower first?"

She shook her head. "After."

"Okay," he said, "so here's what happened." And he told her about the Naughty Angel, about his visit with Perrin, and about the famous Allie Ray Perrin showing up on his doorstep.

"The thing is that both Perrins believe they are being tailed. Allie denies she's having him followed and he denies likewise. Either somebody is lying, or something else is going on. And it just might have to do with the redhead with the gun."

"Miss Naughty Angel," Sunny said. "I'll bet she's gorgeous."

"But not as gorgeous as the famous Allie."

"I leave you alone for a couple of days," she sighed, "and look what you get up to. All these beautiful women running after you."

"Not quite," Mac said with a phony-modest grin, and she snatched up a pillow and whacked him over the head with it.

"No, no," he moaned, pushing out from under the feathers. "No more. I need to take a shower."

She grabbed his hand. "I know a few party games in showers. Oh, and by the way, I hope you haven't forgotten you're taking me out to dinner tonight."

"Right." Mac had been thinking more about room service and sleep but a promise was a promise.

"We're going to Alvaro's," Sunny said, smiling. "Nothing but the best for your girl. Remember?"

CHAPTER 11

Tonight Sunny was all spiffied up in an expensive little slip of black chiffon from one of Rome's famous boutiques, that clung where it should and fluttered around her knees in a very feminine way. She wore black pointy-toe stilettos and Guerlain's Mitsouko perfume. She slicked on Dior's Rouge lipstick, a satisfying brilliant red, then smacked her lips together to smooth it out.

The dresses she had tried on and rejected littered the bed and the bathroom was awash in bubble bath and shampoo. After all, it took a lot of effort for a girl to look her best. She wasn't sure whether Mac had gotten it yet, but she was truly a very girly girl.

Anyhow, there she was now in her new designer black chiffon that had cost an arm and a leg and that, looking in the mirror, she thought was worth every cent. She wore little diamond hoops in her ears and a left hand conspicuously lacking in any sort of ring, be it diamond or gold, large or small. That night she planned to use her left hand pointedly, flaunting its nakedness in front of Mac, who in typical fashion probably wouldn't even notice her perfect manicure, let alone that this was her engagement finger. And that it was empty.

After some persuading Mac had temporarily abandoned his favorite tees in favor of a white linen shirt worn open at the neck and without a tie because he couldn't stand to be buttoned up. His Dolce black leather jacket was a concession to Sunny's beautiful dress and the fact that she'd told him chic Romans congregated at the restaurant she had chosen for its authentic atmosphere, as well as for its fine food. Plus the fact that it was only a couple of blocks' walk from their hotel, so no messing about trying to find one of those elusive and horribly expensive Roman taxis, whose drivers, Sunny had found to her cost, invariably quoted the equivalent of forty bucks even though you were only going the shortest distance.

"Ready?" Mac's eyes smiled at her. He pulled her close, burying his face in her fragrant hair. "Why don't we just get room service?" he whispered, nibbling at her earlobe.

She pushed him away, laughing. "Because

I want to show off my boyfriend. You put the 'cream of Roman manhood' to shame, baby."

"You too," he said, sincerely. "I've never seen you look so beautiful."

To her surprise, Sunny felt herself blush. Mac wasn't given to paying compliments. He was the kind of man who took it for granted that she knew he loved the way she looked. She supposed she did. Still, it was nice to hear him say it.

Linking her arm in his, they descended in the little cage elevator, then walked up the Via Bocca di Leone, named for the pretty lion fountain in the little piazza.

The restaurant had nicotine yellow plaster walls with ancient blackened beams across the ceiling, and white tablecloths with lavish bouquets of scarlet flowers. It was old-world elegant and filled with a chic crowd, there for the food as well as for the "scene." Their table was along the wall near the center and they settled in, pleased with the place and with each other. A tiny amber-shaded lamp lit Sunny's face from below, turning her into a Latina version of a Botticelli Venus. Mac reached out for her hand. The one without the engagement ring.

"I love you, Sonora Sky Coto de Alvarez," he whispered. And lifting her hand to his lips, he turned it palm up and kissed it, then closed her fingers around the kiss.

It was such an intimate gesture that Sunny felt the little answering shiver in the pit of her stomach.

"I love you too, Mac," she whispered, gazing into his eyes.

But then the waiter broke the spell, bustling with importance as he detailed the night's specials.

"Let's share a small Margherita pizza to start," Sunny said, all a-sparkle with love for her man. "Just to go with the first glass of wine."

"No anchovies though," Mac said, remembering she hated them.

Sunny smiled. This was a big concession on Mac's part because she knew how much he loved them.

Mac studied the wine list carefully, finally choosing a Montepulciano. When it came the waiter poured a little into his glass. Mac swirled, breathed its aroma, and sipped.

Sunny liked wine, but Mac was an expert. She saw his face light up and he nodded to the waiter. "Good," he decreed. "Excellent, in fact."

They clinked glasses and toasted each other with their eyes. There was no need for words. This was, Sunny knew, going to be one of the best nights of her life. Here in Rome with her lover, who had just told her he loved her.

"I'm glad you invited me to Rome," Mac said, sipping his wine and nibbling on a piece of the small anchovyless pizza.

"Funny, I thought you'd invited yourself," Sunny said.

But to her surprise she realized Mac was no longer listening. Instead, he was looking at something over her shoulder. She turned and followed his gaze to the door.

"Well, well. Would you just look at that," Mac said, sounding astonished.

Sunny stared at the couple standing at the entrance. You could hardly miss them. Or at least *her*. A redhead, on the arm of a rotund, balding man. There was nothing understated about this woman. Tall, with breasts that defied gravity, her waist was tinier than Scarlett O'Hara's when Mammy had finished tying her corset, and her legs went on forever. She was spectacular in a white silk dress that left no curve unturned. Sunny caught a glimpse of the redhead's ring. A *glimpse?* It almost blinded her. A yellow diamond that must have been all of ten carats. *And* it was on her engagement finger.

"Shoot," she said crossly, turning back to Mac, but he was already on his feet.

80

"Excuse me a moment," he said, then to her astonishment he walked over to the red-head and held out his hand.

"Hi, there," he said to the Naughty Angel. "It's good to see you again. I'm Mac Reilly. Last time we met was in Malibu. Remember?"

The redhead's face turned chalky white. Her hand felt like iced velvet in his.

"Oh, how are you?" she said, in the high, breathy voice Mac remembered calling "sorry" after she'd shot at him.

"Good, thanks. I've gotten over those bruises I acquired tripping over your deck the other night."

"Oh, that wasn't *my* deck." Her voice trailed off uncertainly.

"I believe I have something of yours," Mac said, still with the smile. "You left it in my car."

"Oh, I don't think so," she said too quickly. "I'm not missing anything."

The rotund one gave a discreet cough and she turned her frightened green eyes from Mac to him. "Oh, Renato," she said, "this is Mac Reilly. And this is Renato Manzini. My producer," she added in case Mac might have other thoughts on their relationship.

The two men shook hands. The portly one put a possessive arm around the redhead's waist. "Our table is ready, *carina,*" he said, already edging her away.

She glanced apologetically back at Mac. "Good to see you again," she called as he watched them go.

Sunny was goggle-eyed when he returned. "It was *her,* wasn't it?" she said.

"It was." He took a sip of wine then attacked a plate of antipasto that would have served four.

Sunny stared down at her own little forest of grilled baby artichokes, nonplussed. "How can you just sit there and eat when the woman who tried to kill you is three tables away?"

"I told you she apologized that night. Said it was her mistake." He crunched down the creamy eggplant tart as though he had nothing else on his mind.

"Better watch your waistline," Sunny said.

He glanced up at her, brows raised. "You're the one eating the sugar buns. Two at a time you told me." He winced as her black suede stiletto, Christian Louboutin and *molto* expensivo, caught him on the shin.

"So," she said impatiently. "What's her story?"

"She's with Renato Manzini, her 'producer.' And also, I believe you mentioned,

82

your client Eddie's producer. I still don't know her name."

"That's easy," Sunny said, taking out her cell phone. "I'll call Eddie and find out."

She walked outside to make the call and Mac watched her, smiling at the perfect little twitch of her butt. It was unself-conscious and totally natural and beautiful.

She was back in a flash. "Her name is Marisa Mayne," she said, settling into her chair. "Eddie's seen her around in Hollywood. She's kind of 'a girl around town,' always at the clubs, always on the lookout. He told me she has a walk-on role in the sci-fi movie and that she looks sensational, all bare brown legs and a silver breastplate, with a lacquered silver mask complete with Spock pointed ears.

"Also, apparently at Renato Manzini's insistence, she's been given a couple of lines. Eddie doesn't know where she's staying but assumes, since they appear to be so close, it's with Manzini. His opinion is she's just a girl using her assets to try to improve her status in the movie world. And," Sunny added thoughtfully, "judging by that whopping yellow diamond on her engagement finger, I think she may be succeeding."

Mac took a sip of his wine. "Thanks, babe," he said. "What would I do without you?"

83

"You'd survive," she said.

He met her cool amber gaze. "No, I don't think I would," he said, leaving Sunny breathless, but just then the waiter arrived to serve the lobster fettuccini, interrupting their moment.

Dinner was delicious and the wine got even richer as the night wore on. They were on to dessert — the *dolce* the waiter called it, making them giggle — and a glass of *vin santo,* when Marisa Mayne made her exit. She stopped by their table en route.

"So good to see you again, Mac," she said, offering her hand as though they were old friends. He shook it, waving nonchalantly at Renato Manzini who glowered from the door, waiting for her.

"We have to talk," Marisa whispered urgently. "Call me. *Please,* it's important." Then with a quick apologetic smile at Sunny she was gone.

Mac waited until the couple had finally left. Then he opened his hand and took out the scrap of paper Marisa had palmed him. On it was written her phone number.

"She's not joking," he said thoughtfully. "And this time, I think she might be in real trouble."

CHAPTER 12

The next morning at the hotel, breakfast was a leisurely affair of endless coffee, sweet rolls, and crumbs in the bed, over which Sunny and Mac made love. Marisa Mayne was temporarily forgotten and they were still rolled in each other's arms at noon when Mac said, "Hey, there's all of Rome outside this window. So why are we just lying here?"

"Because this is more fun." Sunny tossed back her long wild hair and snuggled into his armpit.

"Wait a minute." He tilted her chin, rubbing his nose against hers, the silly way lovers do. "We have work to do."

"The Naughty Angel," she sighed.

"Right." Mac unwrapped her from him and reached for the piece of paper with Marisa's number. Grabbing the phone he punched it in.

She answered right away. "Oh, thank God it's you," she said, sounding tense.

"So what can I do for you, Miss Mayne?" Mac asked.

"We need to talk. Please can you meet me, somewhere . . . anywhere *anonymous.* You know what I mean."

"You mean a place nobody will recognize you and see you talking to me?"

This time Marisa sighed. "You're so understanding. But I don't know Rome at all, so tell me, where should we meet?"

Mac looked at Sunny, mouthing "Where?"

"The Tazza d'Oro," she said. "A bar in the Piazza della Rotunda."

Mac told Marisa and arranged to meet her there in an hour.

"Better get up, Miss Coto de Alvarez," he said, grabbing her feet and pulling her the length of the bed. He took her by the shoulders and lifted her up and she wrapped her long legs round his waist.

"Shower?" he suggested.

"Of course," she said.

The umbrella-shaded terrace of the Tazza d'Oro was busy with Romans tossing down espresso so dense that Sunny knew it must hurtle straight to their veins, revving them up to face the rest of the day. It was easy to pick out the tourists because they were drinking cappuccino, something Italians only ever drank at breakfast. She had tied her hair back in a glossy ponytail and wore a cool white shirt and a short white skirt, with her trademark red lipstick. She had two favorite lipsticks: the daytime one was a pure red and the nighttime one had a touch of blue, making it richer. The sun was shining, the air felt warm on her skin and Mac's hand was cool in hers. The glorious dome of the Pantheon seemed to float toward the blue cloud-spotted sky, and weary visitors took their ease on the imposing flight of marble steps leading up to its massive columned portico.

"The Pantheon was built by the Emperor Hadrian in A.D. 118 to 125," she told Mac as they settled at a shady table.

"That's *old.*" He signaled a waiter over.

"*And* it's erected over another, even more ancient temple, built by Marcus Agrippa. Italian kings are buried in there," she added, having done her homework. "As well as Raphael's tomb."

"I want to see it all," Mac said. "But business and a cold Peroni beer come first. What'll you have, sweetheart?"

She gave an exaggerated sigh at his crass dismissal of one of Rome's most important historic monuments.

"Lemonade," she said.

Mac gave the waiter their order, glancing around for Marisa, but as yet there was no sign of her.

"Wait a minute, though." Sunny took off her sunglasses and leaned forward, peering through the crowded square. "There's only one woman here with a body like that."

Mac took another look at the woman with the floppy straw hat pulled over her hair. She wore large dark glasses, jeans, cowboy boots and a loose red linen shirt that barely disguised her assets. It was Marisa all right. He got to his feet and waved her over.

"Oh, thank God you came," she said, sitting down quickly. "I'm so worried."

"Okay, hold on. What would you like to drink?"

"Oh? Campari and soda please."

Sunny was surprised that Marisa was already acting like a Roman, ordering a Roman-style drink. Obviously this woman was adaptable. "Hi." She leaned across the table to shake her hand. "I'm Sunny Alvarez."

"Pleased to meet you." Marisa shook it briefly and Sunny thought for a hot day her hand was exceptionally cold. She really must be frightened.

"You must be wondering who I am," Marisa said to Mac, gulping the Campari and soda as though it were Diet Coke.

"Well, kind of. I mean at least now we know your name."

Marisa took off the dark glasses and took a deep breath. "I'm Ronnie Perrin's fiancée." She held out her left hand with the whopping canary diamond. It caught the light and Sunny quickly put her own sunglasses back on.

"I admired it in Harry Winston's window in New York, so Ronnie bought it for me. But I have to keep our engagement real quiet until after the divorce. He's divorcing Allie Ray you know?" She glanced inquiringly at them and they both nodded.

"Well, anyway when the divorce comes through I will be the next Mrs. Perrin." She beamed at them and Sunny thought how attractive she was with her green eyes and wide sexy mouth. No wonder Perrin had fallen for her. Or had he?

"How did you two meet?" she asked, taking a sip of her lemonade.

"On an Internet chat room," Marisa said, surprising her. "You can go visit people online, ask who they are, what they are, get to know each other before you even meet. I fell in love with Ronnie before I knew who he was," she added defensively. "The fact that he turned out to be rich was a nice surprise.

And Ronnie said the fact that I turned out to be so sexy was a nice surprise too. He loves the Internet, he says you never know who you might meet."

She shrugged, staring down into her pretty pink drink. "The only thing is we can never be seen in public. We never go out together. I go to his Malibu home — he gave me a key. Or else he comes to my place out in the Valley, the suburbs really, where nobody knows what Ron Perrin looks like anyway. To them he's just another guy on a nice Harley."

A Harley girl, Sunny looked interested. "What does he have?"

"Oh, he has a couple, but his favorite is that real old one, not a Harley, the original . . . what's it called?"

Sunny drew an envious breath. "The Indian."

"Yeah, that's it. A man like Ronnie can have anything he wants. Including women," Marisa added, a touch bitterly.

"So you were alone at his Malibu house that night?" Mac prompted her.

She nodded, sending the floppy brim of the straw hat fluttering. "It wasn't quite what it seemed that night though." She hesitated, a little frown between her brows, obviously thinking. "I wonder, have you met Ronnie's partner, Sam Demarco?"

"Haven't had that pleasure."

"Hmm," she said, looking doubtful. "Anyway, Demarco told me Ronnie thought he was being followed. He said Ronnie was real worried about it, afraid some nut was going to shoot him, or else it was Allie Ray on the warpath. Or maybe the FBI keeping tabs on him. I asked Ron about it but he shrugged it off. I wanted him to get a bodyguard but he said that would only make him look like a guilty man."

"And is he? 'A guilty man'?"

Marisa's eyes sparked with anger as she glared at Mac. "Why does everybody have to think that just because someone is rich he's guilty of doing something wrong? It's just not fair."

"Okay," Mac agreed mildly.

Then Marisa stunned them by saying, "Ron likes, kind of to be . . . dominated, y'know."

Mac remembered that look in Perrin's eyes, like a chastised puppy. "Okay, so you are the dominatrix, he's the subject," he said.

"Kinda like that, yeah," she admitted. "But I really hate to hurt him y'know, I try to go as easy as I can on him and . . ."

"Still achieve the end result," Sunny said helpfully.

Marisa hung her head. "It's not really my

scene," she said. "But, y'know, like, I'm an actress, I can play any role."

Watching her, Sunny wondered why, if she was such a good actress, she could tell Marisa was lying.

Marisa took a large gulp of the bitter Campari, shuddering as she swallowed. "I hate this stuff," she said, "but everybody here drinks it."

"So what happened that night?" Mac asked.

"Ronnie had a meeting and I was in the house alone. I went upstairs to wait for him. I had the TV on but I could still hear the surf outside the windows. I had a bottle of champagne waiting in the cooler, the way I always did. Then Ronnie called, said he was running late, he'd be back in an hour. *That's* when I heard the noise downstairs.

"I thought *no,* I'm imagining things, it's just the waves on the rocks, high tide or something. Anyhow, I turned down the TV and listened. I heard the noise again. *A footstep.*" She shivered. "You know those floors, they're some kind of concrete polished until it shines, but they're hard as hell and nothing you can do can soundproof them. You could hear a petal fall from a rose in that house.

"It was a definite footstep. Someone was

moving around downstairs, opening things. *And I was alone.* I was so scared, I grabbed Ronnie's gun from the drawer in the bedside table. I crept to the top of the stairs and peered down."

She stopped with a shudder that this time shook her entire body. She was obviously terrified by the memory. "Jeez, Mr. Reilly — Mac — I wanna tell you, my heart was thudding like a friggin' steam engine. But I'd always told myself that in a pinch, in a situation like this, if it was a 'him-or-me' survival, it would be *me.*

"I still couldn't see anybody so I crept further down the stairs. I was standing at the bottom looking round in the darkness when somebody grabbed me. I screamed and he pushed me to the floor. I dropped the gun and I thought, Oh shit this is it. . . . I was facedown, frantically groping around for the gun. By the time I found it and got to my feet — he was gone."

She looked at Mac. "And then you appeared at the window. I thought you were him, come back again. . . . I didn't recognize you until after I'd shot at you. And now I just want to say I'm sorry."

Mac grinned. "That's okay, it's happened before, and those other times I never got an apology."

"I was terrified I'd hurt you. I thought you'd send for the cops — and that would have meant the end for me and Ronnie. So I just got out of there as fast as I could. I didn't stop to think — except about the gun. I knew I couldn't leave it there in case the cops came, so I wiped it off on my robe — so there'd be no fingerprints y'see. And then I dumped it in your car. I knew the red hybrid was yours, I'd seen you driving by and it was always parked on the street outside your house. Anyway, that's what I did, and then I went back to wait for Ronnie."

"So tell me," Mac said, "why *didn't* you call the cops?"

"No cops. Ronnie wouldn't have liked that."

Mac recalled Perrin saying vehemently, "No police" as he'd left him that morning.

"So what happened later?" he asked. "When Perrin came home?"

"I'd already called him, told him what happened. He agreed it was best not to say anything. But I could tell he was scared. He said it must be the guy who was following him, that he must be some nut who wanted to kill him."

"What happened to the FBI theory?"

"That as well. To tell you the truth, Mac, it was a very paranoid situation. And your showing up didn't help things any."

"Thanks a lot," Mac said. "I'll remember that the next time I hear a woman scream."

"Oh, I didn't mean it like that. Really I didn't." She looked away, embarrassed. "It's just that Ronnie was in trouble and I didn't know how to help him. After the shooting incident he said he had to get me out of the country immediately. Ronnie couldn't afford another scandal, after that divorce and . . . well you know, the court case about mishandling the funds. And with me out of the way no one could ask me any questions. He called Demarco and told him to 'take care of me.' He meant 'get rid of me.' I knew that.

"Anyhow, Demarco called Renato Manzini in Rome, told him he was sending me over right away, and to make things look legit he should give me a small role in his film. Demarco chartered a private jet and got me to Rome that night. He told Renato to put me in the Hotel Eden and to look after me. And Ronnie said he would join me in a couple of days."

"And?" Sunny was hanging on to every word.

Marisa's face fell. "He's never even called," she said. "I'm still waiting for him at the hotel. But now Renato has found me an apartment. I move in tomorrow. Here's the

95

address and the phone number. You already have my cell." She handed Mac the piece of paper with the information. "I daren't try to call Ronnie because he said never to, his lines might be tapped." She looked helplessly at them. "But he just never showed."

"Where do you think he is?"

Marisa shook her head, sending the floppy straw hat fluttering again. "I don't know. Has he dumped me, or what?" She twisted the enormous ring nervously. "I mean, a guy should tell a girl if there are problems. Not that there were. We were happy as two clams. I knew what he liked, he knew what I liked." She glanced meaningfully at Sunny. "Y'know what I mean."

Marisa sighed again. "It doesn't make sense. And that's why I'm worried about him. Somebody was following him. Somebody broke into his house. He told me somebody wanted to kill him. And now I'm afraid they might have. And that's the truth of the matter."

"And what do you want me to do?" Mac asked.

"I need you to find Ronnie. I want to know he's alive. Tell him I'm still here, waiting for him. At least tell him to have Demarco call me and tell me what's up."

Mac said, "So what do you think the intruder was after that night? Obviously he wasn't aware you were there, so it wasn't a rapist or a killer . . ."

Marisa shivered. "Oh, God, don't even say those words. I tremble at the thought of what might have happened. And I really don't know what he wanted."

Mac thought about it. When Perrin had asked him to help he had turned him down, but now Mac needed to know what was really going on. For Allie's sake, as well as for the girlfriend. "Tell you what, Marisa," Mac said. "I'm in Rome for another few days, but I'm gonna call my assistant in L.A., put him onto the case. He'll find out who's following Ron."

"And will he also find Ronnie for me?"

She looked hopefully at him. It was Mac's turn to shrug. "He'll do his best," he said, though he personally felt sure that Perrin had sent Marisa to Rome to get rid of her, and that there was no way she was ever going to see him again.

He thought Perrin's next move would be an offer of a nice little financial settlement. He'd probably also get Manzini to offer her the odd role, keep her here in Rome, out of the way. After all, he had already gotten her an apartment. It would all work out fine for RP. And for Marisa Mayne too. Looking at her, he had no doubt she'd be happy to take the money and run.

Marisa said she had to leave, she was expected on the set. "Just some retakes," she said quickly. "But y'know, it could really lead to something."

She thanked them and Mac promised to call when he had some news. He looked at Sunny who was watching Marisa saunter through the crowd, turning heads along the way, despite the weird hat.

"You're a woman, what d'you think of her story?" he asked.

Sunny looked thoughtful. "It's odd," she said, "but listening to Marisa somehow the word *blackmail* popped into my mind. Y'know the whole S and M dominatrix theme she had going there? I just didn't believe it. Marisa may be lying about what happened that night. Maybe she'd given Perrin an ultimatum, pay up or she'd go to the tabloids and tell her version of 'the truth' about their sexual relationship." Sunny shrugged. "They would have jumped on it."

"But she was onto a good thing," Mac said. "Perrin was generous. Just look at that ring."

"Trust me, I looked." Sunny sighed. "But with playboys like Perrin all good things come to an end. Maybe he was bored with her. On to the next, if you know what I mean. After all, Marisa said he was out that night. I wonder where he was and who with."

Mac took out his cell phone to call Roddy. "Okay, so let's find out where Ron Perrin is. *And* who he's with."

CHAPTER 13

Allie was in her bedroom at the Bel Air house. The same bedroom that used to be hers *and* Ron's, complete with the California king-size bed with the brushed-steel posts Ron had constantly complained about, after getting up in the night to go to the bathroom and cracking his head on them.

"Why can't we have a regular bed? Y'know the kind, with a mattress, box springs, some sheets and a blanket?" he'd yelled. "Why must we have this . . . this *glamazon* of a bed?"

Glamazon was the right word to describe the bed's size and flowing draperies. Silk of course. What wasn't silk in this house? If it wasn't expensive limestone or fossil granite or zebrawood. In fact it was champagne-colored silk with a voile inner lining run through with threads of gold. All in excellent taste, naturally. Ron having chosen the "best" decorator. If you liked that sort of thing. And having finished the house, neither Allie nor Ron had ever admitted to the other that they did not really care for all the opulence.

What the three-thousand-square-foot bedroom suite did have though, were the best closets in the world. His and Hers. They were enormous. Ditto the bathrooms. Hers larger than His, of course, with golden faucets that spilled long flat streams of water into a jetted tub and with towels thicker than Allie could handle. Secretly, disguised in a dark wig and glasses, she and Ampara, the housekeeper, had slipped into Costco and bought a dozen of their super-sale ones so she could dry herself properly. The "good" ones were just for show. Actually, Allie had been pleased to find that the brown wig and glasses were an effective cover. No one had even glanced twice at her.

Today, Fussy, the Maltese, had as usual parked herself right in the middle of the bed. Her favorite place. She had always slept between them, Allie's legs on one side, Ron's on the other. Anyhow, Fussy just sat there now, barking snappily to let Allie know she was bored and that anyhow she'd rather be in the kitchen with Ampara.

The long room with its floor-to-ceiling windows letting in streams of strong California light was filled with people. There was the stylist who'd brought a rack of gowns from which Allie would choose the ones for her Cannes Film Festival appearances, along with the sexy four-inch-heel shoes neatly laid out in a row, and the expensive little bags, and of course the jewelry that came along with a bodyguard, provided, as were the jewels, by Chopard. A seamstress from the design house waited to pin, a hairstylist hovered, and the makeup girl waited to see what she would choose so they could then decide on a "look." Plus there were a couple of gofers, ready to run to the stores or whatever.

The housekeeper had set up a table with coffee, bottled water and soft drinks, as well as her home-baked cookies and chocolate cake, the smell of which was driving Allie crazy. It reminded her of those rare childhood occasions when, with her mother, she had stirred the Betty Crocker chocolate-fudge cake mix then waited, almost dying with anxiety until the oven door was finally opened and the always-sunken cake removed. She had never been able to wait for it to cool, devouring a chunk smiling her pleasure through warm chocolaty lips. It was one of the few highlights of her youth.

She took a large piece now. The stylist

frowned. "Every extra ounce will show in this dress," she warned.

Allie shrugged, uncaring. This was the best she had felt in weeks. Cake was her answer, and maybe about half a pound of M&M's, and how about Starbucks java chip ice cream? Yes! That's exactly what she would have for dinner tonight and the hell with sparkling couture gowns from Valentino and Versace. She needed comfort food.

"Try it," she offered generously. "Ampara makes the best cake you've ever tasted." She put a piece on a plate and gave it to the slender young stylist, who ate it, complaining guiltily she hadn't been this "wicked" in years.

"Go to the gym tonight," Allie said, laughing. "And why should we think it's wicked to enjoy a piece of cake every now and then?"

"I guess it's okay every now and then," the stylist agreed, albeit reluctantly, as she took another guilty slice.

The others crowded round now, all except the bodyguard, who stood stoically, arms folded, next to the large leather boxes holding several million dollars' worth of jewels.

Allie inspected the rack of gowns, all special, all beautiful and all meant for a grand entrance under lights, a photo opportunity for the magazines and television cameras.

"Allie Ray adorable in Valentino and Chopard diamonds at the Cannes Festival," they would say, as she did her job and smiled and waved and stopped to talk to the guy from *Access Hollywood* and the woman from *Entertainment Tonight,* as well as the French TV host, who she always surprised with her ability to speak a little of his language.

"Not fluently," she'd protested, when he'd complimented her last year. "Only enough to get by." It was the compliment that had pleased her the most, though.

She washed the chocolate cake from her fingers and began to try on more gowns, swishing their heavy trains and slinking her thighs together, wondering whether she could even walk. Bored, while they pinned and fussed, she glanced out of the window, thinking of Lev, outside in the black Mustang and probably bored out of his skull too.

What, she wondered, did he do all day to keep himself occupied?

She called him now. "What're you doing?" she murmured into her BlackBerry.

"Isometric exercises," he said, and she heard the smile in his deep rumbling voice. He was the only man she knew whose voice matched his physique.

"I'll bet you're reading the racing form," she countered, having already divined his weakness for the ponies.

"Possibly."

She grinned. "I'm sending you down a little snack. Homemade chocolate cake. You've never tasted better."

"I don't eat cake."

"Today, you're Marie Antoinette," she said, and heard him laugh again.

Pushing the gown pinner away, Allie went to the table and cut him a slab. Wrapping it in a napkin, she handed it to Ampara and told her to deliver it to the paparazzo in the black Mustang. The others stared at her as though she had gone mad.

She said, "And the hell with these gowns. I'm not wearing any one of them."

There were gasps of horror. "But Allie," the stylist protested. "These are gorgeous, they're perfect for Cannes. They're the latest, right off the runway."

"I'll make my own choices from now on," Allie said firmly. "And that goes for the jewels too," she added. "I won't need any."

"But, Allie . . ." The stylist was in a panic now. She had to report back to the producers, the director. The hairdresser and makeup girl waited silently, uncertain of what was expected of them.

"Don't worry," Allie said, giving them that sunny grin. "It'll be all right on the night."

The plan that had been formulating in her mind began to loom as a reality and suddenly she felt light-years better.

Thanking the stylist and her entourage, she sent the team on their way, still protesting her decision.

Allie knew that most women would have died for the choices she had been offered that day. And of course she was aware of her responsibility. She would do her job. But she had her new plan in mind. She had still to figure it out, but she was about to become a different woman and it had nothing to do with the public. She wondered if she should share her plans with Mac Reilly. But Mac was in Rome and anyhow her future was not his business. Only her present.

Worried, she stared out the window. Beyond the thick greenery and the high wall, Lev, or one of his henchmen, kept guard. She was safe now. Wasn't she?

Her thoughts turned to Ron. In her heart she didn't want to believe he was tailing her, but if he was not, that meant it must be the stalker. There had been more of those letters, the last one smeared, the unknown writer said, with his tears. "Next time it will be with blood. Yours? Or mine?" he'd written.

Allie had refused even to look at the letters,

but now she didn't burn them. She'd sent them on to Mac's assistant, Roddy. The game was no longer in her court. They would take care of it.

Feeling guilty about the chocolate cake, she was glad when her trainer arrived to put her through her paces, stretching her body unbearably, urging her on as she sweated on the machines.

"It's worth it, hon," he said, smiling at her. "You've still got the best body in town."

It was the word *still* Allie didn't appreciate. It meant she was no longer eighteen and instead was heading up to forty, a time when actresses were often left in limbo, waiting for those age-appropriate roles that, unfortunately, no matter how good you were, were few and far between.

An hour later, she waved goodbye to him and walked to the window, staring moodily out at the pretty gardens and the deep cobalt blue pool, glinting like a jewel amid the thirsty emerald lawns. She really should think about changing to a desert landscape to save on water, life's most precious commodity. But then, who but she was there to care?

Glancing at her watch, she called her director and canceled their lunch. She would see him later that afternoon at the studio, she said, to do the last of the overdubbing.

After that she changed into jeans, a white shirt and gold flats, put gold hoops in her ears and — after some consideration — slipped her wedding ring back on her finger, and set out for the children's hospital in the Valley, to pay her weekly visit to the young cancer patients.

She called Lev to tell him her plans, watching out for him in her rearview mirror, as he stayed behind her on the freeway. There was no sign of the Sebring convertible.

She'd already spent time at Barnes & Noble choosing picture books and games and had picked up a batch of furry toy animals donated by a caring manufacturer. The kids were always pleased to see her and today they greeted her with smiles and laughter, as though she were Santa on Christmas Day. It made her smile too, and their gaiety in the face of suffering brought her back to her senses and a humble appreciation of the rewards life had offered her.

By four o'clock she was in a darkened Hollywood studio, watching herself on the screen, matching her words to her movements. She was finished by seven, and called Lev again to tell him she was on her way to Giorgio's restaurant on Channel Road in Santa Monica.

She was meeting an old friend there, a woman almost twice her age. Sheila Scott had been good to her when she had first come to town. Sheila was a voice coach and it was she who had taken the Texas twang out of Allie's voice and perfected her sweet gentle way of speaking. And since Giorgio's was also Allie's favorite restaurant, she was looking forward to the evening.

She handed the keys of the Mercedes to the valet parker, who beamed at her, impressed, and said "Good evening, Miss Ray, how are you tonight?" Allie was aware that, as usual, heads turned as she strode through the door and, always conscious of her duty to her fans, even though this was mostly a showbiz crowd, she distributed smiles and stopped to kiss hello to a couple of fellow actors.

She was glad, though, to sink into her chair and share a bottle of Chianti with Sheila, lingering over a plate of fettuccini with langostinos, the house specialty that was her favorite. And also to tell her about her new plan, that was still not a plan. Only an idea.

Sheila Scott, defiantly gray-haired in a town of blondes, her lean face tanned and weather-beaten from years of living near the beach, down-to-earth and motherly, listened in silence.

"I think I've come to the end, Sheila," Allie said quietly. "My new film's no good. I'm about to hit that dreaded — in showbiz anyway — forty. I've lost out in love. Ron has left me, he's found someone else. I have a crazed fan writing scary threatening letters to me, and I'm being stalked. I only feel safe when locked behind my own gates with a bodyguard right outside. I have no privacy, no family. I've reached burnout, Sheila. I need to get a new life."

Sheila nodded, understanding. Allie had been working since she was seventeen and whether a movie lived or died always seemed to depend on her. Not only that, real life had crept up on her. She was a lonely woman trapped by her own fame, abandoned by her husband and stalked by a madman. "Then if that's what you need to do, Allie," she said gently, "go for it."

"There's only one thing — no maybe two — that could stop me," Allie said.

Sheila said shrewdly, "I'm willing to bet that both of those are men. And that one of them is still Ron." Allie gave her a sheepish grin. "So, who's the other?"

"His name's Mac Reilly. The PI. You've probably seen him on TV. But anyhow, like Ron, he's a lost cause. He doesn't seem interested in me, except as a client of course."

Her eyes met Sheila's sympathetic ones. "Do you think it's possible to be in love with two men at the same time?"

Sheila patted her hand across the table. "Only if you're trying too hard, sweetheart," she said.

Just then a couple of fans came over with a request for Allie's autograph, and putting on her best movie-actress face she smiled and chatted with them for a minute.

Then the waiter approached. "A delivery for you, Miss Ray."

Allie's heart jumped into her throat as she took the envelope. For a minute she thought she might faint.

"Allie, are you all right?" Sheila's voice seemed to come from a great distance.

"Who delivered it?" Allie asked. "Where is he?"

"It was just some motorcycle delivery service, Miss Ray. He was still wearing his helmet and I didn't get to see his face."

Allie had recognized the writing. "Oh my God," she whispered. "He's found me."

Horrified, Sheila stared back at her. She'd heard all about the letter writer and the stalker, who were probably the same person. It did not sound good. She said, "Where's your bodyguard?"

"Outside. Waiting for me. He'll follow

me home."

"Call him. Tell him what's happened."

Parked illegally across the street, Lev had kept tabs on the people coming and going at the restaurant. He'd seen the motorcycle arrive and watched the driver go inside. He had thought it odd that the man had not taken off his helmet. It was an automatic reflex: you stopped the bike then took off your helmet. It was that, that had made him take down the bike's license number.

"Don't worry," he told Allie when she called. "Give me the letter and I'll take care of it."

"It's the last straw," Allie said shakily to Sheila. "You see now why I can't go on."

"I understand, sweetheart, but don't let panic send you running away."

"If only it were just that," Allie said, as they kissed good night outside the restaurant.

Lev was right behind her in the Mustang as she drove back along the coast road. At least she knew she was safe. For the moment anyway.

CHAPTER 14

Five days later Sunny and Mac arrived back at LAX. He dropped her off at her place then continued on in the limo, catching up with his phone messages on the way.

Surprisingly, there was one from the supposedly missing Ron Perrin, demanding to know when he was coming back, asking why didn't he pick up his messages anyway and where the hell was he because he needed him. It seemed Marisa was wrong and Perrin wasn't dead after all.

There was also a message from Sam Demarco. "I'm Perrin's right-hand man. I want to talk to you," Demarco said in a voice as crisp and cold as a wedge of iceberg lettuce. "Please call me as soon as possible so we can arrange to meet."

Of course Mac immediately called Sunny to tell her. "Interesting, huh?" he said.

"Which one? Perrin or Demarco?"

"Both. Anyhow, I'm calling them back. I'll

let you know what happens. Meanwhile, babe, get some sleep. You looked exhausted."

"Thanks to you." There was a smile in her voice as she said it.

Mac was smiling too as he rang off.

When Mac finally got home, Pirate greeted him with his usual all-over face lick. He smelled like an old sock. Time for the mobile dog groomers whose blond girl-bather Pirate was in love with, if his goofy expression and complete malleability in her hands was anything to go by.

Mac checked his watch. It was still only noon. He showered and put on a pair of comfortable old khaki shorts. Pirate was glued to his heels as he walked out onto the deck and leaned on the rail, taking grateful gulps of the fresh salty air. After the long flights it felt wonderful. The tide was low and the ocean shimmered, flat and steely under the gray sunless sky. Not a surfer in sight. Of course not, they were waiting for the change of tide, ready for those big green rollers that came crashing onto the shore, riding them like circus performers.

Mac had been in touch with Roddy from Rome and had filled him in on the Marisa situation. Now he called Roddy again and got the news that so far Roddy had no idea who was tailing Allie because, since Lev was on guard, the Sebring had not been seen.

Roddy also told Mac about the latest threatening letter, hand-delivered to the restaurant. "It wasn't written in blood but on a computer. And it was pretty explicit in what he intended to do to Miss Ray. By the way, he never calls her 'Allie,' always the formal 'Miss Ray.' "

"So what are we doing about it?" Mac said.

"Lev got the bike's number and already checked it out. It's a delivery service. Someone hired them and the driver was just doing his job. The person paid in cash. Unusual enough for them to remember, but oddly enough, nobody seems to recall who it was or what he looked like."

Mac sighed. "Par for the course, I guess. I want you to run a check on all Allie's employees, everyone who has intimate contact with her. That means the people who work at the house, gardeners and pool guys included, as well as hairdressers, masseurs, personal shoppers. You know the score."

Roddy did know and it was not a small task.

Mac said, "You ever hear of a guy called Sam Demarco? He's Perrin's partner. Actually his 'right-hand man' is how he described himself in his phone message."

"Not short on ego then," Roddy said. "And yeah, I've heard of him. Kind of a big player around town. Flashier than Perrin. Likes Vegas, the clubs, that kinda thing. Likes to throw big parties at his place. He has a big modern house on one of the 'bird' streets above Sunset, y'know, the ones with the fabulous city views and the 'bird' names, Oriole, Thresher, like that. As well as a new place out in Palm Desert."

"Right. Okay. I'll call Demarco, find out what's so urgent that Perrin's 'right-hand man' needs to talk to me. Marisa wants me to ask him to call her too, she needs to know what to do next."

"Does she want her gun back? If so it's in the bucket under your kitchen sink."

Mac laughed. "Thanks a lot, pal. I'll make sure to get it back to Perrin."

His next call was to RP. Again no answer. He left a message asking him to call him back, then he called Demarco.

After all the hoopla about urgency and the need to talk, Demarco didn't even take his call. Instead his assistant arranged for them to lunch at one o'clock the following day, at the Ivy on Robertson. She also said Demarco needed his help. He wanted to hire him. But she didn't say why.

116

Mac considered calling the "movie star" but that old jet lag was creeping up on him again as well as hunger, so instead he whistled for Pirate, climbed into the Prius and drove to the Malibu Country Mart, all of five minutes away. It was a small complex of chic boutiques and restaurants set around a grassy square with a sandbox and swings where kids played happily. He bought a take-out ham and cheese panini at the Italian restaurant Tra di Noi and with Pirate breathing heavily at his feet in anticipation of his share, sat on a bench outside enjoying his lunch and watching the Malibu world go by.

As usual, the paparazzi were hanging outside Starbucks and Coffee Bean, hoping to get lucky with the "hot" young set, whose appearances there in search of a Frappuccino, with or without babies or small dogs or underwear, could cause chaos. Mac thanked God that though they recognized him they left him alone. His wasn't the kind of fame — or rather notoriety — they were on the lookout for. There was no scandal around Mac Reilly. Unless of course he was to be seen in the company of Allie Ray. Now *that* would be news.

Walking back to the car, he stopped to look in the window of Planet Blue. There was a white T-shirt with the words LOVE IS ALL

YOU NEED in sparkles on the front. Smiling, he went in and bought it for Sunny.

Back home and out on his deck again, he stared at the ocean. The tide was coming in now. Ruffles of white spray fluttered over the rocks and the sun peeked through the clouds. He called Perrin one more time, frowning as one more time he was asked to leave a message. If RP was around he certainly was not answering his phones.

Lulled by the sound of the ocean, Mac lay back on the old metal chaise and in an instant jet lag claimed him and he was fast asleep.

CHAPTER 15

The next morning as Mac drove through the sluggish L.A. traffic on his way to meet Demarco, he was thinking about the "right-hand man's" choice of restaurant.

The Ivy was an L.A. hot spot. It was *the* place to be seen lunching and there were always famous faces there plus the usual wannabes and of course the paparazzi thronging outside with their intrusive cameras. Still the food was pretty good and it was a buzzing little place, a cottage really with an umbrellaed patio surrounded by a picket fence and with various hokey "country" artifacts substituting for décor. Cute, cheerful and expensive.

Mac gave the Prius to the valet, waving to the paparazzi who must have been having a slow day because they bothered to take his picture.

Sam Demarco was already seated at an inside table waiting. Impatiently, Mac observed. Since he was no more than a couple of minutes late, which was a good deal less than par for the course bearing in mind L.A.'s notorious traffic, Mac thought him distinctly ungracious in his greeting.

"I don't like lateness, especially in a man," Demarco snapped, tapping his watch, a whopping gold Breguet that was meant to impress. "I find it discourteous when I made every effort to be here on time."

Mac glanced pointedly at his own watch, a distinctly unimpressive Swiss Army with red numerals and a black rubber strap. "I am exactly three minutes and thirty seconds beyond the appointed hour." He took the seat opposite. "Perhaps in your effort not to be late you got here too early."

Barbed glances shot between them. Demarco was a leonine-looking man, tall, ruggedly built, in his early fifties, lightly tanned and with a mane of thick silver hair. Mac thought he was a little overdressed for lunch in L.A., in a dark blue pin-striped suit from a very good tailor, polished black wingtips, a blue shirt and a yellow Hermès tie. Or maybe it was he who was underdressed, in chinos and a plain denim blue T-shirt from Theodore at the Beach, which happened to be his own personal favorite men's shop. He wore brown suede loafers and no socks. At least *he* was comfortable.

Unlike Perrin, who had eyes like a chastised puppy, Demarco's hard blue eyes told Mac in no uncertain terms to back off.

Demarco offered his hand across the table. Mac reached over and took it, trying not to flinch as it was crushed. Demarco was a physically powerful man and either he was using that power to intimidate, or else he was just unaware of his own strength. Somehow Mac didn't think it was the latter.

The waitress showed up to announce the specials but Demarco waved her away and they ordered quickly, Mac the salmon and Demarco the burger. Demarco asked for a Perrier and though Mac fancied a smooth, round, mouth-filling Cakebread Chardonnay with the fish, he asked for water also.

"Reilly," Demarco said, in his deep sonorous voice, "I asked you to meet me because I know of your reputation, via the television show of course, as well as your exploits off camera."

Mac nodded. He took a sip of Perrier, already regretting the Cakebread.

Demarco waited for a response, looking at him with those better-watch-out blue eyes. When he didn't get one he said, "I take it that anything I say will be in confidence?" Mac nodded of course and Demarco said,

"The fact is, I'm worried about Ron Perrin."

"Your boss," Mac said, letting him know that he knew on just which rung of the ladder of power and fortune Demarco stood.

"First and foremost Ron is my friend," Demarco put him straight. "I started out as his assistant." He shrugged his shoulders, barely wrinkling his immaculate blue pinstripes. "Now we are partners."

"You worked your way up," Mac offered helpfully. Then he thought the hell with it, summoned the waiter and ordered a glass of the house white.

"You might say that." Demarco sat back. His face was a mask but Mac got the feeling he didn't like him. He wondered why Demarco was even bothering. After all, he could hire any PI in town. Anyway he wasn't sure he wanted to work for the guy. In fact he could easily do without both him and RP.

"Reilly," Demarco said again, without benefit of a Mister or even a may-I-call-you-Mac. "I'm worried about Ron. He's been behaving very oddly, claims he's being followed, that somebody wants to kill him."

"So? Is he? And do they?"

"How should I know? That's your department. Didn't he ask you to work for him? Find out what was really going on?"

Mac wondered how Demarco knew that.

Perrin must have told him. But then if he had, wouldn't he also have told him why?

He took a sip of the house white. It was okay but he regretted the Cakebread. "Perrin did and I turned him down," he said.

The food came. Mac stared at his poached salmon, artfully presented in a nest of chopped tomatoes and basil with a delicious vinaigrette. He no longer fancied it.

He heard Demarco sigh. Then Demarco said, "I seem to have offended you. I'm sorry, I didn't mean to do that. It's just that I'm upset. I'm concerned for Ron. He's my friend. He's been more than good to me, I can't let him down in what might be his hour of need. I am asking you to work for me and when I tell you why, you'll understand."

"Okay," Mac agreed. "I know these situations can be stressful."

"The truth is I think Ron's going a bit nuts," Demarco said suddenly. "And you know why? He thinks the FBI is investigating his business dealings."

Mac glanced up from tasting his salmon. This was the second person to mention the FBI. "And is that true?"

"The FBI is always interested in men with multinational billion-dollar businesses, but whether it's true in Ron's case or not, I don't yet know."

"And you want me to find out?"

"As discreetly as possible, of course."

Mac remembered Perrin shredding documents and the overturned shredder box. He'd thought Perrin had simply been hiding financial evidence from his wife's lawyers, but it seemed there was more to it. "Okay," he agreed. "My assistant will let you know about fees and expenses. Meanwhile, have you any idea where Perrin might be?"

Demarco shrugged again, spreading his hands. "I haven't heard from him in over a week. And nor has Allie. Of course, her lawyers are going crazy, calling me at all hours demanding I tell them where he's hiding. Ron was supposed to appear in divorce court this week but he never showed up. Then they tried to serve a subpoena on him and couldn't find him. He's not at any of his houses. My guess is he's hiding out somewhere until he can work things out to his better advantage."

"Or else he's on the run from the FBI."

"It's possible, but I think Allie is the more immediate problem. He doesn't want to part with another cent."

"Tell me, does Perrin have a girlfriend?"

"A rich man always has lady friends."

"Yeah. Anyone special, though?"

"You might as well know that, women-wise, my partner does not have a good reputation. Don't quote me on it."

Mac nodded. "You want me to find Perrin? And find out what the FBI is after?"

"You got it, Reilly. But one thing for sure, no police. Perrin wouldn't like me setting the cops on him. Absolutely no police." He stood up. "I'm late for my next meeting." He shook hands again.

"Let me know how it goes," Demarco said. He paused then added, "I'm really worried about Ron. He's a good guy. *Decent,* y'know. I'm afraid he might do something . . . well, foolish. Y'know what I mean?"

Mac did know, but hardly thought Perrin had looked suicidal. In fact quite the opposite. RP definitely did not want to be dead.

"I'll do my best," he reassured him.

He watched Demarco stalk through the tiny room. The man dwarfed everyone in sight. A lion on the prowl was the image that came to mind. Though maybe a kindly lion, deeply concerned for his friend and partner.

He looked at Demarco's plate. He had not touched the burger.

CHAPTER 16

Allie was in her garden when her BlackBerry burbled a tune. She looked at the display to check who it was, but all it said was "Wireless caller."

"Hi, Allie," a familiar voice said.

A smile lit her face. She walked along the terrace overlooking the parterre garden modeled after the ones at Versailles. Not as big but certainly as sculpted. Not a leaf out of place.

"Hi, Mac," she said, her soft voice conveying her smile. "Are you calling from Rome?" She crossed her fingers, eyes raised to heaven, praying he was home.

"I'm back. I wanted to speak to you, make sure you're okay and that Lev and his friends are doing a good job."

"Perfect. Except for the biker with the letter."

"I heard about that," Mac said. "Look, we need to get together. I'd take you out to dinner if it wouldn't cause a scandal in the tabloids."

She laughed. "I could come over to your place," she said, thinking how it would be, just the two of them in his cozy little home. "We could send out for pizza."

"You got it," he said. "Seven okay with you?"

"I'll be there," she promised.

"Oh, just one thing . . ."

"Yes?"

"You like anchovies on your pizza?"

"Love them," she said, laughing again.

"Seven it is then."

Mac had called Sunny to tell her what was going on. He'd asked her to join them and now she was sitting on his deck. Her hair was pulled back and tied with a scrap of black ribbon and she was wearing the Planet Blue T-shirt with LOVE IS ALL YOU NEED written in sparkles across the front, topped with a cute little orange and hot pink striped cardigan against evening beach chill. She was holding Tesoro on her lap. Despite the balmy evening, the little dog shivered the way Chihuahuas often did, in what Mac always claimed was a deliberate play for attention.

"It's not cold for God's sakes," he grumbled, keeping a keen eye on Pirate, who was lurking near the steps at the very edge of the deck, ready to run if Tesoro jumped him.

Sunny threw him an exasperated glare. "She's sensitive, that's all." She'd hoped that bringing Tesoro might thaw the cold war between the two dogs, but the Chihuahua wasn't having any, and nor apparently was Mac.

Even though it was not grand like its neighbors, she loved Mac's little house, especially on soft sunlit evenings like this. The house had wood siding painted pistachio green with those cheap aluminum sliding windows that were original and maybe qualified as antiques by now, and there was a touch of gingerbread trim — an afterthought, she guessed, by some previous owner who'd wanted to jazz it up a little. Inside was a tiny wood-floored living room with a picture window overlooking the ocean, a small kitchen, mostly taken up by a large wine cooler; a bathroom at the back; and tacked on at one end and separated by a narrow corridor, the master — and only — bedroom.

She glanced at her watch. Ten minutes before seven. Ten minutes before the fabulously beautiful Allie Ray got here. She wondered what she was like. Mac certainly liked her, and besides she was paying him well to

128

take care of her problems. Of course he couldn't possibly be interested in the beautiful movie star, other than as a client who was in trouble that is. Still, as any woman knew, you couldn't take that kind of high-wattage beauty and fame lightly.

The captain's bell clanged unmusically, making her wince. Nothing she could say could get Mac to part with that cracked old bell. She took another sip of the good red he'd broken out from his best stash, kept in the properly refrigerated cooler that took up a good part of his tiny kitchen. For a wine buff like Mac, wine had priority over food, which after all, could be ordered in and delivered to his door. Just the way the pizzas were being delivered right this minute.

Smiling, she watched him through the window as he turned the oven to high, ready to reheat the pizzas, reaching into the cupboard for the hot peppers and shuffling plates around. He was looking particularly cute tonight in the baggy white linen pants she had chosen for him at the expensive little boutique on the Via Condotti, and an old black T-shirt faded through many washings to carbon gray. His hair was still wet from the shower and she knew exactly how his skin would smell, spicy and sexy and . . . well, she wouldn't go there right now.

The captain's bell rang again and she saw Mac hurry to answer it. Pirate was right behind him, barking enthusiastically as he opened the door and welcomed Allie Ray, one more time, into his home.

"Hi, good to see you again," he said as the petite beauty smiled up at him.

"Good to see you too," she murmured, standing on tiptoe putting an arm around his neck and kissing him on either cheek.

"I brought Fussy along." She held up a small bundle of white fur whose black button eyes were half-hidden behind a long fringe. "It's my housekeeper's night off and I couldn't leave her alone. I hope you don't mind?"

"Err . . . no. No, of course not." Mac eyed his own dog doubtfully, but Pirate just stared up at the Maltese, seeming stunned to have yet another female invading his home.

"Okay," Mac said. "Let me get you a glass of wine. White or red?"

"Since it's pizza how about red?"

"Perfect. I already opened one of my favorites."

Allie walked with him to the kitchen, still clutching Fussy and followed by Pirate.

"A Nobile — Antinori 'ninety-six," Mac said, pouring her a glass. "I hope you'll like it."

She took a sip. "Hmm, delicious. I didn't know you were a wine expert."

"That's probably because you don't know very much about me, other than the guy you see on TV attempting to solve a few old crimes. Closed cases. With the generous help of the LAPD, of course, without whom none of it could take place."

"I suppose not." Leaning against the kitchen counter she took another sip. "But you know something, Mac. Somehow I feel I *really* know you anyway."

Their eyes met. Surprised, he wondered if he was reading more in her words than she had meant. "There's someone I'd like you to meet," he said quickly. And taking her arm, he walked her out onto the deck.

"Oh, hi." Sunny got gracefully to her feet and found that she stood a good foot taller than the petite movie star. But my lord, she was beautiful. The long straight blond hair fell like a curtain over eyes that were bluer than any she had ever seen. She saw a flicker of surprise cross Allie's face, quickly covered with a smile. It was obvious she hadn't known there would be anyone else here.

Mac introduced them. Clutching their dogs the two women shook hands and said hello. Tesoro extended an aristocratic nose toward Fussy who sniffed back then let out a sudden abused yelp.

"Oh my God," Allie said to Sunny. "Did your dog just nip Fussy's nose?"

"She certainly did not. Tesoro wouldn't stoop to such behavior." Sunny's eyes met Mac's over the top of Allie's head and he grinned. "Your dog just barked, that's all," she added defensively.

"That's all she ever does." Allie sighed.

"Tesoro too." Sunny was suddenly all sympathy. "We should have dogs like Pirate. He's such a good boy, look how well behaved he is."

They both turned to look at Pirate, back at his post near the beach steps, ears down, his one eye wary. Now he had two smart-ass bitches to run from.

"Sunny runs her own PR company," Mac told Allie. "But sometimes she helps me out on my cases." Brows raised, he grinned again at Sunny who looked amazed but pleased.

Allie was taking in Sunny's exotic looks; her glorious body and long slender legs. She didn't see how any man could resist her, especially a sexy guy like Mac Reilly.

She sat on the edge of the metal chair near Sunny, put the Maltese down and took a sip of her wine. "Are you working on my case then, Miss Alvarez?"

"Please, call me Sunny. And actually

no, though I do know about your being followed."

Sunny wasn't about to tell the movie star she'd heard all about her storm of tears and confessions. No woman would care to have her life exposed to a stranger like that. Besides when Mac confided his business secrets to her he trusted her to keep them.

The captain's bell jingled again and both women glanced inquiringly at Mac.

"That'll be Roddy," he said, heading for the door, leaving them alone.

"It must be awful, being stalked," Sunny said with a little shudder. "I sympathize with you."

"It's certainly not comfortable," Allie said.

From the house they heard Mac's laugh and then Roddy's voice, lighter than Mac's and excited.

Allie looked expectantly toward the two men as they stepped out onto the deck. She'd thought she would be alone with Mac. Now not only was there another woman, there was also another guy, Mac's assistant, who she knew had been taking care of her problem when Mac was in Rome.

Roddy's spiky hair was bleached platinum. He sported a spray-on tan, white linen shorts, a tight red Gaultier T-shirt and Havaianas flip-flops.

He hugged Sunny enthusiastically. "Missed you when you were in Rome," he said, holding her away from him and smiling into her eyes, and Allie suddenly understood that Sunny had been in Rome with Mac. *And* for an entire week.

Sunny said, "Allie, this is Roddy Kruger, Mac's assistant. And good friend."

He was beaming at Allie, revealing the whitest teeth she had ever seen.

"Allie Ray! Oh . . . My . . . God!" He sank to his knees in front of her then took her hand and kissed it reverently. "I've loved you since forever," he said dramatically. "You're even more beautiful in person. I'm thrilled to meet you."

He was so genuinely excited Allie smiled. He certainly wasn't Sunny Alvarez's boyfriend though. "I'm glad to meet you, Roddy." She leaned over and kissed his cheek. "There," she added, "now we really know each other."

He scrambled to his feet, beaming back at her, his hand on his cheek. "I swear I'll never wash that spot again," he said, and they both laughed.

"Oh my God," Roddy said again, as Mac handed him a glass of wine. "Will you just look at that dog. It is a *dog,* isn't it?" He glanced at Allie for confirmation and they all stared at her Maltese.

Looking like a white kitchen mop, Fussy was on her belly edging slowly toward Pirate, who sat frozen to the spot, his one wild eye glued on her.

"Oh my God!" Sunny whispered, repeating Roddy's words. "We should do something."

She meant Allie should do something, but Allie simply stared, and Mac held up a hand, watching silently.

Fussy inched closer. A quiver ran through Pirate's body as she drew nearer. There was nowhere for him to run because the tide was already over the rocks.

Everybody held their breath. Fussy was twelve inches away and still Pirate sat transfixed. The Maltese lifted her head, button eyes peering at him through her fringe. A moment passed in silence. Then she rolled over onto her back, paws waving in the air, and peeked flirtatiously up at him.

"Well, the shameless little hussy," Roddy said, breaking the silence. "Will you just look at her, *flirting* with Pirate."

"And take a look a Pirate," Mac added. There was a bewildered look on Pirate's face as he bent his head and the two dogs sniffed, nose to nose. Then he wagged his tail and flopped down beside her.

"How about that? I believe Pirate's in love," Mac said, beaming at Allie.

Sunny's heart sank. In the two years they had been together Tesoro had never as much as acknowledged Pirate's presence, except for the occasional snarl and swipe at his nose. Now Allie Ray's Maltese had Pirate wrapped around her paws like a dog in love. All her theories about Mac not wanting to marry because their dogs were incompatible went out the window.

"I'll take care of the pizza," Roddy said, heading indoors. "We eating out there?" he called back to Mac.

"We are," Mac said. "Unless you're cold," he added, looking, concerned, at Allie.

"Nothing a sweater couldn't fix," she replied. "I enjoy eating outdoors. Californians don't do it often enough. Take advantage of our climate, I mean. I've learned to enjoy Malibu's mists and winter storms as much as our beautiful sunny days, when you wonder why you would ever want to live anywhere else."

"And would you? Ever want to live anywhere else, I mean?" Sunny asked.

Allie gave her a surprised look. "Sometimes I think about it," she said. "Sometimes, I dream of another life." Then she gave a quick shrug and added briskly, "But

136

this is the life I created for myself. I'm a very lucky woman, I know. Millions of women would want to change places with me. Wouldn't they?"

Her turquoise blue eyes fixed on Sunny, who said, surprised, "Yes, I'm quite sure they would. Though of course there would be no way to replace the real Allie Ray."

"I need a magic wand to wave over my life," Allie said softly, as though she were voicing private thoughts, not meant for another's ears. "All I need is that magic wand to make me disappear."

Mac pushed open the rattling screen doors. "Pizza. Come and get it," he said, as he and Roddy put two enormous pies on the redwood trestle table, plonking down bottles of wine and Pellegrino, as well as a container of hot chili peppers. Roddy added a bunch of paper napkins which blew away in the wind, sending him dancing after them. Mac offered Allie his old dark green cashmere sweater and received a smiling thank-you.

Sunny recognized that sweater. She had bought it for Mac a couple of Christmases ago. Buttoning her orange and pink striped cardigan, she took the chair next to Allie.

When they were all seated and their glasses filled Mac said, "Okay, first we have news, Allie. *Good* news," he added. "Well, let me amend that. *In a way* it's good news. The Sebring has not been seen since Lev has been guarding you."

Her shoulders slumped. "But what about when he leaves? And anyway who is it? It's so scary knowing someone is watching you, living your every moment. It's as though they're stealing your life."

Sunny could see she was genuinely frightened. "You'll be all right now," she said gently. "Mac'll sort it out."

"It'll be okay," Mac told her. "We're still working on it." He was slicing the pizza with the expertise of decades of experience. "Allie, I know you like anchovies. Sunny no. And Roddy just a little. You'll have to scrape them off," he added, putting loaded plates in front of Sunny and Roddy.

Sunny was worried. First the Maltese and Pirate and now the anchovies. Maybe she and Mac were not compatible after all. And Allie was so cute, drowning in Mac's big green sweater, looking like a fragile mermaid ready to swim off into the ocean. Even Roddy was under her spell. From under her lashes, she saw Mac smile at Allie.

"So, when do you leave for Cannes?" he asked.

"In a couple of days. Not that I'm ready for it. In fact I'd rather not go, especially knowing what I know about the movie." She shrugged and bit into her pizza. "But it's my job. I'll show up and I'll do my best for them. After all, that's why they pay me."

Sunny took a bite of the pizza. It still tasted of the anchovies. In her lap the shivering Tesoro whinnied like a mini-pony. She got up, went inside, snatched a chenille throw from the sofa, wrapped the Chihuahua in it and went back out again. The wind was definitely chilly now, and like her dog, she shivered. Nobody, meaning Mac, took any notice. They were talking about Allie's new movie, and about Cannes.

"You should come with me." Allie was looking directly into Mac's eyes. "I could use a friend. And we could talk about the situation with Ron, work out what to do. And about the stalker, and the letters. I simply don't have time before I leave," she added.

"Maybe I will," Mac said.

Sunny's heart sank. She was losing a battle she hadn't even known existed. Scrambling to her feet again, she said, "Sorry, the jet lag is killing me. I have to go. I have an early start tomorrow." Grabbing the whining Tesoro, she dropped quick kisses and called out goodbyes. She was surprised when Allie got to her feet to give her a hug. There was a look of distress on the movie star's familiar face that spelled out her loneliness.

"Listen," Sunny said, suddenly concerned, "if you need someone to talk to, call me. My number's in the book." Their eyes met. "I mean it," she said gently. "Sometimes another woman's opinion can help sort things out."

Allie smiled and hugged her again. "I'm not used to women liking me," she said. "Usually they're jealous."

Sunny smiled, guilty but absolved. Her jealousy had been temporary, even though she was leaving her lover in the company of the famous beauty. Call her crazy, she hoped she was doing the right thing. And anyhow the jet lag was true. She could hardly keep her eyes open.

Mac walked her to the door. He grabbed her shoulder, swung her round to face him. "You okay, honey?" he said.

She pulled the ribbon from her hair and tossed her head, sending her dark curls cascading in a J.Lo gesture that made her look even more beautiful and made Mac smile.

"Hey, babe, come on," he said persuasively. "You're not jealous of Allie, are you? She's just a lost soul, Sunny. It's nothing personal."

"I know," she said, then added with a grin, "Just be careful, okay?"

Still, with Tesoro zipped safely inside her leather jacket, as she roared the Harley along PCH, heading toward Santa Monica, and Marina del Rey, she did wonder if she had made the right move.

CHAPTER 17

The sound of Sunny's bike faded and Roddy held Fussy on his lap, feeding her morsels of pizza. "It's the anchovies she's after," he said, offering her another. Only this time Fussy bit his finger instead.

He glanced aggrieved at Allie, who sighed and said she was sorry and that Fussy had her problems. "She only has one love and that's Ampara, my housekeeper," she said. "And I can't tell you how many times she's bitten Ron."

Mac took the chair next to her. "Speaking of Ron, have you seen him lately?"

"No, I have not seen him. He was due in court last week, something to do with the divorce proceedings, but he never showed. They issued a subpoena but haven't been able to serve it because nobody can find him."

"Think he's avoiding your divorce issues?"

"Of course he is. Why else would he do a

disappearing act? That is unless he's run off with the other woman."

Mac shook his head. "I can reassure you on that. He's definitely not with the other woman. And she doesn't know where he is either."

"Am I allowed to ask how you know that?" Allie said.

"Better not. But I can tell you you don't have to worry on that score."

"This isn't Ron's first extramarital affair. He even hired one as his 'secretary,' and he bought her expensive gifts. The bills were supposed to go to the office but occasionally they would come to me by mistake. He was a generous man," she added drily. "I remember a diamond watch that cost more than the one he'd given me." She smiled. "That hurt. Though to tell you the truth by then I would have been happy with a bunch of daisies."

"You're worth more than that."

She shrugged. "Now I can afford to buy my own diamonds, but you know how it is? It's just not the same."

"And that's how things stand now?"

"That's how things stood as of two weeks ago. Anyway, *where is* Ron?"

"That's exactly what Demarco asked me today."

"Sam Demarco?" Her tone was angry. "He

143

calls himself a friend, but he muscled his way into Ron's good books by persuading him to take it easy. 'Take time off, play with the girls, go on vacation,' he said to him. I put part of the blame for our breaking up on Demarco."

"So Ron seems to have left town with no forwarding address. And since the mistress is eliminated, can you think of any other reason he might want to disappear?"

She shrugged. "Business, you mean? I don't know anything about that."

"If Ron were to die, wouldn't you inherit?"

Allie's long blond hair shifted in the breeze. In a gesture graceful as a ballet dancer's she wafted it out of her eyes.

"All I know is what I'm supposed to get in the prenup."

"Plus whatever else your attorneys can negotiate over and beyond that. Given Ron's rep with women."

"Exactly."

Roddy threw a glance at Mac. It was time for him to leave. "Okay," he said. "I'll call it a night." He took a business card from his wallet and handed it to Allie. "Anything you need, just call. I'll be there."

She smiled as he bent to kiss her cheek. "Thank you again, Roddy. It was lovely meeting you."

"My honor." He backed away, bowing like a commoner before royalty and almost crashing into the plate-glass door, making them laugh.

CHAPTER 18

"He's so nice," Allie said.

"He is," Mac agreed.

Silence fell. The ocean roared in the background as the waves hit the rocks and a final lonely pelican raced home in the dark, wings whirring overhead.

"And so is Sunny," Allie said into the silence. "She's beautiful."

"She is," Mac agreed again.

Their eyes linked and the silence grew deeper. A world of possibilities stretched between them.

It was Mac who finally broke the spell, pouring wine into their glasses, throwing some leftover pizza to Pirate who wolfed it eagerly while Fussy watched from Allie's knee, tossing her fringe like a mini–movie star posing for the cameras.

"So how did you find Pirate?" Allie said.

"I saved him from a fate worse than death. Actually" — Mac glanced at Allie, who was sipping her wine, watching him — "I saved him *from* death."

"Tell me," she said.

"Well, as they say in the old potboilers, 'it was a dark and stormy night.' I was driving over Malibu Canyon when I saw this body in the road. A frowsy little mutt, just lying there, his head all bloody and one leg crushed so bad it was hanging off. I bent down to stroke his mangy fur, thinking what a way to die, hit by some speedster on a canyon road. But then the mutt opened an eye and looked at me. I'll tell you it gave me quite a shock since I'd imagined he was dead. But there was a kinda hopeful look in that eye." Mac shrugged. "What could I do? Of course I took off my shirt, wrapped him in it, put him in the back of the car and drove to the emergency vet in Santa Monica. I paid the necessary, told them to do their best and went on my way, glad that I'd at least given the mutt a chance. I went out of town and I didn't pick up the vet's message until a week later. The vet said, 'We had to amputate your dog's left hind leg and re-move one eye. He's on the mend. You can come and get him. He's ready to go home.'

" 'What d'ya mean, get him?' I said when I called him back. 'He's not *my* dog. I just scooped him up off the canyon blacktop, gave him another shot at life.'

"And this is what the vet told me: 'There's an old Chinese saying, that if you save a life you are responsible for that soul forever. He's all yours Mr. Reilly. So come and get him.'"

Mac shook his head, remembering. "So of course I did. And despite his sorry state that dog greeted me as though we had known each other forever. And believe me, now it seems as though we have. I wouldn't be without him."

Allie reached for his hand "And tell me, Mac Reilly," she said, surprising him. "Would you ever be without Sunny?"

Mac took a deep breath. He was looking at one of the most beautiful, the most famous, the most desirable women in the world. Temptation hovered between them, soft as silk.

"I could never be without Sunny," he said quietly.

Allie sighed. Rejection was not sweet. "I like your honesty," she said, gathering her bag and her dog. "It's getting late. I must be on my way."

He walked with her to her car.

"I asked you to come to Cannes with me," she said. "You told me 'maybe.' Is that 'maybe' a promise?"

Mac put his hands on her shoulders.

"A lot depends on your answer," she said. "More than you'll ever know."

He kissed her gently on each velvety peach cheek. "Maybe," he said again.

She turned for the door, then turned back again. "Mac," she said. "Rich men don't just go missing, do they?"

He shook his head. "Particularly ones who owe their wives."

She nodded. "That's what I thought. You will find Ron for me, though?"

"I'll do my best."

She smiled, that heavenly smile that had made moviegoers fall in love with her. "That's all I can ask," she said.

As Allie drove away Mac took Pirate and went back into their little house and closed the door. It was sad, Mac thought, that golden opportunities sometimes missed their mark. It just went to show you, timing was everything.

He went out onto the deck, listening to the surge of the tide and thinking about the missing husband, and about the weird letters and the real stalker. He did not like the scenario. Not one bit.

He dialed Lev Orenstein's number and discussed it with him. "Allie won't bring the police in," he said, "so it's up to us."

"I'll keep her covered," Lev replied, "and keep my eye out for the crazy guy. But the rest is up to you."

Worried, Mac knew he was right. Now, he had three people anxious to find Ron Perrin. His wife, who wanted a divorce; his mistress, who wanted marriage; and his business partner, who wanted his friend back, and possible absolution in case of any financial misdoings. RP was an interesting man.

Mac's thoughts returned to Sunny. Hadn't he said in front of the love of his life that he might consider going off to the South of France with a world-famous beauty? Just him and her? He was lucky she was still speaking to him. It was late and he didn't want to call her because of her jet lag. He knew she was probably already sleeping.

He sent her a text message instead. "HOPE U SLEPT OFF THE JET LAG. MISS U LIKE CRAZY. THE MARINA'S NOT AS FAR AS ROME. ANY CHANCE OF U INVITING ME OVER FOR DINNER TOMORROW NIGHT? TAMALES WOULD BE GOOD. AND NO, I'M NOT GOING TO CANNES WITH THE GORGEOUS MOVIE STAR. I WAS JUST BEING 'POLITE.' LOVE U, BABE."

Her return message was waiting for him

the next morning.

"MY PLACE. SEVEN O'CLOCK. FORGET THE TAMALES. I CAN COOK OTHER THINGS YOU KNOW."

CHAPTER 19

Sunny had it all planned. Mac was an old-fashioned guy at heart. And she decided to start with butternut squash soup that tasted like sweet velvet scattered with crumbled almond cookies and cinnamon. No one would guess how easy it was to make. Then his favorite chicken cacciatore, which, like the soup, could be made in advance, so she wouldn't be racing around the kitchen at the last minute. For dessert, her own personal favorite, a light-as-a-feather lemon cake from Mrs. Gooch's on Melrose, served with low-fat Dreyer's chocolate fudge ice cream.

She'd thrown in the low-fat as a concession to her own conscience, though when you were romancing a guy you aimed to please. Plus she had bought a good champagne, Henriot, a lesser-known marc but famous in France, and a very splurgy bottle of Bordeaux, a Ducru-Beaucaillou that had cost far too much and which practically had to be opened the day before in order for it to breathe.

Anyhow, that was her man-pleaser menu, everything carefully planned, the table beautifully set, in honor of her man, with very masculine graphite-gray table mats, plain square white dishes and streamlined silverware. Not a curl or a flourish in sight. Except for a crystal vase of Sterling roses in that off-beat grayish lilac color.

And she was at her most girlish in a sweet-but-naughty silky little dress in the same color as the roses, spaghetti-strapped and with a touch of lace at the dipping neckline. Towering-heeled sandals — just a sliver of silver — Jimmy Choo of course and four years old but she loved them to pieces.

Her dark hair was brushed loose and arranged to fall sexily over one eye, and she wore her favorite Dior Rouge lipstick — the evening one — and of course, the Mitsouko perfume. The fire (gas logs only, but still effective) sparkled in the grate even though the night was mild. Neil Young was singing "Harvest Moon," very softly, and votive candles glimmered in the shadows.

For once the place was tidy, because knowing how Mac hated her chaotic housekeeping, Sunny had shoved everything into cupboards and drawers. Of course this meant she wouldn't be able to find anything for weeks, but tonight would be worth it.

She took a chocolate-covered fig from the refrigerator and bit into it, wondering, at the same time, why she had no self-control when it came to sweet things, when she knew perfectly well that her butt would get bigger by the minute. She sighed, and told herself that after all life was made up of small pleasures. Sweet ones especially. And besides, those figs were heaven.

Looking round, she thought somehow her home didn't seem to lend itself to tidiness. The Shabby Chic butter-colored linen sofas had permanent dents in them where she sprawled with Tesoro on her lap watching TV, or just catching up on paperwork. The glass desk with the brushed nickel trestle legs, a recent purchase from Williams-Sonoma Home in an attempt to add a little modern chic, was perpetually covered in papers as well as Tesoro's paw marks. The Santa Fe–style rug her mother had given her didn't really go with the rest of her things, but its vivid colors reminded her of home, though anyhow it kept skidding across the floor because she'd never gotten around to finding one of those pads to anchor it.

The wooden coffee table in front of the fireplace was one of her flea market finds that she was hoping would turn out to be an authentic antique, and was now covered in assorted vases filled with flowers picked up at Gelson's supermarket, and made more beautiful with the addition of smooth gray pebbles to hold them in place. Photos lined another wall; of her *abuela,* her family and her friends. But in the place of honor was a steel-framed photograph of her first Harley and, right next to it, the other love of her youth, her horse, a chestnut filly by the name of Jupiter. Sunny's heart still filled with emotion when she looked at her. Jupiter had shared her life for fourteen exciting years, and she still missed her. She missed the old Harley too, but for other reasons, mostly because it reminded her of her youth and all the fun she'd had misspending it.

Only the huge many-branched silver candelabra on the coffee table was a genuine antique, given to her by her grandmother. Her sister, Summer, had its twin, and it meant more to Sunny than anything else in her home. She lit the candles now, faintly scented with beeswax and honey, and let the aroma drift slowly through the room.

Books were piled on the shelves that covered the rear wall, but best of all, and the reason she had bought the condo, was the wall of windows facing out onto a perfect view of the marina, a-sparkle with lights. It was where she stood, every evening, watching the sun set and then dusk fall and the lights gradually come on. It was like Christmas Eve every night — though at real Christmastime it was even better, because then the boats were trimmed and decorated and lit like colorful chandeliers. She and Mac would watch the parade go by from her deck, shivering in the December night, hot toddies in hand to stave off the cold.

The place did clean up nicely, though, she thought, glancing round approvingly. And of course the one room that was always immaculate was her kitchen. She couldn't abide mess in there. She had to know where everything was: her knives, her spices, her dishes. She was a cook and tonight she would tempt Mac Reilly all over again with her dinner and the sexy dress. He would forget about Allie Ray for a while, though somehow she knew she could not.

Loneliness was the thing that had struck her most about Allie. Yet with all her success, with all the people she knew, and everyone wanting to know her, how could the woman be lonely?

But this was her night, not Allie's. And if there was ever a setup for seduction, this was it. Sunny shrugged. Hey, a girl could only try, right? She might even have to pop the question herself tonight if Mac didn't come through.

When Mac walked in her door, though, her heart did triple turns and her knees turned to Jell-O. He was just such a heartbreaker, even though he claimed to be quite ordinary.

"Ordinary" he most certainly was not. She loved his lean, muscular body, she wanted to run her fingers through his dark hair, and when his blue eyes looked deeply into hers, she could swear he saw her soul. And his hands were beautiful, thin and kind of bony, firm when they needed to be, gentle and tender when you wanted it most. "Romantic hands," she would call them. Yes, definitely "romantic."

They stood looking at each other for a long minute. Then Sunny stepped into his arms. She breathed in the familiar scent of his skin, smelled his aftershave, his freshly washed hair, his old cashmere sweater that had a tang of Pirate about it — and now also a hint of another woman's perfume. Chanel, she thought.

They staggered backward into the living room, still entwined, lips still clinging.

"Don't ever leave me," he murmured, sending a thrill through her.

Then she glanced down and saw Tesoro

standing next to them, her cute little Chihuahua face upturned.

"Uh-uh," she said, suddenly apprehensive. Mac removed his lips from her hair to look. Tesoro bent her head slightly to sniff his shoes. Then, daintily of course, because she was a very well-bred dog, she threw up all over them.

Sunny yelled and Mac groaned, and Tesoro retreated to the sofa, where she propped one leg heavenward and proceeded nonchalantly to lick her more intimate parts.

"That dog is uncivilized," Mac said as Sunny ran to fetch paper towels. "Pirate would never do anything like that, and he's just a mutt from the L.A. streets, for God's sakes. What is it with these aristocrats? They think they don't need manners?"

"So much for good breeding," Sunny agreed humbly.

Tesoro had succeeded in bringing the romantic tone of the evening down a notch or two, so Sunny quickly poured the champagne, hoping it might soothe Mac's ruffled nerves. It seemed to do the trick and they sat, holding hands and talking about Demarco and the Perrins.

"I'm not sure exactly what Demarco's role is yet," Mac said. "If he really is RP's full partner I'll bet it was a recent thing. And if a takeover is what they were planning he'll need his friend Perrin around. So I don't think he has anything to do with his disappearance. Especially as he's asked me to find him. As for Allie, I'm not sure she's happy to be rid of him. Anyhow, now all three of them are looking for RP and it's up to us to find him."

Sunny took a sip of champagne, eyeing him over the top of the glass. "You getting paid for all this?"

"Financial arrangements are being made with Demarco. Allie Ray is paying more than generously to have me on call twenty-four/seven, not only to find her missing husband but also to nail the real stalker, who might have become a major threat. As for Marisa, well I guess she just goes along for the ride."

"Hah!" Sunny sniffed.

"Hah — what?"

"You'll never get rich."

He grinned at her. "But just think how I'm enjoying myself."

Mac changed the subject, admiring the flowers and her dress. He even commented on how neat the place was, then he went onto the terrace and took in the view, while Sunny turned the music up a little and served the soup. As she had hoped, it knocked his socks off, taste-wise, and they were back on romantic course once again. She smiled happily. She definitely knew the way to a man's heart.

The special Bordeaux with the main course mellowed Mac even further. Forgetting about dessert, they took their still-full glasses and, with Mac's arm around her shoulders, hers around his waist, wandered into the bedroom, already artfully arranged with piled up pillows and the old-but-good Frette sheets her mom had given her, as well as a soft cashmere throw. Oh all right, she'd admit it, it was only cashmere *blend.* Still she knew it would warm their naked bodies in its soft folds in the event they felt chilled. Not much chance of that though.

The lamps threw soft golden pools of light, the music was to make love by, and her bed — their destiny — awaited.

She allowed Mac to undress her. Not that there was much to take off, but she loved the way he lingered over the important bits. Then he picked her up and lay her on the bed.

They were both being transported to heaven, or at least that part of paradise you can achieve while still on this earth, when she heard a menacing growl. Opening her eyes she was just in time to see Tesoro launch herself at Mac. All claws extended.

Mac's yell was not the one of passion she had expected. He flung himself upright, cursing the dog, who gave him a contemptuous look then jumped off the bed and stalked out.

Sunny ran to get the Bactine, dabbing it onto the scratch marks that furrowed neatly in four places down his back. He yelled again as it stung.

"Think of the neighbors," she reminded him. "They'll say it's domestic abuse and call the cops."

"Tell them to call Wildlife Control instead," he said, still hurting.

Sunny thought how odd life was. Here was a man who could dodge bullets and killers with impunity, but put him up against a cute little Chihuahua and he was no match. She sighed. So much for her romantic evening.

"Tell you what," she said brightly. "How about a grappa?"

"Sure." Mac was already flinging on his clothes. "At my place. It's safer there."

And so they left poor Tesoro to lick her paws and consider repentance.

Malibu worked its old charm though, and soon they were in Mac's bed. Sunny was carefully avoiding putting her hands on his wounds, and Pirate was a discreet bump in his basket by the window. All was sweetness and light again. And oh God, it was good, she thought, as she fell asleep wrapped in Mac's arms.

No proposal tonight, though. What could a girl do when her love life was sabotaged by her own Chihuahua?

CHAPTER 20

Sunny departed early the next morning to pick up her strained relationship with her dog, so Mac decided to wander down the road to Coogies for coffee and pancakes. *Blueberry* pancakes, he thought, whistling for Pirate. Plus enough coffee to float a battleship. His TV show was on hold, with no decision yet made and for once his time was his own.

Feeling good, he waved to the guard at the gate and was strolling out onto PCH when he noticed the car parked a little to the left in the sandy area just off the highway. An old Cadillac. Deep burgundy color. Dusty. It looked as though it had maybe been dumped there, but then he saw a man sitting in the driver's seat. Thin-faced, olive-skinned with a beard that looked like it might be a disguise, and of course, dark glasses.

It flashed through his mind that he'd seen that face, that *man* before, strolling slowly back and forth on the beach, along the surf line.

Calling Pirate to heel Mac walked over to the car. The windows were down and he stuck his head in. "Hi," he said. "What's your problem, buddy?"

The man looked silently at him. The beard was real. He heaved a deep sigh. "I might have known you'd catch on to me," he said. "Of course I know who *you* are." He took a card from the dash and handed it to Mac. "Sandy Lipski," he said. "Private investigator."

Some private eye, Mac thought. He couldn't even conduct a surveillance. He stuck out like a sore thumb.

"We need to talk," Lipski said.

"What about?" Mac said.

"Ronald Perrin."

Mac was surprised but he didn't show it. Of course he could have just gotten in Lipski's car and said okay so talk but he preferred to see people in their own habitats. He found it gave him a clue as to who they *really* were.

"Okay, so we'll go to your office," he said. "I'll get my car and follow you."

Who Lipski was became obvious when Mac saw his office. Small, on a Santa Monica side street. Tired file cabinets; a battered

164

desk with an old leather high-back chair for Lipski and a dingy airport-style chair for his client. Grimy windows; a Sparkletts water-cooler with a stack of Dixie cups; a scratched wooden floor and torn screens. No AC but that was usual at the beach where everybody swore you didn't need it because of the sea breezes. It wasn't strictly true, especially today in Lipski's office, but Mac steeled himself against the nicotine-tainted air and got down to business.

"Before we get to Perrin, first tell me who you are," he said, taking the airport chair and making himself approximately comfortable.

Lipski's story was all too familiar: an ex-cop drummed out of the force for drugs, he'd drifted into a seedy underworld life. A few years down the road he'd found a 12-step program, gotten a life back and taken up the investigating business.

"Nothing fancy," he said, lighting up a Marlboro. "Just spying on fiancés for women who want to know if their future husbands have a past. Or else 'a present' they don't know about — like for instance another wife. Following erring husbands to motels. That kinda thing."

He took a long drag on the cigarette. The ash dropped down his shirtfront. Mac waited. His fly-on-the-wall technique never let him down.

"I met Ruby Pearl in rehab," Lipski said. "We kept each other going, encouraged each other, y'know. She was cute, blond, full of life. She always had me laughing. I really fell for her. Then she met another guy. She told me he was really rich. She'd met him on a chat room on the Internet. She started seeing him and soon dropped me. He gave her presents — a diamond watch, and like that. Stuff I could never have afforded even if I'd worked twenty-four/seven." He shrugged miserably. "How could I compete?"

This was the second time Mac had heard about a diamond watch.

"But I still loved her," Lipski said. "Y'know how it is? Sometimes there's a woman you can never get out of your system? I would have taken her back in a heartbeat. But then she disappeared. Just like they say, 'into thin air.' I didn't know then who the guy was she had been seeing, only that he was rich.

"Months went by. The police put her in the missing persons file. And you know what that means. Nobody was even looking for her. I couldn't rest, I had to know what happened. A girl like that doesn't just disappear. Somebody had to have something to do with it. And then I found out from an old friend in the LAPD it was Ron Perrin she had been seeing. He'd even given her a job as his secretary.

"I know she's dead," Lipski added quietly. "It's that old gut feeling. Y'know how it is? It just doesn't sit right."

"I understand," Mac said. He was thinking that Ron Perrin was in deep trouble. No wonder he'd done a disappearing act.

"It was me that night in Perrin's house," Lipski said, startling him.

"How'd you get in?"

Lipski shrugged. "Sometimes it's just who you know. An employee with a grudge, a stolen key . . . y'know how it goes. I made it my business to find out who . . . why . . . Anyhow, I got the key and the alarm code."

"Jesus," Mac said. "It was that easy?"

Lipski shook his head. "No. I'm that clever."

Mac laughed. There was more to this man than met the eye.

"I saw Perrin drive out the gate. I thought the house was empty. It was quite a shock when the woman came downstairs. And with a gun, for chrissakes. I didn't want to hurt her — and hey, I didn't want to get hurt either and then shoved in jail for burglarizing Perrin's place. I just wanted to get away. So

I pushed her to the floor and took off as fast as I could. I figured she'd be too shocked and frightened to come after me with the gun." He shrugged. "I was right. I got away easy."

"So what were you looking for at Perrin's house anyway?"

"Evidence," Lipski said simply. "He killed my girlfriend. There has to be something there, doesn't there?"

Lipski's weary eyes, deep-set like two black coals in his thin bearded face, looked directly into Mac's. "You have to help me, Mr. Reilly," he said. "I'm beggin' you. Please."

CHAPTER 21

Loneliness had Allie in its grip, that kind of strangling sensation when, staring out the window, she felt that the rest of the world was out having a good time, while she was trapped in a prison of her own making. It was exactly the way Sunday afternoons had felt back in that small Texas town when she was a teenager and life was rushing past her and she knew she would never get to participate in it.

She thought about Sunny Alvarez. Now there was a woman who would never allow life to pass her by. Sunny was a driver in the fast lane on her Harley chopper, her black hair crammed under a silver helmet, like the god Hermes in full flight. Allie remembered Sunny's eyes looking directly into hers, that evening at Mac's place, and her saying, "Listen, if you need someone to talk to, call me. . . . Sometimes another woman's opinion can help sort things out."

Of course Sunny had not meant it. Sunny had a life and she was enjoying it too much to take time out to listen to Allie's selfish tirade of woes. And to any other woman Allie knew her complaints must indeed sound trivial. After all, she was a woman who supposedly had everything.

Still, remembering the concerned look in Sunny's eyes, her hand hovered over the BlackBerry. She called Inquiries and got her number. After all, she had nothing to lose but her dignity. She punched it in.

Sunny answered immediately. "Hi, Allie," she said, sounding surprised. "I'm glad you called."

"Really?" Allie was also surprised.

"Of course *really*. Listen, you want a cup of coffee? We could meet at Starbucks. . . . Uh-uh, wait a minute, in a rash moment I forgot you can't do that kind of thing, go to a Starbucks and not get mobbed, I mean. So why don't you come over here instead? Then we can talk."

"You're sure I'm not interrupting your day?" Allie said cautiously.

"Of course you're not," Sunny lied. She gave her the address and said she was putting the coffee on to brew right that minute. "You like espresso?"

"Love it," Allie said, casting aside any worries about caffeine.

Sunny immediately called the potential client with whom she had a meeting and told him she'd been unexpectedly delayed and would get back to him later. So what if it cost her a job? Allie was a lonely woman who needed to talk and she had promised she would be available.

Half an hour later Allie walked in Sunny's front door and stood looking round, taking in the offhand furnishings, the photos of the Harley and the horse and the general chaos.

"Sorry about the mess," Sunny said. "Somehow it's always like this, I don't know why."

Allie saw that Sunny was wearing jeans with the shirt that said LOVE IS ALL YOU NEED, in sparkles. "Do you believe that?" she asked, perching on the edge of the sofa.

Sunny came and shifted a pile of papers from behind her. "Make yourself comfortable," she said. "And no, I don't believe it's all you need, but it's a nice thought."

Allie got up and followed her into the kitchen that she saw was surprisingly neat. "What else do you believe in then?" she asked, taking a seat on a bar stool at the center island, while Sunny poured espresso into two small dark green French café cups.

Sunny thought about it. "Honesty. Loyalty. . . ." Then she grinned. "A good chopper, fun . . ."

"And love."

"That goes along with the rest." She passed Allie the bowl of sugar and another of sweeteners. "So where are you at, Allie Ray?" she said, leaning her elbows on the black granite island, opposite her, and taking a sip of the deep rich coffee.

"It's not so much where I'm at, I suppose." Allie shrugged. "It's more about the loneliness. It's one of the worst feelings in the world. I was remembering, just before I called you, that this was exactly the way I used to feel when I was a teenager, doomed to be locked into that stifling gray life forever. And now, after all my success, I seem somehow to be right back where I came from."

"But you escaped then," Sunny said.

Allie raised her coffee cup in salute. "I did," she agreed. "I picked myself up and got out of there. I left no trace. I never wanted anybody to find me and drag me back again."

"Kicking and screaming," Sunny said.

"Something like that. Anyhow, I went to Vegas. Where else would a girl who looked like me end up? I worked two jobs, cocktail waitress and like that, skipping from casino to casino. Then I got a job as hostess at a steak house. That's where I met my first husband."

Her eyes met Sunny's. "He was a nice man, y'know. He treated me like a lady — and believe me, by then I was far from feeling like a lady. But I won't go into that now."

"I understand," Sunny said, wondering if she did.

"Those awful tight revealing costumes we had to wear." Allie's eyes were half-closed, remembering. "After that, the steak house felt like a slice of heaven. Safe, you know?"

Sunny nodded, and she went on. "He wasn't really rich, the man I met," she said, "but to me he was. I'd never known anyone who could take you out to dinner at a good restaurant and say, Have whatever y'want, babe. And he bought me presents, flowers, and a bracelet. He was older and I leaned on him. I needed him I guess. So when he asked me to marry him, I did. And I ended up trapped again.

"Oh, I was a good wife. He asked me to give up my job. I stayed home, fixed dinner, watched TV with him. He was in his fifties and I was eighteen, trying to seem older and lying that I was twenty-one. Nobody cared except me. I guess I just wanted to belong."

"And did you?" Sunny poured more coffee. She pushed the sugar bowl toward Allie and went to sit on the bar stool next to her.

"Not enough." Allie sighed. "I left him, went to another city . . . New York. We divorced a year later. I've never heard from him since."

"Married at eighteen, divorced by twenty," Sunny said, astonished.

"I wasn't lucky enough to get an education like you." Allie was looking at her with that wistful expression in her eyes again. "There was no money for college and anyhow my father wouldn't have allowed it. I was his 'whipping boy' . . . me and my mom both." She shuddered, remembering the ugliness of that life.

"I was twenty-two and working as a hostess again, in a fancy brasserie in Manhattan, and living in a cockroach-infested walk-up in Greenwich Village." She shrugged. "It was like history repeating itself, only I was older and a little bit wiser, and I knew I was beautiful now. I wasn't about to throw that asset away again, and when I met the man who became my second husband I made sure he was rich."

She turned her head to look at Sunny. "It didn't make any difference. It ended up the same way. Only this time he was nice about it and settled a little money on me, enough to fly to L.A. and get a small apartment and try to become an actress. Like a thousand other girls just like me."

She shrugged. "I was like all the rest of the pretty girls in town, creating a confident image when I'd never really acted in my life. I went to acting classes, learned my craft, went on auditions, played a little theater here and there, small roles, nothing important. After a while, a couple of years, I was doing okay, getting jobs, minor roles in movies, and TV . . . but never really breaking through. And still fending off the Hollywood Romeos that came courting. And then, a couple of years later, I met Ron."

She stopped her confessional monologue and looked at Sunny. "You'll know what I mean when I say it was meant to be. The thing between me and Ron. It was like we had known each other in some other life."

"You were head over heels," Sunny said, and Allie laughed.

"And so was he. We met at a New Year's Eve party in Aspen. We left them all to their Happy New Years and champagne and went back to Ron's log cabin in the woods with the snow piled outside the door. I ruined my expensive shoes that I'd gone without lunches to afford but I didn't care. I just wanted to be with him. And I was lucky, he wanted to be with me.

"Ron was a superb skier," Allie said as Sunny brewed more coffee. "His body's compact but it's hard from all the weight lifting and his legs are strong. In his black Bogner ski suit he skimmed down those mountains, looking like a dark bird of prey. I wasn't nearly as good, in my fur-hooded movie-actress ice-white suit that he made me promise never to wear again.

"I remember him saying, 'Don't you realize that if you're dressed in white and anything went wrong, an accident, a fall, or God forbid an avalanche, the rescue team would never be able to see you in all that snow?' The next day he took me to the store and bought me a bright red suit and boots to match. Then he took me to a jewelers and bought me a diamond ring. An eternity band.

" 'It's only the beginning but I feel in my bones, in my heart, this is for eternity,' he told me, holding me by my scarlet shoulders and looking into my eyes."

She looked at Sunny. "It was then I felt our souls connect," she said. "And I knew he was right. We were meant for each other.

"We took the cable car to the top of Ajax Mountain where we celebrated over a mug of hot chocolate, with many soft and gentle kisses. Few people realize that at heart, Ron is a very tender man, because he never allows that side of him to show. He was trained in business and the boardrooms of the world to be poker-faced, unemotional, a hard man who never gives up.

"Ron got me my first really big role, starring in the movie that made my name. I was already well known, of course, but the sexy, glamorous *Good Heavens, Miss Mary,* established me in the kind of fresh comedic romantic role that became my trademark. That role 'branded' me and made my fortune. After ten years I was an overnight success.

"We had celebrated with a trip to Europe, stopping in Paris to shop, and in Saint-Tropez to lie on the beach and drink rosé wine over long lunches. Life was sweet then.

"I owe a lot to Ron," she said. "And that's why I'll never forget him."

"And now you've lost him," Sunny said gently.

Allie's eyes met hers. "Have you ever been heartbroken?" she asked.

Sunny considered. "Once or twice I thought I was. But now I know, never in the way you mean."

"So now you understand why I'm unhappy."

"Because you're lonely," Sunny said, putting her arm around Allie's shoulder. "And you are hurting."

"It's just that I thought I had it made, all my ducks in a row — and now it's all falling apart," Allie said.

She wasn't crying but Sunny could see the held back tears glittering silver in her blue eyes.

"Thank you for telling me," she said gently. "I'll never betray your confidence."

"I'm sorry I used your shoulder to cry on," Allie said wistfully.

Sunny leaned across and kissed her cheek. "That's what girlfriends are for," she said.

But she thought Allie seemed almost embarrassed when she left, as if she had revealed too much of herself. Sunny only hoped that spilling it all out had helped her in some way.

"Call me again, let's get together soon," she said when they kissed goodbye. And though Allie promised she would, somehow Sunny knew she would not.

Minutes after Allie had left, Mac called. Sunny was dying to tell him all about Allie's visit, but he cut her off.

"Tell me later," he said, sounding urgent. "I've got something more important to tell you."

And then he told her the Lipski story. And

about RP buying an expensive diamond watch as a gift for a woman who was supposed to be his secretary and who was now missing, and that Lipski believed she had been murdered.

"Interesting," Sunny said, thinking of poor Allie, who despite everything, still wanted to believe her husband was a good guy.

"What's even more interesting is I'm on my way to break into Perrin's house myself, just the way Lipski did."

"Why?" she demanded, shocked.

"To look for evidence, of course."

"You can't do that. You'll be committing a crime —"

"No I won't," Mac said calmly. "I'm not breaking in. I have the keys."

"*Technically* it's a crime . . ."

Mac was laughing. "I'll let you know how it goes."

"No. Mac, wait. I'm coming right over. Promise me you'll wait."

"Okay, I promise," he agreed.

CHAPTER 22

As usual the traffic was hell. Stalled off Surfrider Beach on the throbbing bike, Sunny thought that if you were not in a hurry sometimes driving PCH could be quite a turn-on, what with all those bronzed muscular young surfers stripping off their skintight wet suits behind parked SUVs, or else simply gift-wrapped in skimpy towels.

But she had no time for such erotic thoughts now. Immediately after she'd spoken to Mac, she had called Roddy to enlist his aid in stopping Mac from breaking and entering, but Roddy was in Cape Cod for a long weekend. So now it was up to her to stop him.

You didn't need to be an expert to know that housebreaking was a criminal offense. And besides, she had that gut feeling that not one of the people involved — not Marisa, not Demarco, not even Ruby Pearl, and especially not Ron Perrin — was worth it. It was up to her to use all her arts of female persuasion to stop Mac from making a fool of himself, and maybe ending up in handcuffs, photographed by the paparazzi, disheveled and looking guilty with a two-day growth of beard, en route to the Malibu courthouse.

The traffic unraveled and she coasted to the Colony, waving to the guard as she drove in. She didn't bother to ring the unmusical captain's bell; as usual Mac's door was unlocked.

There was one of those red sunsets going on where the sun looks like the ball of fire it really is, painting the neon blue sky with a coral and orange glow. Pirate glanced up, no doubt checking to see if the dreaded Tesoro was with her. Satisfied she was not, he gave Sunny a welcoming grin then went back to his snooze.

Mac got up to give her a little more than a mere welcoming grin. "Hello, fellow housebreaker," he said, kissing her soundly.

"What d'you mean?" she gasped, coming up for air. *Fellow housebreaker?*

"Okay, so technically, it's not housebreaking. I have the key."

"Where'd you get that?"

"Lipski. I have to give him credit, he even got the alarm code."

"Jesus!" She dropped onto the metal lounger and felt it sink a little even under her modest weight. "You need new chairs," she reminded him while she thought of it.

"Okay. Now listen. So far no police are involved. There's no way anyone will even know we're in the house."

"We?"

"We," he said firmly. "I'll need help. Anyway, there's nothing for the police to be involved in yet. Perrin's just a rich man who's taken off by himself somewhere. Rich men do that all the time, y'know."

"I didn't." She gave him a withering look.

The wind blew her long hair and Mac leaned over and gently brushed it back again. The last of the sun's glow lit her face and he bent again to kiss her. "I love you, Sonora Sky Coto de Alvarez," he murmured, dropping his lips from her hair into that favorite place of his where the pulse beat at the base of her throat.

"You're just smooth-talking me," she said warily, but she was softening. She gave him a smile.

"I am," he agreed. "But maybe I'll save that for later."

"After the break-in."

"Right. Meanwhile, why has Perrin disappeared? Is it because the FBI are after him for fraud? Or perhaps for money laundering? And maybe he killed Ruby Pearl who knew too much? Who knows, maybe Marisa and Allie are next on his list."

"Jesus," Sunny said again. She was a little scared by this time.

Mac took a seat on the end of the lounger. Leaning forward, hands clasped between his knees, he gazed earnestly at her. "I don't know if Perrin is really a killer, but I do believe Lipski is right about one thing. There has to be something, some evidence, a clue at least in that house. Remember I told you RP was shredding documents that morning? I'm hoping he didn't get around to all of them. I'm asking you to help because it'll be faster having two of us go through the place." He looked at her, one dark eyebrow raised. "So? Are you with me or not?"

Sunny sighed. It was a foregone conclusion.

They waited until it was dark, then, leaving Pirate at home this time, they walked down the empty beach. The tide was receding and Sunny worried out loud that their footprints in the wet sand would lead incriminatingly to the Perrin house but Mac said to stop being Sherlock Holmes, nobody was looking.

She hurried up the beach steps after him,

glancing nervously over her shoulder as he opened the side door. The alarm pinged and, twisting her hands in an agony of fear that they would be discovered, she waited until he'd switched it off.

"Oh my God," she said, shivering. "Tell me, Mac Reilly, why am I doing this? Am I crazy or what?"

"Crazy," he agreed. He was standing by the window almost exactly in the place he had been that night when Marisa had pointed the gun at him. "There's got to be something," he whispered to Sunny. "Some evidence of wrongdoing, something Perrin forgot."

"Okay." She had stopped shivering but was still distinctly nervous.

In an alcove of the enormous living room was a computer. Mac went over and switched it on. Its start page was an Internet chat room with pictures of young women with short bios and even shorter skirts, and messages urging you to get in touch.

Mac whistled. So Perrin really was into chat rooms. Marisa had admitted that's where she'd met him, and Lipski had said the same about Ruby. He guessed that for Perrin it was anonymous and better than going the old-fashioned route via a Holly-wood madam who might one day get arrested and spill the beans about your sexual activities and preferences.

The two of them began to go systematically through the house. Mac took the upstairs, Sunny down, grumbling when they found nothing. There were no files and the shredder was gone.

Sunny was in the kitchen when she heard the steel gate leading to the street sliding back.

"Oh my God, Mac," she called in a piercing whisper. *"Somebody's coming."*

Mac raced down the stairs. He grabbed her hand and rushed her through the kitchen door into the garage. He stood for a second until his eyes adjusted to the darkness. He'd wondered what car Perrin would drive and now he knew. A Hummer. Silver. RP's color, it seemed, with windows tinted black so you could not see inside. Next to it was a silver Porsche. Plus a red Harley. And in a corner, the pièce de résistance, an original Indian motorcycle.

"Oh — My — God." Sunny stared reverently at the bike.

Mac was peering through a crack in the door to the kitchen. He could make out someone moving around. A man. He had not put on the lights, so like them, he obviously didn't want it known he was there. It definitely was not Lipski though. The man turned and looked his way.

Mac grabbed Sunny, opened the Hummer's back door and shoved her inside. "Get down on the floor," he said. "Don't say a word, whatever happens."

"Oh my God," she said again, but this time with a panicked wobble in her voice. "Is it the FBI?"

Mac climbed into the front seat. He got down on the floor and stretched out as best he could. Then he locked the doors from the inside.

"It's Demarco."

"*Ohh . . .*" Her agonized moan almost made him laugh.

His shoulders were cramped and his head was stuck under the steering wheel. Above him dangled not just the Hummer's key but a whole set of keys and an electronic opener. He reached up and slipped them into his pocket.

"It's hot as hell," Sunny whispered. "I think I'm dying."

"No you're not," he said confidently. But lying on the floor of the squat Hummer was like being in a hearse without the benefit of a coffin. He was sweating.

In the back of the car, Sunny gripped the side pockets, hauling herself into a more

186

comfortable position. Her fingers encountered a piece of paper. She took it and stuffed it in the pocket of her shorts.

Demarco was standing in the garage, staring around, a baffled look on his face. He was still wearing a pin-striped suit. Not the usual uniform for housebreaking. And unlike most people's garages, this one was not full of stored junk. It was clean as a whistle. There was nowhere to hide anything. Except in the two cars.

Mac ducked as Demarco made for the Hummer, hearing Sunny's little whispered whinny of fear.

Demarco cursed as he tried the Hummer's door and found it locked. Next he tried the Porsche. He opened the door and looked inside it, but obviously did not find what he was looking for. Mac watched him stalk back into the house, allowing the door to slam behind him.

Mac felt in his pockets for the keys.

"Guess what," he said to Sunny. "I left the house keys in the kitchen."

She shot up from the floor in back of him. *"You mean we're locked in this garage?"* she said in a loud whisper.

"Shh." He gestured toward the house. "Actually, yes. That's exactly what I mean."

Sunny moaned. "I'm gonna die in here. I'll

187

leave a note asking that the Indian bike be buried with me. We've got to get out," she added. "Who's gonna give Pirate his dinner?"

"You are." Getting up, Mac helped her out of the car. "See that?" He pointed to the locked side door with the dog flap.

"You mean that *doggie door?*"

"Must have been for Allie's Maltese." He glanced encouragingly at Sunny. "Think you can make it?"

She groaned. "Explain to me why we don't just use the garage door opener and get out the usual way."

"Because we don't want to announce our illegal presence to all and sundry passing by on the street. Especially if Demarco is still around, though I doubt that. Tell you what, why don't you get out, then go around to the beach side. The window should still be open. Then come back through the house and let me out."

Sunny glared at him. "I've a good mind to leave you here."

"Aw come on. You've got to admit it was worth a try."

Sighing, Sunny eyed the doggie door. "I'll never forgive you for this," she said.

In a few minutes she was standing outside breathing the fresh salty night air.

"Oh thank God," she whispered to herself. But she had to be quick and rescue Mac before somebody got suspicious or Demarco came back.

"Thanks," Mac said, when she finally opened the kitchen door and let him out. "Tell you what," he said as he reset the alarm and they exited quickly, locking the door behind them. "I've got a bottle of good champagne chilling just for you. How about it, babe?"

"I hate you," she said, smiling.

After the second glass of champagne, when Sunny's nerves had stopped twitching and she had agreed not to keep looking back down the beach at Perrin's house, she told Mac about Allie's visit, and how she had revealed her very personal life story.

"She just wanted to talk," Sunny said. "And I was the anonymous person who would listen. A 'girlfriend.' "

"So what do you think of Allie, now?" Mac asked.

Sunny took a thoughtful sip. "I like her. I think she's had — is having — a hard time. And I admire her. She came from a tough background and fought her way to the top, even though she says it was Ron who in the end gave her that final leg up to movie-biz

189

stardom. But I get the feeling she's at a crisis point. I don't think she knows which way to turn. And besides, she misses her husband. I got the impression that she depended on him for everything. Ron Perrin was her rock in a very craggy business."

"You sure she misses him?" Mac sounded surprised.

"I'm sure of it," Sunny said firmly. "In fact I'd be willing to bet she still loves Ron Perrin."

CHAPTER 23

It was a few days later and Allie was in the South of France, alone on the terrace of her luxurious suite at the Hôtel du Cap. She was clutching a glass of champagne, taking a gulp from it every now and again to steady her nerves, thinking of what she was about to do. Snatches of conversation and laughter drifted from the gardens below, along with the faint slurp of the Mediterranean hitting the shore. Umbrella pines straggled across the skyline as the horizon turned a neon blue to match the sea, and the air felt soft against her skin.

She thought of Malibu, where the Pacific Ocean always let you know who was master, curling in high iced-green waves that slammed against the shore in a torrent of white foam, then receded with a whisper over the rocks. She thought of Ron and their home at the water's very edge, of how, when they had first bought it, they would lie awake listening to the ocean that somehow soothed them into sleep with its noise. And she remembered Mac's humble little place, perched precariously on its wooden pilings, and as charming and casual as the man himself.

A glance at her gold Cartier watch told her it was almost time. She just had one last call to make. She punched in the numbers, hoping Sheila would be there. Luckily, she answered right away.

"Sheila, this is it," Allie said softly. "I have nothing to return for."

"Sweetheart, are you sure?" There was a hint of panic in Sheila's voice.

"I've never been so sure of anything since I was a teenager and wanted to get out of that deadly little town in Texas. Sheila, it's what I have to do. What I *need* to do. I don't know where it will end, but I have to be on my own. I have to try to create a new life."

"But how, what will you do, Allie?"

Sheila was worried, but there was a new lift to Allie's voice as she answered. "I have no idea. I guess I'll find something. I'll let you know, my friend. But you promise to say nothing to anyone?"

"Not even Ron? If he should return that is?"

"Especially not Ron."

"And what about the detective? Reilly?"

Allie hesitated, but she decided quickly she had to do this alone. "Not even Mac Reilly," she said firmly.

Sheila wished her luck, said she would be thinking of her, and Allie promised to call before too long. Then she walked back into the suite, checked her appearance in the full-length mirror and called for the bellboy to carry her small suitcase to the waiting limo.

Taking a deep breath, she walked to the door. She turned for one last look at the charming room with its view of the sea; at the silver ice bucket with the open bottle of excellent champagne and the massed bouquets of scented flowers; at the piles of expensive clothes that the maid would straighten out for her. At the life of a movie star on her way to her premiere. And then she closed that door and took the elevator downstairs to the lobby where her director was waiting for her.

She caught his slight frown of disapproval at her plain outfit, but still he smiled and said how lovely she looked.

"Sometimes simplicity is better," she said "It's just a pity we didn't keep that in mind when we made this movie."

They sat in silence for the almost forty minutes it took to drive what usually took only twenty. The traffic was hell and the director was biting his fingernails, afraid they were going to be late. But Allie knew they would wait for her. Everyone always did.

As the limo drew up at the Palais des Festivals, she stepped out and posed smiling for the photographers. Compared to all the glitter and the gowns and the glamour, she caused a sensation. So simple, so different in her narrow black silk pants and plain white taffeta shirt with the sleeves rolled up, her only jewels a pair of gold hoops and, oddly, since it was known her marriage was on the rocks, her gold wedding band. Her blond hair was pulled back into a chignon and tied with a black satin bow, the way Grace Kelly used to wear hers in the sixties, and in fact more than one person commented on her resemblance to Monaco's late princess.

She walked to the enormous red-carpeted flight of steps leading into the Palais, holding hands with her director, smiling and waving to the crowd, posing some more, the complete professional, making sure the photographers had what they needed. Then she strode up the steps, turning at the top for one final wave. No one would have guessed that at that moment she felt she was the loneliest woman in the world.

She sat through the screening of her

movie, with its title, *Midsummer's Dream,* half-stolen from Shakespeare. Despite some drastic last-minute cutting it was as slow and emotionally unmoving as she'd suspected it would be from the first week's shooting, when the script had begun to be changed. From then on it had been changed on a daily basis until nothing was left of the charming little love story, which was how it had started out. But anyway, her thoughts were not on the film. That was the past.

She was thinking about Mac Reilly and his phone call before she'd left for France, wishing her good luck. "Sure you won't change your mind and join me?" she had asked wistfully, already knowing his answer. Loneliness had made her try, and despite her brave words to Sheila, she was scared by what she was about to do.

The movie was over. Time to face the press, do the interviews, pose for the photographers one more time. Then on to the cocktail party given by the studio on the enormous yacht moored in the bay, and then to dinner at the famous Moulin de Mougins restaurant, where once again the beautiful Sharon Stone was conducting a live auction to benefit AIDS.

And after that? After that Allie's time was her own.

At the auction, she bid on a luxury cruise for two. Surprised when she won, she generously donated it back to be reauctioned. Then whispering to her director that she was tired, she said good night and slipped from the darkened room.

Her small suitcase was already stashed in the back of the limo. Allie asked the driver to take her to Nice airport. She opened the case, took out a long cardigan and slipped it on. She pulled a straw hat over her hair and adjusted the brim so it shaded her face. A pair of square-framed glasses hid her eyes and she wiped off her lipstick.

When they arrived, she tipped the driver a generous couple of hundred dollars, said she did not need any help, hefted her suitcase, then walked into the departures terminal and headed for the restroom.

In a stall, she quickly changed into jeans and a sweatshirt. Then she picked up the suitcase again and walked out to the car rental facility.

This was the test. Would they recognize her? Or would they not?

The woman at Euro-Car was tired and disinterested. Yes, Madam's car was waiting. She just needed to see her driver's license, passport, credit card, and she should sign here.

Allie gave her the new passport with her real name, Mary Allison Raycheck, and the new credit card. She held her breath. Would the woman look at her to check?

"Row C, number 42. Left out of the door. *Et bon voyage.*"

Bon voyage, Allie thought, elated, throwing her suitcase into the trunk of the small baby blue Renault then climbing into the driver's seat. Little did the woman know this was to be the *"voyage"* of a lifetime.

She shut the car door with a solid thud, then sat for a moment, suddenly overwhelmed by fear. Desperate, she took out her BlackBerry and called Ron at the Malibu house. There was no reply. She tried Palm Springs. The same. She called Mac Reilly. Again no reply, only the request to leave a name and number, which she did not.

Tears glittered in her eyes but she brushed them away. There was nobody to even care what she did. Still, there would be no more threatening letters, no more crazy stalkers, no more complicated love life — or rather, lack of one. And no more movie star. She was free. She was Mary Allison Raycheck.

Back to that again.

CHAPTER 24

Roddy was having a busy day and one not quite to his liking. First he paid Allie's housekeeper a visit to question her about the staff and the extra cleaners, the pool service, the gardeners. Allie was in Cannes and Ampara was alone.

She was holding Allie's little white dog on her knee while they talked over a glass of iced tea in a rich man's kitchen that could have graced the cover of *Architectural Digest,* and that Roddy thought was big enough to double as a ballroom. *Cozy* was not exactly the adjective that sprang to mind, but he smiled and said how great it was and how happy Ampara must be to be working in such elegant surroundings.

Ampara was from El Salvador and knew what poverty looked like. She had worked for Allie and Ron Perrin for four years and sent money home regularly to her family. Comfortable-looking was how Roddy would have described her, small and round and sort of grandmotherly, though in fact she was only forty-five. She spoke excellent English, which she told him was essential if you wanted a good job like this one because it meant you could take messages.

"And have there been any messages for Miss Ray since she left for Cannes?" Roddy asked.

Ampara shook her head, looking sorrowful. "No, sir. There's been no messages for Mr. Ron either. And with both of them gone and maybe splitting up, I'm not sure where I stand anymore, job-wise."

Roddy looked concerned. "Even though they're away, you still get paid though, right?"

"Oh yes sir, the accountants have always taken care of that. But it gets lonely here, in this big house all by myself. Especially at night. I have my own apartment over the garage and I don't mind admitting that me and Fussy lock ourselves in there and bolt the door. I'm glad of the little dog's company," she added, lifting Fussy to kiss her on her nose.

Roddy could have sworn he saw the dog smile, quite different from the snappy little creature that evening at the beach. And he didn't blame Ampara for being intimidated.

This was a huge house that demanded to be filled with people, a party house meant to be exploited and shown off. A bit like Allie herself, he thought.

Ampara told him that all the indoor and outdoor staff had worked there for years. The only casualty was Allie's assistant, Jessie Whitworth, who had worked for Allie for almost a year and who she had "let go" a couple of months back.

"I think Jessie was surprised when Miss Allie told her she didn't need an assistant anymore. Miss Allie said she was cutting back on work and personal appearances, and anyway she wanted to take over her own life," Ampara added.

Roddy's ears perked up. A sacked personal assistant sounded likely to be an angry ex-assistant, one who might want revenge. But when he asked Ampara she said no, Jessie wasn't like that. She was a nice quiet young woman, always polite and with a smile.

So were some serial killers, Roddy thought, writing down Jessie Whitworth's name, address and phone number.

"More iced tea, sir?" Ampara asked.

"Thanks, no. I'll be on my way. You've been more than helpful."

The housekeeper's round face looked doleful. "I surely hope Miss Allie and Mr.

Ron gets back together, sir. I'm not happy being here alone. I need someone to look after. That's why I'm so fond of Fussy here."

She saw Roddy out through the massive front hall with its double sweeping staircase and crystal chandelier, standing on the steps, the dog in her arms, watching as he got in his car. She waved as he drove off.

Roddy dialed Jessie Whitworth's number. She answered right away. A pleasant low voice, precise and businesslike. She sounded like the perfect secretary as she agreed to meet him at the Starbucks near Wilshire and Third in Santa Monica.

She was already there when Roddy made his way in. He could have picked her out even if she hadn't waved hello from a corner table. In contrast to the young clientele, who were mostly in lowrider jeans and cropped T-shirts with their hair bubbling down their backs in blond extensions, she was tall and very neat looking in a buttoned-to-the-neck blouse and well-cut brown pants, worn with Gucci loafers. Miss Whitworth's hair was cut in a neat black bob, and she was pretty in an unobtrusive sort of way. Roddy got the feeling she had spent a lifetime trying to appear unobtrusive. He guessed it was the only way to survive as an assistant to important people who sometimes acted as though they were more important than they were.

"That's what I'm paid for," Jessie Whitworth told him with a smile when Roddy mentioned it, surprising him with a set of expensively veneered, dazzlingly white teeth.

"It's not me up there center stage," she added, and Roddy saw the smile disappear.

"Is that where you'd like to be then?" he said, looking interested as they sat over their low-fat macchiatos.

"About a dozen or so years ago, when I was young and foolish enough to believe I had talent. Hollywood soon robbed me of that illusion," she added. "Not that I minded. I realized I wasn't cut out for the game that had to be played. All I wanted was to act. And you know what, Mr. Kruger? I simply wasn't good enough."

Roddy knew exactly what she meant. He himself had never had eyes for the acting profession, but he had friends who still cherished the hope of that one role that would change their lives, meanwhile struggling on with a bit part here, a nonspeaking part there, even willing to work as extras.

"It's a tough life," he said, all sympathy. He was liking Miss Whitworth and her honesty about herself.

"Allie didn't exactly *fire* me," Jessie said. "She told me her life was changing and she needed to do things herself. Become more independent. Lose all the trappings. That sort of thing. I knew she was unhappy and on the verge of splitting up with Ron and at first I thought it was just a reaction to that. I believed she would get over it, move on. But it seems she hasn't."

"You think she was still in love with Ron?"

Jessie's cool gray eyes met his. "I would say so. Yes, definitely. I mean they fought because . . . Well, you know how it is with couples. I don't want to talk about my employers' private lives," she added. "It's not right."

"I understand."

Roddy was on the verge of eliminating her as stalker material when she said, "Of course Allie paid me three months' wages in lieu of notice and gave me glowing references, but the fact is I still don't have a proper job. I'm temping right now, out at Mentor Studios. Just another assistant's assistant. A jumped-up secretary. Quite a comedown," she added bitterly, and Roddy caught the quick flash of anger that crossed her face.

So, he thought surprised, poker-faced Little Miss Goody Two-shoes has emotions after all. And not all of them are good.

"Thanks for talking to me, Miss Whitworth," he said. "You helped clear up a few points."

"Like that Allie was planning on making changes to her life?"

"You believe that, do you?"

She nodded. "I think she might have become tired of being America's darling with no privacy."

Roddy thought she was right. He shook her hand, waving from the door as he left. He thought Miss Whitworth was a dark horse and that an eye should definitely be kept on her.

He waited in the parking lot until he saw her leave. She was driving a bright blue Porsche Carrera. A fancy car for an ex–personal assistant turned jumped-up secretary, he thought, surprised.

He had already checked the other staff on Mac's list, the hairdressers, stylists, et cetera. Now he drove to the studio to snoop around there, see who was what, and what, if anything, anyone had to say about Allie.

He'd gotten personal approval to be on the lot from the producer of her last film, and the guard at the gate checked him out on the computer, made out a visitor pass for him then waved him on. "Visitor parking to the left, sir," he said, so Roddy swung a left and drove down a line of densely packed cars. The studios must be really busy, he thought, turning down another aisle looking for a spot.

He slammed his foot on the brake, threw the car into reverse, backed up, then stopped. He was looking at a black Sebring convertible with very dark windows.

CHAPTER 25

Roddy drove back to the gate. From the procedure he had just gone through he knew the guard would have a list of all visitors and their car numbers. He explained who he was, told him what he wanted to know, annoyed but understanding when the guard explained he was not at liberty to give out that information.

Roddy had the number of the production office of Allie's latest movie. He called, explained what he wanted and asked for help. Within minutes the guard had been given permission and Roddy had the information. Or at least some of it.

"That car has been on the lot for a couple of weeks, sir," the guard told him. "It's a rental and the woman driving it told me it had broken down. She said the rental company would come and tow it." He shrugged. "So far they've not shown up. Beats me how they keep in business," he added.

"So who was the driver?" Roddy asked, as casually as he could because he was excited to think he might be on the brink of identifying Allie's stalker.

"A woman by the name of Elizabeth Windsor, sir. I remember because like it's the Queen of England's name."

"The Queen of England right here in your parking lot?" Roddy said. "Imagine that."

The two men laughed together, then Roddy said, "You remember what she looked like?"

"Couldn't forget. Tall, blond and long legs." He made a curving gesture with his hands and they grinned at each other.

"How old?" Roddy said.

"Oh, I dunno — twenties, I guess. Come to think of it, she looked a bit like Allie Ray."

"Better than the Queen of England," Roddy said. "Younger, anyway."

Bristling with excitement, Roddy called Mac to tell him the news.

"I'm sitting here right now, looking at the Sebring," he said. "Of course I know there's probably hundreds like it, but this one was parked on the Mentor lot where Allie was filming. Anyhow, it's a rental in the name of Elizabeth Windsor."

"Like the British Queen?"

"The same, only this one's in her twenties,

a tall blond who looks a bit like Allie."

"Check out 'the Queen,' " Mac said. "See if anyone remembers her, and what she was doing on the Mentor lot anyway."

Roddy was weary. It was almost seven and he'd been thinking more along the lines of an iced vodka martini in the bar at the Hotel Casa del Mar in Santa Monica with a soothing view of the ocean. It was close to where he lived in an all-white condo — dogless, thank God — and with a more prosaic view of other people's backyards.

He told Mac about his conversation with Jessie Whitworth, said that he thought she was a dark horse, that she drove a bright blue Porsche, and that she had been fired a couple of months ago. "By coincidence she's working as a temp right here at Mentor Studios," he added.

"She's not tall and blond?"

Roddy laughed. "Quite the opposite. A bit of a plain Jane really, in a nice efficient way. Knows her place and sticks to it."

"So how does the prim Miss Whitworth afford to drive an expensive car?"

Roddy said, "You know L.A. First and last month down and you too can look like a big shot and drive a Porsche."

"Yeah, but that doesn't gel with what you're telling me about the rest of Jessie's image," Mac said thoughtfully. "I wouldn't bet she's the stalker either. Our stalker is really crazy. I'm talking potentially violent crazy and I'm not getting that message about Miss Whitworth."

"Right," Roddy said. "I just thought with her being fired and all . . ."

"And you're right. We can't dismiss her out of hand. I'll do a little background search on Miss Whitworth myself, okay?"

"Right," Roddy said, relieved that he didn't have to work all night. That martini was getting closer.

"I'll also check out the Sebring," Mac said. "Thanks, Roddy." There was a grin in his voice as he added, "You're the greatest, y'know that?"

Roddy smiled, showing a beguiling pair of dimples. "I always kind of thought so myself," he said, preening in the rearview mirror.

CHAPTER 26

It didn't take long for Mac to find out that the rented Sebring had been reported as stolen. The driver's license in the name of Elizabeth Windsor was a fake, and the credit card with which she had paid had a false address in another city.

Driving to the rental facility at LAX, Mac thought about Elizabeth Windsor. Could this really be the stalker? The writer of those filthy abusive letters, filled with explicit threats? Most often a stalker was a man, but this one might also be an expert in using women. Elizabeth Windsor might merely be a pawn in his game.

Planes zoomed low over his head as he parked, merging with bewildered-looking travelers hurrying to pick up cars, only to struggle to find their way out of the traffic into the City of Dreams.

He got lucky. The young guy in charge of the rental place remembered Miss Windsor.

"Tall," he told Mac, "with long blond hair. Kind of a looker, if that's your taste."

It was obviously his, which was why he remembered. Mac said, "So where was she from?"

"Wait a sec and I'll look it up for you." He drummed on his computer and within a minute had what Mac had asked for. It tallied with the information the cops had given him. The address did not exist.

"She seemed okay," the young man added. "Pleasant, y'know. Not the sort to steal a car."

"She didn't," Mac said. "Apparently she just forgot to return it."

At eleven o'clock that night the phone rang. Mac was in bed but not yet sleeping. Lying at his feet, Pirate opened his eye. Mac knew it couldn't be Sunny because she was in New York for a couple of nights on business.

He grabbed the phone. "Yeah?"

"Mr. Reilly, sir, it's me, Ampara. Miss Allie's housekeeper."

Mac sat up quickly. "Yes, Ampara?"

"I'm scared, Mr. Reilly. I just noticed that the alarm system has been turned off. Now only I know where that switch is, sir, and I haven't been up in the house for the past five hours. I'm here, all alone, just me and

Miss Allie's dog, and I'm scared, sir. I want to call the police, but I know Mr. Ron doesn't want no police here. I don't know what to do, sir."

"Are you securely locked in, Ampara?"

"The door is locked and bolted, and so are all the windows . . ."

Mac thought quickly. He knew Ampara was safe for the moment and that he needed to get there before he called the cops. If someone really had entered the main house, that person had a key and knew exactly where to turn off the alarm. He might just have his stalker.

"Stay right where you are," he told Ampara. "I'll go into the house and check it myself. You won't see me from the windows because I'll turn off my lights and leave my car at the end of the driveway. I'll call you immediately I get there. Okay?"

"Okay," Ampara said doubtfully.

"And Ampara."

"Yes, sir?"

"If you hear anything at all, anyone near your apartment, if you're scared, call Security and the cops immediately."

"Yes, Mr. Reilly. I'll do that," she said, sounding relieved.

Next Mac called Lev Orenstein. He told him what was going on and arranged to meet him at the house.

Lev was there before him, his black Mustang half-hidden under an overhanging tree. Mac pulled in behind. Lev got out to meet him. He looked every inch the movie-style tough bodyguard, lean and mean and intent on business in a black turtleneck, black jeans and with a weapon tucked handily in a holster under his arm.

Mac was in his usual T-shirt with a pair of shorts he'd pulled on quickly as he leaped out of bed. Pirate sat in the car looking disconsolate at being left but was well trained enough to know not to make a song and dance about it.

"The gate is locked," Lev said. "No car parked anywhere within sight, though he may have parked near the house."

"That would be a very confident move," Mac said, on the phone again with Ampara.

"Oh, Mr. Reilly, it's you," she said, sounding relieved.

"Lev Orenstein and I are outside the gates. Can you open them for us, please, Ampara?"

The gates slid smoothly to the sides and Mac and Lev jogged along the grassy verge by the driveway leading to the house. Again, no car was parked outside. Except for the twin lamps illuminating the steps, the house was in complete darkness.

Mac got the housekeeper on the phone again. "Don't you usually leave lights on?"

"Yes, sir, Mr. Reilly. There's always a light in the front hall, as well as in the kitchen, and in the master bedroom. Always."

Eyebrows raised, Mac looked at Lev. "You ready?"

The front door was not locked and no alarm sounded when Mac pushed it open. He slid through, keeping to the wall with Lev right behind him. They paused for a few seconds, waiting for their eyes to adjust to the gloom, listening. There was no sound but Mac felt that someone was here. He thought of Allie, alone at night, unaware that someone had the key and the alarm number. He got the feeling she had left for France just in time.

Lev knew the house and he led the way through the vast rooms to the kitchen. The refrigerator purred and half a dozen green lights flashed the time on various appliances, enough Mac thought to cook for a restaurant. There seemed to be two of everything and sometimes three or four.

Back again in the front hall, they crept silently up the curved staircase. The big double doors at the top obviously led to the master bedroom. Mac turned the handles and pushed them open.

A sudden crash sent them spinning, weapons in hand. It had come from behind a second closed door on the right. They ran toward it, turning quickly, backs against the wall, cop-fashion, as Lev flung it open.

No one was there. Mac switched on the light. The noise they had heard was of a pile of empty hangers crashing from a shelf where they had been clumsily thrown. The clothes they had held lay in a heap on the floor. All Allie's beautiful expensive gowns, slashed to pieces. The perpetrator had done a good job. Nothing had escaped his knife.

"I guess we're too late," Mac said, switching on the lights in the bedroom. "This is an example of what he could have done to Allie."

"Jesus," Lev said, shaken because he knew and cared about her. "I'll make the bastard pay when I find him."

Mac could no longer rule out the police. The crazy guy had to be caught before he did real damage. He called the Beverly Hills PD and informed his contact there what had taken place.

Next he called Ampara, told her that the cops were on the way and that he wanted her to pack her things and move to somewhere safer.

"You have a friend? Someplace to stay?" he said. "If not, we'll get you a rental apartment for the time being."

Ampara said she had friends, she was calling them now and she and the dog would go there.

The police arrived in minutes; they didn't mess around in Beverly Hills and Bel Air. They surveyed the mess, inspected the alarm system, brushed everything for prints, took pictures and agreed it would be better not to let this leak to the media. "Don't want to scare everybody," was what the detective in charge said to Mac, who knew him well from his guest appearances on his TV show. "What we need to do now is get in touch with Mr. Perrin and Miss Ray."

"Doesn't everybody," Mac said. "Trouble is, she's in France and nobody knows where he is."

"Probably on some tropical island drinking mai tais and getting an expensive tan," the cop said.

Mac didn't think so but he wasn't about to tell the cop that. He was just glad that Allie was away and that Ron Perrin had not been around for the main event.

Or maybe he had? It gave him something to think about.

CHAPTER 27

The next morning, Mac drove to the address Jessie Whitworth had given, on Doheny in West Hollywood. If she was home, he'd ask to speak to her about Allie. If not it would give him a chance to check her out with the apartment manager, find out if she was a good tenant, what the manager thought of her.

The apartment building was not a bad one though he guessed the apartments themselves would be on the small side, studios and one-bedrooms, most likely. Still, it looked well maintained and there was a smart new canopy over the entrance. He pressed the button that said "J. Whitworth."

There was no reply and he pressed again. When he got no answer he rang the bell for the apartment manager.

A woman answered. He told her he had seen the For Rent sign outside and was interested.

"Wait a minute, I'll be right there,"

she said.

She arrived in a hurry, all dressed up in a fluffy top, cropped jeans and strappy heeled sandals.

"I'm Mila. Gotta be quick," she added with a cheerful grin. "I'm late for my date."

"Sorry to bother you then," Mac said. "It's just that a friend of mine lives here. She told me how much she liked it, said she thought I would too. Jessie Whitworth's her name."

"Jessie?" Eager to show the apartment, she was already unlocking the door to 3J. She stepped aside, waving Mac through. "Wait a minute though, don't I know you from somewhere?" She looked straight at him for the first time. "Oh my God, it's *you,*" she said, stunned. "Mac Reilly from the TV show."

Mac smiled. "Got it in one," he said.

"But hey, what would you — I mean a famous man like you — be wanting an apartment like this for? I mean they're nice but not in your league." She caught Mac's rueful glance and said, "Uh-uh, have I put my foot in it?"

"Well, it's really for a woman I know," Mac said. "I just don't want it broadcast around. Right?"

"Right. I mean, of course. I'm the soul of discretion, anyone here will tell you that."

"So how's Jessie anyway?" Mac asked casually. "I haven't seen her in a while."

"Jessie? Oh she and her friend left early this morning. Off on vacation. Cancún, in Mexico she told me. Lucky things."

"Yeah," Mac agreed, walking round the small apartment, opening cupboards, checking out the bathroom. "So who's her friend?"

"Elizabeth, you mean?"

"I don't think I know Elizabeth."

Mila turned out to be far from discreet; in fact she was full of chat and ready to diss anyone and everything.

"A bit haughty, I thought," she said. "Though I guess she's nice enough. Tall and blond, kinda thought she was gorgeous and heading for Hollywood's hot spots. In fact she looked a bit like Allie Ray, the movie star. You know who I mean?"

Mac agreed that he did. "Cancún, eh?" he commented thoughtfully as they walked back outside. "Well, thanks for showing me the apartment, Mila. I think it's a bit small for what I wanted, but if I change my mind I'll let you know."

"Thank *you*, Mac Reilly," she said, giving him a hopeful flirtatious smile. "You can call me any time."

Back in the car, Mac got Roddy and Lev

219

on the Bluetooth in a conference call and filled them in. "Elizabeth Windsor's a blond Allie Ray look-alike," he told them. "A copycat."

"Maybe a *jealous* copycat," Roddy said.

"And I'll bet a *dangerous* copycat," Lev agreed.

"Anyhow, she's Jessie Whitworth's roommate, and Jessie had access to the Bel Air house keys and I'll bet she also knew the code for the alarm."

"Bingo," Roddy said.

"Anyhow, the two of them left for Cancún this morning. Taking a little vacation."

"Are you sure that's their destination?" Roddy said.

"I know what you're thinking," Mac said. "But Allie's at the Hôtel du Cap. And nobody will get in there, not with all the extra security for the Film Festival. In fact, I'll call her right now and tell her what's going on."

But when he called she did not answer. And when he called the Hôtel du Cap he was told Miss Ray had already left.

CHAPTER 28

Mac heard about Allie's disappearance on the car radio, driving north on PCH, on his way to the Malibu Fish Company for a quick lunch with Roddy.

The news reporter said the movie star had left the gala dinner in Mougins alone and been driven to Nice airport. She had not boarded a flight and she had not been seen since.

Since her new movie had received scathing reviews, they said perhaps Allie was simply avoiding the press. But there were also rumors about her marriage being on the rocks, and about Perrin's penchant for another woman, and after all, Allie was about to hit forty. They said there was more trouble in Allie Ray's life than the average fan knew about.

"But hey," the newscaster added with a smile in his voice, "that's Hollywood for ya. On top one minute, down the next. So? Is it goodbye, Allie Ray?"

Mac made a quick right into the parking lot then walked to the wooden shack that sold fresh fish to take out, if you were lucky enough to have a woman at home who could actually cook, that is. Otherwise you sat outdoors at scarred wooden benches and they cooked huge platters of fresh snapper or halibut, or almost any other type of sea creature, with mounds of fries that would satisfy any carb addict's soul.

He ordered the Cajun salmon sandwich, took his number and went and sat on the upper deck with a view of the beach and the pushy gulls begging for scraps and squadrons of brown pelicans zooming past, and the sun casting a golden glow. It was another beautiful day in California. Sheltered from the breeze by a clear plastic awning, he drank his Diet Coke, waiting for Roddy and thinking what to do about Allie.

He wished now that he had said yes when she'd asked him to accompany her to Cannes. But that would have been wrong, it wasn't his place to intrude on her life. Besides, it would have caused havoc with his relationship with Sunny.

He checked his watch. Twelve noon. Nine p.m. in France. He dialed Allie's cell. It rang but there was no reply. Still, her phone worked and if he knew women she wasn't going anywhere without it, so she had to be somewhere around.

He drummed his fingers on the table worried that she had not at least gotten in touch with *him*. She knew whatever happened, he was on her side.

He checked his watch again. Roddy was late. His order number was called and he loped down the wooden steps to pick it up. Biting into his Cajun salmon sandwich he called Sunny.

She was at the spa having a massage but she took his call anyway, to the annoyance of her masseur who grumbled that it ruined the whole aura. Mac could hear new age music whining in the background. He didn't understand why they always had to play Enya to soothe your soul. What was wrong with a little Bach?

"Allie's gone missing," he said, taking another bite of the salmon.

"Where's she gone?" Sunny said.

"If I knew she wouldn't be missing, would she?"

"Ohh. Right. Well, after those reviews I'm not surprised. She probably wanted to get away for a bit of peace and quiet."

"She hasn't contacted anybody."

"Not even you?"

"Right." He waited a minute then he said, "If she doesn't show up soon I might have to go and find her."

Sunny's groan rang in his ear.

"It's not what you think," he said quickly.

"That's what they all say."

"Jesus, Sunny, give me a break. The woman is missing. Her husband is missing. And his girlfriend is still waiting for him in Rome."

"How d'you know that?"

"She e-mailed me yesterday. She's panicked, doesn't know what to do."

"Isn't Demarco taking care of her? Or the Italian producer?"

"I guess so. I might have to go to Rome too. Find out what's going on there."

"I'm coming with you."

"Okay."

"Oh." Sunny had thought he would object. "Well, maybe I'll let you go on your own this time," she said. "I've got a job that needs my presence."

"I'll miss you," Mac said, grinning.

"Huh. Right. Of course you will. Anyway, where are you now?"

"Waiting for Roddy. Eating a Cajun salmon sandwich at the fish place on PCH. With the wind in my hair and loneliness in my heart . . ."

"You bastard," Sunny said. But he could tell she was laughing.

CHAPTER 29

Roddy pranced up the steps, holding a giant-size Coke and a shrimp cocktail.

Mac said, "I keep telling you, you should get Diet. You'll get fat drinking that."

"Fat? Me? I'd poison myself first!" Roddy tossed his blond head, ever the drama queen. "Anyhow you're the one who'll put on weight. Just look what you're eating." He waved a shrimp, minus cocktail sauce, in front of Mac's eyes. "Stick with me, Mac Reilly. I know what's good for you."

"I'm hoping you know something else," Mac said.

"Like who broke in and slashed all Allie's frocks? I wish I knew."

"What else?" Mac asked.

"I've got all the dirt on Perrin. And trust me, it's juicy."

"Just tell me what he's been up to, businesswise."

"Okay! So. Perrin's business dealings are a

very tangled web indeed. Multiple companies, multiple transfers, multiple everything. There's no way to trace where money comes from, or where it goes, though the Caymans and the Bahamas are strongly suspected. Perrin flies to both places frequently, on his own jet, always with the excuse of a fishing or gambling trip, and always in the company of half a dozen or so male friends. No women this time."

"So I guess we can assume Perrin took more with him than just his male buddies."

Roddy nodded. "Anyways, the FBI caught on to it, and Perrin found out. That's why he's a frightened man."

Mac said, "Plus Allie's attorneys are demanding a full accounting. Which obviously Perrin couldn't give. *And* he had Marisa lobbying either for marriage or money. Who really knows which? And then there's the case of the missing woman, Ruby Pearl."

"Ruby Pearl?" Roddy cocked his head inquiringly. "Sounds like the poor girl's mother was into jewelry! Am I wrong, or did I miss something?"

Mac told him the Lipski story and about his break-in at Perrin's, and that he'd seen Demarco also looking around the house.

"I can't find anything much on Demarco," Roddy said. "Except like Perrin he seems to have an awful lot of spare cash. He just built a large house in the desert y'know. A very classy area, movie folk and celebs and just plain rich folk. Our Mr. Demarco is mingling with the best."

"I'm not surprised. Which reminds me . . ." Mac took out his cell and placed a call to Lipski.

Lipski said, "I hope you have some satisfactory news for me, Mr. Reilly."

Mac sighed. The poor guy was desperate. "Not yet. I'm sorry, Lipski. I checked the Malibu house — absolutely zero of any interest. I'm off to France for a few days, but I'll tell you what, when I get back I'll also check out Perrin's Palm Springs place. You got the address?"

He wrote it down, said goodbye and looked at Roddy, who was looking back at him, a question in his widened eyes.

" 'Scuse me?" Roddy said. *France?*"

Mac glanced at his watch. "If I move my butt I can be on the late afternoon Air France to Paris."

"Paris?"

"Then the flight to Nice. Be back in a couple of days."

"Thanks very much for telling me." Roddy turned his head away huffily. "I don't suppose you're gonna tell me *why*

228

you're off to France."

"Allie Ray. She's gone missing."

Roddy's head swung back. His eyes opened wide. "No shit. And you're setting off on your charger. The cavalry to the rescue of the poor missing maiden?"

Mac grinned. "Darn right I am," he said.

CHAPTER 30

Sunny paced her condo with Tesoro nipping at her heels, demanding to be picked up, but for once her mind was not on the Chihuahua. She was wishing she had gone to France with Mac. She had recognized Allie's despair and her isolation, and was concerned about her. Women — especially famous ones — didn't just go missing. But wasn't that exactly what Allie had said about her husband, Ron? And he was missing too.

Before he left Mac had brought Sunny up-to-date on Perrin's possible money-laundering activities. Mac said Perrin had done it before and now it looked as though he was doing it again.

"Once a thief," Sunny had said and Mac had agreed she was probably right.

She stared out at the expensive boats in the marina. Maybe she would go out with a couple of her girlfriends tonight, sink a couple of martinis at one of the local bars, try a little "fine dining" . . .

She picked up Tesoro, who bit her hand just for the hell of it. Scowling, she walked into the bedroom to check her closet in search of something to wear.

She picked up the pair of shorts lying on the floor and put them into the laundry hamper. As she did so, a piece of paper fell out of the pocket. Impatient, she threw it into the wastebasket.

Wait a minute though. Those were the shorts she'd worn when they'd broken into Perrin's house. And that must be the piece of paper she'd grabbed from the Hummer's side pocket.

Retrieving it, she smoothed it out. She was looking at a receipt for property taxes received from Ronald Perrin on a dwelling named the Villa des Pescadores, Nuevo Mazatlán, Mexico.

She stood for a minute, letting it sink in. Then with a yelp of joy she grabbed the Chihuahua and swung her high in the air, dancing around the room.

"I think I've got it, Tesoro," she yelled gleefully. "Like Professor Higgins in *My Fair Lady*. . . . I think I've got it."

If she were quick she might just get ahold of Mac at the airport. She punched in his number but his phone was turned off. He must already be in the air. She tried Roddy next, relieved when he answered. Excited, she quickly spilled out her story to him.

"Sweetie," he said patiently, "I'm in Napa Valley, wine tasting with my boyfriend. Mazatlán's a long shot, not worth ditching my weekend for. I'll be back Monday. I'll check it out then. If RP's really in Mexico we'll go down together, have a few margaritas and 'bring him back alive' as they always used to say in the western movies."

Sunny got the strong feeling he wasn't taking her seriously. "But he could disappear between now and then . . ."

"If he's really *there. Sweetie, we just don't know.* . . ."

She was getting nowhere. Frustrated, she said goodbye, then paced the floor some more. Time was ticking by. Anything might happen. This was too good a clue to miss. It was up to her to take the initiative, and besides she quite fancied herself in the role of girl detective. Make Mac proud of her, she thought with a grin.

Dumping the dog on the sofa she made a few quick calls, threw some clothes into a carry-on, and packed Tesoro, wailing, into her doggie travel bag. She dropped her at the posh kennel near the airport where the staff knew all Tesoro's likes and dislikes and where she knew the dog would be treated like the princess she was.

In under an hour Sunny was at the airport. She was on the trail of Ron Perrin. *She* would be the one to find him. She would find out what made RP tick and see what he knew about his missing wife.

CHAPTER 31

STAR CHECKS OUT ON HER NEW MOVIE — AND SO WILL THE PUBLIC.

Allie read the headline sitting in a corner of an autoroute café near the medieval walled town of Carcassonne, looming on the horizon, turreted and battlemented like the façade of a Disney theme park.

She was on her third cup of coffee and her second croissant with a Michelin map spread across the table in front of her, trying to work out a route. But since she had no clear idea of where she wanted to go, other than simply disappear, it was proving difficult.

She glanced round the café. Nobody was taking any notice of her, everyone caught up in their own breakfast coffee. Anyhow they would not have recognized her. That first night in the small motel off the Autoroute du Soleil she had taken the scissors to her long blond hair, cutting until all that was left was a short untidy mess that she had dyed brown. Unmade-up and with her dark cropped hair and in square-framed eyeglasses she looked like a child's idea of a schoolteacher. And in jeans and a baggy T-shirt, she was just another eccentric woman with a bad haircut, driving on her own through France. Nobody had even looked twice at her. It was a first in Allie's life and she liked it.

Draining the last of her coffee she went back outside to the parking lot and the baby blue Renault. Lying next to it in the shade of a tree was a dog. A large dog. It had a German shepherd–type head and a stocky Labrador body, and was shaggy furred and muddy. It looked up as she approached but did not move.

"Hi, dog," Allie said, nervously. "Or should I say, *Bonjour, chien?*"

The dog's eyes were fixed wearily on her. His tongue lolled and he panted even though he was in the shade. She wondered if somebody had dumped him. Thrown him out of their car? No longer wanted? Make it on your own, buddy, they'd probably said.

"Okay . . . So . . . okay," Allie said. She always said, "so okay," when she was thinking. "I guess you're hungry. You wait here. I'll be right back."

In the café only coffee was available. Lunch was served between noon and two and the rules were the rules. Since she couldn't buy the dog a steak she bought a couple of ham and cheese sandwiches from the dispenser and a bottle of water, then went into the boutique and purchased a pottery bowl with BIENVENUE À CARCASSONNE written on it.

When she got back the dog was still there, his head sunk between his paws. He lifted his eyes and looked at her. He obviously expected nothing and was used to it being that way.

Allie knelt beside him. She poured the water into the bowl and put it in front of his nose. The dog scrambled to his feet and began to lap.

He had not, Allie guessed, had a drink for hours. Maybe even days. Unwrapping the sandwiches, she pulled out the ham and tore it into pieces. She put it on the ground in front of him. The dog took one sniff then, delicately for such a large animal, began to eat. Watching him, Allie marveled. He didn't snatch or wolf the food down. Even under stress, this was a civilized dog.

When he'd finished he sat back on his haunches, looking at her. She thought she caught a glimmer of gratitude in his soft eyes. She put out a hand and lightly touched his fur. It was harsh with ingrained dirt.

"Good boy," she said, refilling his water bowl. She pulled the rest of the sandwich apart and put it down in front of him, watching as he devoured it.

"So. Okay then . . ." She gave him a farewell salute. "We're on our own, you and I, boy," she said. "I wish you good luck."

The dog regarded her gravely.

"Okay, so okay. *Au revoir, et bonne chance, chien.*" She climbed quickly into the Renault, glancing at him in the rearview mirror as she drove out of the parking lot and onto the autoroute heading north. The dog was still sitting there, his eyes fixed on the departing car.

"Okay. So it's okay," she told herself nervously for the umpteenth time. "He'll be all right. I mean, he's just a dog, somebody will find him, look after him . . ."

She slowed down. Cars flashed past her, their drivers hooting angrily. She was remembering Mac Reilly's story of how he'd found Pirate almost dead, and how the vet had told him that once you had saved someone's life you were responsible for their soul forever. She told herself she was a fool, crazy, mad. She was having a hard enough time getting her own life together and nobody

237

was coming along to save *her* soul. Not Ron. Or even Mac Reilly.

She swung the Renault onto the off ramp, crossed under the motorway then reentered going south.

Back at the autoroute café the dog was still sitting where she had left him, next to his water bowl with BIENVENUE À CARCAS-SONNE painted on it.

The brakes squealed as she pulled up, got out and opened the passenger door. Her eyes met the dog's.

"So okay," she sighed. "Get in."

The big dog was firmly planted on the front passenger seat, eyes fixed on the road in front of him as she drove down the autoroute.

"Sit down and make yourself comfortable," Allie said, giving him a little pat.

His gaze shifted sideways. He looked at her then he slid wearily down until most of him was on the seat, the rest just sort of spilling over.

"You speak English?" she asked, surprised. He gave a little whimper.

"Ah, at least you have a voice," she said, smiling. "I don't know where we're going, you and I," she added, "but I guess from now on we're in it together."

She stopped at the next small town where she found a tiny auberge which, like many places in France, accepted dogs. The owner recommended a *salon de beauté* for *les chiens,* where they gave Allie black looks, exclaiming over the dog's condition. "Madame must take more care," the owner said frostily, until Allie explained that she had just adopted him from the side of the road.

The woman recommended a vet, and Allie told him the same story. He gave the dog a thorough examination, said he was badly nourished and had also been mistreated, probably beaten; there were wounds on his back. He gave the dog some shots, gave her an antibiotic gel for the wounds and pills for nutrition, and she bought enough food for a month.

Her dog was now a different animal from the one she had picked up only hours ago. His big shepherd head lifted at a more confident angle; his coat had emerged from the bath as a silky golden brown that almost matched the color of his eyes. His big paws were manicured. His ears were clean. He had his own water bowl and a metal dish that she filled with food which he ate like a perfect gentleman, then lay quietly beside her in a nearby restaurant while she devoured her own bowl of cassoulet, the local specialty, feeding him chunks of the duck meat and sausage, which he accepted with a grateful wag of his bushy tail.

Back at the auberge her dog accompanied her upstairs to her room, nails clattering on the wooden steps, and when she had showered and climbed into bed he lay on the floor, his eyes turned her way.

"Just checking, huh?" Allie said, pleased with him.

When she woke in the night she heard him breathing quietly as he slept. She smelled the clean dog smell of him and loneliness retreated a notch. This dog was all hers and nobody would ever take him away from her. Even though he still did not have a name. She smiled. But now, neither did she.

CHAPTER 32

Ron Perrin, hungover from last night's tequila, sat on the deck of the small beach villa, if such a grand word could be used to describe the desiccated concrete building built cheaply many moons ago, and that only he knew about. No one else. Not even Demarco. Or Allie. *Especially* Demarco and Allie.

The Villa des Pescadores, so called because it had once been owned by a Mexican fisherman, was his secret. It wasn't his *only* secret, but certainly now it seemed to be his best kept one.

Sometimes he thought the square ugly little house, with its yellow walls faded to a patchy butter color, was his true home. A place where a man could be alone with his thoughts, where he could be himself, and where, should he need company, all he had to do was drive into Mazatlán town and find solace in a bottle of tequila, while enjoying a local spiny lobster with a decent mariachi band to sing along with. Here, life was simple. Nobody knew him except as the eccentric gringo who kept to himself and liked his tequila.

A CD played in the background. He turned it up louder over the crash of the surf. "Will you still love me tomorrow?" Roxy Music's Bryan Ferry sang hauntingly. Sighing, he closed his eyes to listen.

In all the years Ron had owned this place he had never brought a girl here. Or, for that matter, brought anyone. Battered and beat up, its saltillo-tiled floors cracked, its plumbing unreliable — sometimes there was water, sometimes not — and with electrical wires sticking here and there out of the walls, it was just a shack on the beach. Forget Malibu and the miniature rolling stock and the expensive art; forget Palm Springs and Bel Air and all the grand rooms and rich décor and the knickknacks and the gardens. Here there was a terrace just wide enough for a chair and a table, on which he propped his bottle of Corona and his feet. It was the only place that soothed his soul.

He took another slug of the Corona and contemplated the waves pounding in a wall of white foam not too many yards away, thinking of the Malibu home he had shared with Allie. His love. Maybe his one true love. Maybe the love of his life.

Why was it then, he could not tell her that? Why, for fuck's sake, had he treated her the way he did? Why did he always have to show off his power and his masculinity with a series of women he didn't care for? Something was wrong with him, and it was a disease he did not know how to cure.

Sitting here, alone, in a way he never could be in Malibu, where his next-door neighbors' decks overlooked his, with phones ringing constantly, barraged with endless problems, not the least of which was the reason he was here, he tried to get that feeling of peace again. But this evening it wasn't happening.

Why couldn't it all just go away? he thought gloomily. Why couldn't he simply wipe the slate clean and start all over again? How differently he would do it now. How different he would be with Allie. He would eliminate the bitter memories and begin again, crazy in love the way he'd been when he first met her.

The sound of the waves roared in his ears and hunger made his belly rumble. Leticia usually left him food. She came in a couple of times a week to clean the place, if you could call a somnolent flicking of a broom across the tiles and a quick swigging down with a bucket of water that. Still, she did change the sheets every now and again, and she did do his personal laundry, taking it home with her and returning it the following week. Ron figured that since it took her a whole week to wash his simple shirts, her husband was probably wearing them four days while he got the other three. No matter, it wasn't worth worrying about, and people had to live, didn't they?

He put the Bryan Ferry CD on again, scrunching his eyes as if in pain, mouthing the words. "Will You Still Love Me Tomorrow."

His thoughts returned to Allie. He could see her clearly, in his mind's eye, the way she'd been when he first met her. Almost fifteen years ago now. Could it really be that long?

Anyhow, there she was, the petite blond beauty he'd seen on the screen with a skin that glowed like a ripe peach with the pink bloom still on it, and Mediterranean blue eyes that reached into his soul. He'd "recognized" not only who she was, but the person she was, in that way that happens so rarely, when a man instantly understands this woman is for him. But at the back of his mind, even though Allie had always

244

protested that she loved him, was always the nagging worry that how could a beautiful woman like her really love a man like him: unattractive and rough around the edges?

Yet, Allie had said she did, right from that first night together. They had spent it in each other's arms in a forested snowbound log cabin in Aspen, away from all the other New Year's Eve revelers. The Bryan Ferry record had been playing then, as it played in his head now. They had been together ever since. Until lately that is, when the bitterness had risen like bile in his throat, and he had lost her and was verging on losing all his money.

A "shit" was what many called him, and Ron finally believed it was true. And that was the reason he was here, on a lonely stretch of beach outside the funky Mexican tourist town of Mazatlán, hiding from the law and from his own emotions.

This was no time to be alone. His morbid thoughts were getting the better of him. Setting down the empty beer bottle, he walked into the apology of a bathroom, took a sparse shower — water pressure was bad tonight — combed his hair back from his sunburned forehead, put on a faded old T-shirt with MAZATLÁN printed on the front, flowered beach shorts and flip-flops.

He was ready to take on the Mazatlán nightlife.

He knew just the place: a joint on the beach where the booze flowed, the mariachi music blasted to the rafters and nobody knew him. He could get drunk there. Again. In peace.

CHAPTER 33

Sunny was glad the Alaska Airlines L.A./Mazatlán flight was a quick two and a half hours. She was itching to get there and see if her theory about Ron Perrin was right. Arriving at the airport along with the planeload of holidaymakers already in their shorts and tank tops, she picked up the rental car and asked the way to Nuevo Mazatlán.

The sun blazed from a sky as blue as any Raffaello fresco, turning the interior of the little Seat into a furnace. Hitching her skirt to her upper thighs, and gasping for breath, she turned up the air-conditioning. Sweat trickled between her shoulder blades as she drove the airport road, past humble little ranches with thin black cows in sparse fields and lumbering dogs, sniffing for scraps. Past run-down houses and auto repair shops, and cheap street cafés with plastic gauze slung overhead to keep off the sun offering *mariscos* and tacos. And past a yellow-painted school, and lavender and pink and turquoise houses.

The road rambled into the outskirts of Mazatlán, threading through congested traffic out to the marina area, bristling with new condominiums. Then over the new bridge and, at last, into the quiet countryside of Nuevo Mazatlán.

Here development petered out and the scenery reminded Sunny of Provence: stony, dotted with scrub and low-growing shrubs, and with a kind of singing silence in the air.

Following the signs, she came at last to the entrance to the Emerald Bay Hotel. A guard checked her at the gate and she drove slowly down a long avenue, turning finally in to the approach to the hotel, guarded this time by a tall aviary where bright macaws and parrots shrieked a welcome.

The lobby was centered with a large stone fountain set beneath a lofty dome, and its water music immediately cooled her. Slipping off her sandals she walked gratefully across the travertine floor and was checked in by a smiling young woman.

Ten minutes later, in a gauzy white skirt that flowed around her pretty knees as she walked, a black T-shirt and sandals, water bottle in hand, she hesitated at the door to her room. Should she call Mac? Or Roddy? Tell them what was up? She grinned. Nah! Let them wait. They'd find out later what a clever girl she had been.

The door slammed behind her as she hurried along the jungly path leading back to the entrance where her car was already waiting. Pausing only to ask the concierge the names of the most popular bar-restaurants in town and directions, if any, to the Villa des Pescadores, she was on her way, she hoped, to meet the infamous Ron Perrin.

The empty road followed the long curve of the beach as she left civilization behind, or at least the resort version of it, passing small shacklike homes and makeshift outdoor bars. Children stopped their play to wave and a couple of thin brown dogs chased her tires, making her think wistfully of Tesoro, pouting, no doubt, in the fancy kennel with the sofas and cushions and doggie TV. But right now her mind was on bigger things.

As dusk settled over the empty landscape, the road seemed to wind on forever. Loneliness crept around her, almost tangible in the warm air, and for the first time she was nervous. Then suddenly, a house popped into view, right at the edge of the beach. A small square yellow box, very much the worse for wear. She slowed down to read the name painted on a rock by the sandy path leading to it. VILLA DES PESCADORES.

She took another doubtful look. Could this *really* be a place the worldly, flamboyant billionaire Ron Perrin would stay? Remembering his glamorous Malibu home, she knew she must be on the wrong track. A man like that would never live here. He wouldn't even spend the night here. And neither, she thought, stumbling up the rutted path to the house, would she.

The slatted unpainted wooden door was closed. There was no bell, so she rapped, then waited, glancing anxiously round. It was getting dark and she was in the middle of nowhere. Alone. She was doing everything a woman in a foreign country was not supposed to do. *And* she was on the trail of a criminal. Who knew how he would react when she confronted him?

Uneasy, she rapped again, harder this time, waiting again, sucking on her bruised knuckles. Still no answer. She tried the door. It was not locked. In fact there was no lock.

Calling hello, she stepped inside and found herself in a single small shabby room. There was a narrow unmade bed in one corner. A lamp stood next to it. A couple of cheap woven Mexican chairs fronted the adobe corner fireplace that looked, from the blackened areas around it, as though it smoked badly. The sink and a tiled counter were piled with empty beer bottles and plates, while an open cupboard held a few shirts. Other garments spilled out of drawers in untidy heaps. An old CD player and a Bryan Ferry disc were on the table.

"Is anyone here?" she called hopefully, though she didn't see how anyone could be since there was nowhere else to go, except the primitive bathroom. She checked it, shuddering, then the terrace, if it could fancifully be called that. It overlooked the sea and, in a way, she thought wistfully, had a lot in common with Mac's own place.

Still, out on the terrace she felt the appeal of Perrin's little beach shack. The wind blew coolly through her hair and the only sounds were the thud of the sea and the cries of seabirds. Hard green waves, framed by a couple of blackened mesquite trees, pounded the long crescent of beach, then slurped noisily back again. A squadron of pelicans flew past at eye level, and hanging high in the sky were seabirds with wings like stealth bombers that reminded her of Batman.

Suddenly she understood that Perrin came here in search of peace and simplicity. And to her surprise, she found herself almost hoping he had found it.

Getting a grip, she reminded herself of her mission and also *exactly* why Perrin was here. He must have *something* to do with his wife's disappearance. She was certain he knew where Allie was. Anyhow, he was dodging the law and she was going to find out what was going on and hopefully bring him back.

Tripping over the stones on the path and wishing she had put on sneakers instead of flimsy sandals, she went back to the car. In Mazatlán town, she parked, then hailed a passing mini-moke open-air taxi and asked to be taken to the Bar La Costa Marinera.

The bar was down a side street leading to the beach and it was jumping. Impressed, Sunny counted half a dozen Harleys parked outside, and the mariachi trumpets blasted her eardrums before she'd even stepped through the door.

The café was crowded with locals and holidaymakers dressed for a Saturday night out, most already into their second or third margaritas, served in massive goblets. The wooden tables were packed and brawny waiters in yellow shirts lofted enormous platters of seafood over their heads. The aromas of deep-fried red snapper, of shrimp and hot chilies and cheese nachos filled the air, and those seated on the terrace reaped the benefit of the evening breeze while devouring guacamole and spicy salsa. Telling the waiter she was looking for a friend, Sunny wended her way between tables, fending off bantering offers to come take a seat.

A mariachi troupe in tight black pants and short jackets, glittering with silver studs and sequins, were playing familiar Mexican songs very loudly. There were two girls on fiddles, their long hair swaying as they played, warm dark eyes smiling, while an old man hefted the traditional huge Mexican bass. Plus there were four guitars and two trumpets. Sunny stopped to listen, applauding with the others.

In a far corner of the terrace the Harley bikers sat at a big round table, arms around each other's shoulders. Sunny watched disapprovingly as they poured tequila shots down their throats. Bottles of Tecate and Corona littered the table and huge platters of lobsters and rice and beans were being delivered. Sunny hoped they were not riding those bikes home.

She wasn't the only one watching. At his lonely corner table, Ron Perrin sank another tequila shot followed by a slug of beer from a bottle topped with a crescent of fresh lime.

The bikers yelled to the mariachis to play "Guadalajara."

"Guadalajara," Perrin yelled with them. It was his favorite.

Sunny turned to see where the American voice came from. And there he was. The rugged, beetle-browed billionaire mogul in a pair of flowered bathing shorts and an old Mazatlán T-shirt.

"Guadalajara," Perrin began to sing along in Spanish, in a firm tenor voice, and the waiters crowded round, joining on the chorus. Ron knew every word and brought it to a fine Mexican-style yipping finale, taking a bow at the applause and whoops of acclaim when he'd finished.

Sunny threaded her way toward him. His head was lowered now and he stared somberly into his glass.

"Ron Perrin," she said.

It was not a question, it was a statement, and he knew it. He lifted his head and looked at her with those molten brown puppy-dog eyes.

"Aw fuck," he groaned, reaching for the bottle. "I'm busted."

CHAPTER 34

Perrin did not invite her to join him, but Sunny pulled back a chair anyway and sat down. Her soft white skirt ruffled around her pretty knees and she noticed that Perrin noticed that. He was still the ladies' man, even in a moment of crisis.

Perrin raised his hand to summon a waiter. "What would you like?" he asked, polite despite the circumstances.

"Mango margarita, rocks, no salt," Sunny told the waiter.

"Are you crazy? Whoever heard of anyone drinking a *mango* margarita?"

"Obviously you haven't traveled in the right circles, Mr. Perrin. Mango margaritas are very popular."

"Hah! I prefer my tequila straight."

"So I notice."

They stared silently across the table, taking each other in. Even though Perrin was drunk, Sunny thought he had an oddly magnetic personality, forceful, drawing you toward him with soft brown eyes that were in complete contradiction to his supposedly tough character. Yet she knew this was a man of steel, a man who did battle in the boardrooms of the world, a dangerous man nobody ever said no to. Except his wife.

"So who the hell are you anyway?" Perrin demanded. "You're not the cops, not the FBI . . ."

"Is that who you were expecting?"

The waiter brought Sunny's margarita. Her throat was parched by the hot dusty drive and she sipped it through a straw, eyes raised to watch Perrin.

"You're a beautiful woman," he said, avoiding her question. "I came here to get away from that."

"Seems you succeeded." She took another long sip. The trouble with mango margaritas was you could almost forget there was tequila in there. Two and it crept up on you like rocket fuel. Suddenly hungry, she waved the waiter over again and ordered nachos.

"You don't need to worry about your figure then," Perrin said.

Sunny tried to assess if he was coming on to her, but decided he wasn't, though his eyes still admired her. He seemed suddenly more sober.

"And you speak perfect Spanish," he added.

"My father's Mexican. Other than my grandmother, I make the best tamales you've ever tasted. And since Abuelita wasn't fat, I'm hoping I inherited her genes as well as her recipes."

The nachos arrived: steaming-hot refried beans and cheese on crispy tortilla chips. Sunny dived right in. "Go ahead," she said to Perrin. "Try them, they're good." She stopped in midbite. *What was she doing? This was the enemy, her "prey," and she was acting like they were friends out on a dinner date.* She took another slurp of the margarita and said, "So, anyway, what are *you* doing here, Mr. Perrin?"

He grinned at her, brows beetling. "Well, darn it, I thought you must know. Otherwise why are you here?"

"We're going round in circles," she said, and found herself smiling back at him. She had to admit again that Ronald Perrin had a certain charm. She was meant to be the interrogator, doing Reilly's job, and here she was schmoozing. "Anyway," she said briskly, "I came here specifically to find you. We need to talk."

"So how *did* you find me?"

"The tax receipt in your Hummer for the

Villa des Pescadores."

"Should have shredded that," he said with a sigh. "I just couldn't get around to everything, there wasn't time. Anyhow, what were you doing in my Hummer?"

Sunny blushed. "Oh God, now I have to confess," she said, lowering her eyes to avoid his gaze.

"So?" He was waiting for an answer.

"I guess you'd call it 'breaking and entering,' " she said with a sigh. "We were looking for clues."

"Clues to what?"

She glanced up, uncertain. This conversation had gone all wrong. Wasn't she meant to be the interrogator?

She avoided his question. "Don't worry, we didn't find any, except the tax receipt."

They stared silently at each other across the table. Perrin downed his tequila.

"How about that train set?" she said.

He shrugged. "I grew up too poor to have a train set. Call it the child in me."

Sunny smiled back at him, understanding.

"You still have not told me who you are," he said.

"My name is Sunny Alvarez," she said. And to her astonishment he offered his hand across the table.

He held it fractionally longer than necessary, looking deep into her eyes. Then he said, "And exactly who *is* Sunny Alvarez? With the Mexican father and a grandmother who makes the best tamales in town?"

"I'm Mac Reilly's —" Sunny stopped herself quickly from saying *fiancée,* because of course she wasn't, and *girlfriend* sounded too cute. "Assistant," she substituted at the last second.

"I should have known. I asked him to work for me. He turned me down. If he hadn't, maybe I wouldn't be in the fix I'm in now," he added bitterly.

"And exactly what *fix* is that?" Sunny knew she had him nailed. He would tell her everything now.

Instead he summoned the waiter to bring a second mango margarita.

"So, what *really* brings you here?" He gave her that beetling glance.

"Allie Ray," she said simply.

Perrin stared silently into his glass. "Don't tell me you're here to serve me with a subpoena," he said, suddenly turned to ice.

She shook her head. "And don't *you* tell me, Ron Perrin, that you don't know Allie is missing."

Perrin lifted his eyes. He stared at her, his face suddenly devoid of expression.

"You must know she disappeared from the

Cannes Film Festival," Sunny prompted. "I thought you'd know where she's gone."

He seemed to pull himself together. "Why should I know where Allie goes these days? She's her own woman."

He was playing it cool but Sunny sensed a thread of despair in the slump of his shoulders, in the suddenly-tired face and the blank eyes that were not willing to show his true feelings. Could he really not know where his wife had gone?

"You care though, don't you?" she said softly.

He lifted a dismissive shoulder again. If he was shocked by the news of his wife's disappearance he wasn't going to talk about it.

"What does it matter? She's going her way, I'm going mine." He called for another tequila and sank it down, then poured the Corona down his throat. "Anyhow, you're not here to talk about caring. You're here to make some kind of deal."

"She's not the only woman missing," Sunny said. "What about Ruby Pearl?"

He stared at her, obviously surprised. "Ruby was a secretary. She helped me out for a couple of weeks. I don't know where she went after that."

"And then there's Marisa. She's still in Rome and still waiting to hear from you. She

260

showed us the canary diamond engagement ring you gave her — no small diamond either, my friend. She told us you'd promised to marry her." She looked hard at him. "She's worried, Ron."

Frowning, he ran a hand through his receding brown hair. "I never talked 'love' or marriage to Marisa. That ring was a token. She admired it in a shop window so I bought it for her. It gave her pleasure, but it was not an engagement ring. Hey, I'm still a married man," he added.

"Pity you didn't think of that before you started an affair."

Perrin leaned in to her across the table. "Let's get this straight, Sunny Alvarez," he said. "Marisa was under no illusions, despite what she said to you about an 'engagement' ring. She knew the score and I wasn't the first rich guy she'd hung out with. Marisa is one savvy chick."

"She also said you were into S and M sex."

"She said *what?*" His face was blank with shock. "That's crap."

"I thought she was lying. I even thought she might be planning a bit of blackmail. You know the kind of thing — she'd sell her story to the tabloids unless you came through. And anyway, what happened that night in Malibu when Mac heard her scream? She *was* there, even though you denied it."

"Okay, so she was there. I just didn't want the world to know about it. Can you blame me? I'm a married man. What happened was, I was at a meeting. She'd gone to the Malibu house to wait for me, said she heard footsteps, got scared. Then Reilly came bursting through the door."

"So she shot him."

"Not exactly."

"She tried."

He lifted his shoulder in a dismissive shrug. "A strange guy comes through your window at midnight you'd shoot him too."

"So why were you with Marisa anyway?" Sunny was suddenly curious. There was more to this strange man than met her eyes and she wanted to dig deeper, find out what he was all about beneath that tough-guy layer, stripped of the glamour of power in his ridiculous flowered shorts and old T-shirt.

He gave her a long dark look. "You want the truth?" he said quietly. "It's love gone wrong. That's all, Sunny Alvarez. *It's love gone wrong.*"

He slumped over the table and put his head in his hands. "Allie doesn't love me anymore. That is if she ever did," he added. "Can I blame her? Of course not. And wouldn't most men in the world have liked to be Mr. Allie Ray? You bet they would. It was what I wanted more than anything, more than all the money, the houses, the possessions, the power. I wanted Allie and now I've lost her. And in answer to your question, Miss Detective, I'm here at my little hangout trying to find my soul again."

Sunny stared at him, stunned into sympathy. She gulped back the tears that threatened. "I'm so sorry," she whispered.

"I've never had a woman feel sorry for me before," Perrin said. "And I don't know that I like it." He lumbered to his feet, swaying slightly. "I'm tired," he said. "That's why I'm talking too much. I've gotta go."

He thrust his hand in his pocket and gave the waiter a sheaf of pesos. "That'll take care of it," he said. And ignoring Sunny, he made for the exit. She followed him out into the street. "How did you get here?"

He searched in his pocket for the car keys. "Drove of course."

"Hah! Well you're certainly not driving now. Not with all that tequila in you."

"Are you implying I'm drunk?" He asked the question with all the pompousness of the inebriated.

"I sure am, Ron Perrin." Sunny hailed a cruising cab. She opened the door and pushed him inside. "Take him home," she told the driver, and she gave him directions.

Slumped in the backseat, Perrin looked at her from under drooping lids. "I'll bet you're in love with Mac Reilly," he said. "A woman like you, you could twist any man round your finger."

"I wish," she replied, smiling. Then, "Look, I'm coming over to see you tomorrow morning. We'll talk some more."

"What about?"

"Well, you know, like about money laundering," she said, then immediately wished she hadn't. "And maybe about love," she added, softening the blow. She slammed the door shut and stood, watching the cab drive down the side street then turn left at the corner, out of sight.

There was no doubt in her mind that Ron Perrin was still in love with his wife. And no doubt he was in trouble. Tomorrow she would see what she could do to help.

The next morning, Sunny was up early. She took a walk along the beach, enjoying the fresh clean air and the early warmth of the sun, watching pelicans drop like dive-bombers into the sea, emerging with glittering silvery fish in their beaks.

She took her own breakfast on the terrace: deep dark coffee with sweet rolls that this morning tasted as good as any she had ever eaten. Then she ordered a thermos of coffee to go and she drove slowly over to Perrin's place.

She knocked on the door but there was no reply. She turned the handle and pushed it open. "Hi, it's me, Sunny," she called. "I brought you some coffee, thought you might need it."

Putting the thermos on the table, she glanced around. The room was empty. The terrace was bare. And the Bryan Ferry CD lay broken in two on the table.

Her heart sank. She wasn't bringing Perrin home in triumph to Mac after all.

CHAPTER 35

It was hot in Cannes and Mac drove with the top down on the rental Peugeot, hoping the slight breeze would blow away the cobwebs of jet travel. He checked in at the Hôtel Martinez and went directly to the concierge to ask about limo services.

Armed with a complete list of those in the area, he went to his room, showered and ordered room service breakfast, even though it was one in the afternoon. He pulled on a pair of shorts and sat on his little terrace enjoying the sea view and thinking about Allie.

He had no doubt she had run away from life as she knew it. She'd had enough of being a movie star, enough of being second woman to her philandering husband, enough of the glamour and the riches of the Hollywood lifestyle. Basically, she was a simple woman who, as she had confessed to him, had never felt at home in her skin, and never felt part of the scene in which she played such a major role. Allie wanted her privacy back. She wanted to be anonymous.

He'd had Sunny e-mail the last pictures of Allie, taken at the Festival. She looked beautiful and serene, doing her job, posing for the press and waving to the fans. He thought there was nowhere she could go and not be recognized. She would have to have some sort of disguise, somehow change her look.

He drank the freshly squeezed orange juice that tasted like sunshine in a glass and ate the eggs scrambled with wild mushrooms, still thinking about her.

The most obvious thing would be to cut off her hair. He winced at the thought. Allie's hair was one of her signature features: thick, shiny and naturally blond. Of course then she would need to dye it. She'd go for a simpler look, jeans and T-shirts probably, and she would have to wear glasses. At least if he were in charge of her disguise, that's what she would do. And even then he wasn't sure it would work. Most certainly she would have to leave the glitzy parts of the South of France behind, probably head north, into the countryside.

Taking out the Michelin map he'd picked up at Nice airport, he studied the main routes. One led to Provence and the Luberon, an area filled with movie people, writers and vacationing socialites. The others led toward Toulouse and the west, or north via Agen. Sighing, he folded up the map. France was even bigger than he had thought.

Taking out his list of limo services he began systematically to call each one, trying to locate the service assigned to take Allie to the Festival. He got lucky on the fourth try. Telling the limo company manager he would be right over, Mac threw on a shirt and sandals and drove to the outskirts of Cannes.

The manager was suspicious. "We have already been hounded by too many reporters, monsieur," he said coldly. "We never talk to them about our clientele."

"I work for Madame Ray, I am involved with her security." Mac showed him his card. "I would like to speak to the driver who took her to the airport."

As it happened the man was about to show up for work, and Mac waited under the manager's frowning gaze for him to arrive. When he did, Mac stood up and greeted him warmly, shaking his hand and calling him *"mon ami."* It didn't go down too well. The driver was definitely not his "friend"; he was standoffish, worried that he might be in some kind of trouble. He was a short man with a big nose and beady eyes that stared sullenly at Mac.

"All I need to know is where you dropped Madame Ray, and what she did next," Mac said.

The driver, whose name was Claude, said he had no authority to talk to anyone about Madame Ray and had not said a word to any reporters. Mac had to go through his spiel all over again and though still reluctant the driver finally spoke.

"Madame Ray is not a missing person," he said nervously. "There has been no report to the police of her being 'missing.' The cops are not involved."

"True," Mac agreed, knowing he was probably the only person now who might raise questions about Allie's whereabouts. But he still thought she was a runaway and he didn't want the police on her trail.

"I dropped her at Departures," Claude said. "She was dressed up for the Festival, but when she got out she had on a sweater and a hat pulled over her hair. She wore dark glasses. She didn't even let me get her bag — just one small valise. She took it herself. She gave me a good tip, thanked me . . ." He shrugged. "And that, monsieur, was that."

"Did you watch her go?"

"For a minute, yes. She went straight inside, heading I supposed for check-in. I don't know where she was going, monsieur.

That's all I know."

Mac sighed. It had not gotten him much further. All he had was Allie in place, in semi-disguise, at Nice airport on the night she disappeared. She might have done anything after leaving the limo, taken a taxi to the train station for instance, or rented a car.

Thanking Claude, he drove from Cannes back to the airport in Nice, where he found all the flights that had left late that night and their destinations. Next he went to Arrivals and checked out the rental car agencies. None of them had any record of Allie Ray renting a car.

Stymied, and knowing he had wasted his time, he drove back to the hotel. There was an e-mail from Sunny. "Off on a little trip of my own," it said. "I'll call you when I get back."

Of course he called her right away. There was no reply. Mac sighed. The women in his life were giving him trouble. He wondered what Sunny had been up to now.

CHAPTER 36

Allie circumnavigated Toulouse, panicked by the spaghetti junction of motorways with signs reading BARCELONA, BORDEAUX, PARIS. She didn't want to go to any of those places.

Sandwiched between cars whose confident French drivers were mostly on the phone, she honked at them furiously, then slid the baby blue car between them, putting her foot down and scattering them honking behind her.

The road sign now said AGEN so she guessed that's where she was going. It took longer than she had thought and she was weary of driving. She wanted somewhere rural with fields and trees and maybe even little bubbling streams, the tourist's dream of the French countryside.

The dog snuffled gently in his sleep as she headed north toward Bergerac, through the small town of Castillonnès, where she stopped for a cup of coffee and to walk the dog, and also to ask directions to the scenic route.

Back in the car, she drifted farther into the countryside. Soon emerald green fields rolled away on either side, with caramel-colored cows browsing in the shade of ancient chestnut trees. A small river slid lazily through, making eventually for the big splashy Dordogne. Rocky paths led into hills topped with mysterious turreted castles. Horses grazed in verdant paddocks and long white avenues lined with fluttering poplars led to small peaceful châteaux, while farms with slatted wooden barns looked as though they had been there for centuries.

There was a sudden loud squawking and Allie stomped on the brakes. A bunch of rusty-feathered chickens stood at the gate to a farm, clucking aggrievedly at her. The dog stuck his head out the window to watch as a gigantic rooster stepped into the road. The rooster glanced both ways. Then, squawking and flapping his wings, he sent his harem scurrying back to the gate, out of the way of a small car rushing down the lane in a cloud of dust. The rooster waited till the car was gone, then stepped out into the road again. Again he looked both ways. Getting the all clear, he marched his fluttering flock safely across the road and into the field beyond.

Allie grinned, amazed. She hadn't realized French chickens had road sense. And now her mouth was watering at the thought of a fresh omelet. She pressed on, looking for a place to stay, hopefully near a restaurant.

Quite by accident she found exactly what she was looking for in the back roads around the medieval stone village of Issigeac, where she'd stopped to fill up the car. A red-haired woman was the only other customer and, gathering her courage and adjusting her hat and her dark glasses, Allie asked in nervous French whether she knew a place to stay.

"Why yes," the woman told her, speaking in English. "My friend Petra's B & B is just down the road." She gave her directions, told her it was reasonable and that Petra had also recently opened a small restaurant.

"It sounds like heaven," Allie exclaimed, smiling.

The woman gave her a penetrating look. "Well, not exactly heaven, but if you've come on a long and difficult journey, it might be exactly what you're looking for."

"Thank you." Allie turned away.

"By the way," the woman said, "my name's Red Shoup."

"Hi." Allie hesitated. Then for the first time she said, "I'm Mary Raycheck."

The woman nodded, still looking quizzically at her. "Be sure to tell Petra I sent you," she said, getting into her car. She took a card from her bag and gave it to her. "Here's my number," she said. "If I can be of any help, give me a call."

Allie watched as she drove away. Did Red Shoup suspect who she was? Or was she just seeing a troubled-looking woman with an unnaturally quiet and far too big a dog, roaming around France on her own?

She got back in the car and following Red's directions found herself crunching up a gravel driveway to a small manor house surrounded by clumps of shade trees, under which more cows lazed. A couple of black and white Border collies raced toward the car, and her own no-name dog cowered worriedly back. The front door was flung open and a tall, plump blonde dressed in what appeared to be a red satin nightgown flew down the short flight of steps toward her.

"*Bonsoir, ma chérie,*" she called in French strongly laced with Brit. "Are you the one looking for a room? Red Shoup just called to say you were on your way. And your dog too. No need to be afraid of this couple of old collies, all they want is to chase sheep. Not that there's many of them here, but we had them on the farm back in Wales, y'know. So come on, darlin', let's have you out of there. I have the best room in the house for you."

Petra stopped her onslaught of words and beamed at Allie, still sitting in the car, stunned by the fiftyish blond and scarlet vision peering shortsightedly back at her.

"I still don't know your name, love," Petra said.

"Oh. Right. It's Mary. Mary Raycheck."

"That's unusual. Polish, is it?"

"Originally, I think so. And you must be Petra."

Allie got out of the car and they shook hands. She put the dog on a lead then walked with her new landlady into the house.

"I'm Petra Devonshire. Posh name for a bit of all right like me, eh?" She nudged Allie, laughing. "Used to be a dancer, TV variety shows, touring musicals, that sort of thing, though everybody said I was more the Benny Hill type. Y'know the one always being chased around the garden in fishnets and a garter belt. Anyway, somehow I ended up here. Remind me to tell you the story, love, one of these nights when we've nothing better to do than natter over a glass of wine."

Petra had not yet stopped for breath and Allie had no desire to stop her. She was charmed with Petra's free-flowing barrage of information, enchanted by her free-form lifestyle, amazed by her uninhibited red satin nightie at four in the afternoon, and by her invitation to hear her life story over a glass of wine.

"Follow me, darlin'." Petra twitched her way up a broad staircase lined with family portraits. "None of 'em's mine," she explained. "They all came with the house." She flung open a door at the top of the stairs. "Best room in the place, love. How does this suit you?"

Dazzled, Allie looked at the gilded Empire bed upholstered in threadbare blue damask; at the battered old pine armoire and the ornate dressing table with three blotchy mirrors and a pair of silver candle sconces; at the fluffy white sheepskin rug and the lavish once-red satin curtains now faded to a pale pink. A tiled fireplace dominated one wall, fronted by a sagging sofa whose flowered chintz bulged at the seams, and a massive gilt mirror looked down over all. On a red tray on the table in front of the big window was a coffeepot and the fixings, and a door led into a blindingly white new bathroom with a plastic shower cubicle barely big enough to hold a grown-up.

All memories of Allie's lavish three-thousand-square-foot boudoir in Bel Air disappeared in an instant. "Petra," she breathed, a hand clutched to her chest. "I love it. You may never get rid of me."

Petra's raucous laugh mingled with her own. "That's good," she said. "Because I could use the money. Now make yourself at home, then come down and join me for a cup of tea. And please, love, take off that awful hat. It doesn't suit you. You'll find plenty of straw hats hanging on the rack in the front hall. I'll be in the kitchen when you're ready."

She disappeared in a flurry of red satin, completely uninhibited about being caught in her lingerie.

Allie walked to the window, opened it and stuck her head out. There was no traffic noise; no sirens; no stalkers. A breeze set the leaves of nearby poplars rustling softly, a cow mooed in the distance and the two Border collies chased each other around the meadow at the bottom of the drive. She could see a horse, black and glossy, nibbling the grass, and a couple of old bicycles lay cast to one side by the front steps, where twin urns held a tumble of geraniums in every color of red, and the fragrance of jasmine twining up the old stone walls mingled with the faint tang of hay.

She sighed happily. Could she have found Paradise at last?

Chapter 37

After Allie had showered and changed into a pair of black jeans and a white linen shirt, she walked downstairs, smiling as she passed a suit of armor standing guard draped with a magenta feather boa.

She found Petra in the kitchen, a long room with the kind of plasterwork known as *columbage,* creamy plaster set between a pattern of small beams. Cupboards lined two of the walls while tall windows, inset here and there with stained glass, occupied another. A planked wooden table took up most of the room, big enough to seat at least twenty. On it was piled folded laundry and a basket of knitting wools complete with two sleeping orange kittens. There was an enormous half-finished jigsaw puzzle of the Dordogne countryside and an old radio blasting the latest French hits, several mixing bowls and chopping boards, pottery mugs and plates, together with various magazines and a scattering of papers.

Rickety side tables held tottering piles of books, and coats were thrown over armchairs in front of the enormous fireplace, as if the owners had come in and simply left them there until the next time they were needed. The kitchen door stood open, letting in the breeze and the smell of the countryside.

In the middle of the chaos, Petra, serene, with a pink negligee over her red satin nightie, was fixing tea in a large serviceable brown pot.

"Somehow it always tastes better from a brown pot," she told Allie, pouring boiling water over the tea leaves then clearing a space at the table by leaning her arm along it and shifting the nearest pile to one side.

"There you are, love, you'll be gasping, I'll bet. And here's good English biscuits. McVitie's chocolate digestives. And what about the doggie? What's his name anyway?"

"He doesn't have a name."

Petra's shocked blue eyes were ringed with black and fringed with heavy mascara. "How can that be? We *all* have names, love, even if it's just to differentiate one from the other. I mean, what do *you* call him?"

Allie thought. "Dog dear," she said. "That's all I've called him. So far anyway."

"Well there you are then. Dog Dear he is. Dearie for short. How's that name suit you, old love?" she added, bending to take the newly named Dearie's muzzle in her hand. "I can tell, you're a good dog," she said gently. "You take care of your mum, then. That's your job. All right?" Aiming a critical look at Allie, she added, "Besides, she looks as though she needs it."

Allie took a gulp of the hot tea, burning her throat. She met Petra's mascaraed blue gaze. "Do I look that bad?"

"That haircut's a killer. Whatever possessed you?"

Allie hid her face behind the big blue mug of tea. "Necessity."

Petra nodded. "I've been there myself. A couple of times." She poured milk into her tea from a cow-shaped white jug then added three heaped spoons of sugar, stirring vigorously. "Man trouble was it?"

Allie put her head in her hands, suddenly filled again with despair.

"I left my husband," she said in a whisper so that Petra had to lean toward her to hear. "Or rather he left me. I'm getting a divorce."

Petra sucked in her cheeks and her breath noisily. "Ah, that's bad."

"There was another woman."

"That's even worse!"

"Then he disappeared. Nobody knows where he is."

"Trying to skip out on the alimony, is that it?" Petra heaved a sigh that sent her ample bosom a-tremble. "Men," she groaned. "They never know how good they've got it until it's too late. Trust me, love, if he's got any sense — and since he married you in the first place, I have to credit him with that — he'll come running back, begging you on his knees to let him come home."

"Is that what happened to you then?"

Petra took a biscuit and dunked it into the mug. "I love it when the chocolate melts," she said. "And yes. It did. Twice, in fact."

They were silent as she ate the English cookie, obviously enjoying it.

"So did you take him — both of them — back then?"

Petra's laugh filled the kitchen. "Of course I didn't. I was already on to the next bastard by then. Never could pick 'em, love. Just wasn't in my genetic makeup. Ah well, four husbands and numerous lovers later, here I am on my own again. And I have to admit I'm kind of enjoying it. Though there is a gent I have my eye on. The local squire in fact. Owns a big vineyard hereabouts; tall, dark, good looking. You should see him astride a horse. It turns a girl's knees to jelly."

She threw a penetrating glance Allie's way. "Actually, you're probably more his type than I am. Classier, you know what I mean? Except for the haircut. We have to let our local hairdresser loose on you, see what she can do to tidy it up. Give yourself a break, Mary, just because your marriage is on the rocks doesn't mean you have to let yourself go, now does it?"

Allie recalled standing on the red carpet in Cannes being photographed for the world's press, and she smiled. Her disguise seemed to have worked, even if she did look like hell.

Petra pushed both hands down on the table and heaved herself to her feet, startling the two sleeping orange kittens, who tumbled meowing out of the knitting basket. Rummaging in the table's chaos, Petra found a bowl and filled it with milk from the cow jug.

Just then the two collies bounded through the open door, leaping up to sniff the kittens, who cowered back for a second, then went on calmly lapping. Dearie watched the other dogs nervously, looking ready to run. He turned his big head to look at Allie, and she smiled and gave him a pat and said it was okay.

"Peace reigns," Petra said. "And now, my

love, I have to get dressed. My restaurant opens at six and I haven't even started the pastry for the beef Wellingtons." She raised a questioning eyebrow at Allie. "You ever make a beef Wellington? No? Well then don't bother. I don't know what inspired me to try it but I can assure you it's a bitch to get right. Still, nothing ventured, nothing won is what I always say. And this is a fairly new business so I have to offer something a bit different, don't I?"

She made for the door, turning to look again at Allie. "You want to come along, Mary? You must be hungry, though you don't have to order the Wellington. There's simpler stuff on the menu." She hesitated. "Unless you have other plans, of course."

"No! Oh, no. I have no plans." Allie had no plans for her entire life. She was free-falling and right now she wanted to do it in the company of Petra Devonshire.

"You wanna help then? Or are you going to be a customer?"

"Oh, I'll help. I can be a waitress, washer up, anything you like."

"Right. Here's an apron. Put it on, love, and let's get going."

"Can Dearie come too?"

Petra eyed the dog, who was now up on his

feet and practically glued to Allie's side.

"Doesn't look like we have much choice, does it?" she said with a cheerful grin.

CHAPTER 38

The Bistro du Manoir turned out to be a small converted stone barn. A graveled pathway led to it, just wide enough for two cars to pass, with a sandy parking area to one side. The front entrance was protected from the elements by a gorgeous fluted glass canopy similar to the art nouveau ones found at some Paris Métro stations, and which Petra told Allie she'd had copied by a local glassmaker.

A spacious flagged pergola shaded by a hundred-year-old wisteria, dripping with purple blossoms, wrapped around the back, where tall French doors stood open to the warm night air. Candles flickered on rosy tablecloths, and the zinc bar was already being propped up by a few die-hard locals who looked very much at home.

"Evening, all," Petra said, floating fast through the room, like a comet with Allie at her sparkling tail. "Jean-Philippe," she called to the bartender. "Don't send those boys out of here drunk, okay? We don't want any accidents."

"Nobody's drunk," grumbled one of the younger customers.

"Fine. Great. Just don't do it here." Petra continued on to the kitchen, grinning as she heard him ask aggrievedly why they even bothered coming here when she didn't want them to drink.

Petra rammed the small bunches of flowers she'd picked from the garden into low glass vases, handed them to Allie and told her to put them on the tables. Then she put her to work brewing up a ratatouille, chopping courgettes and eggplant, onions, garlic and tomatoes, after which she showed her how to make salad dressing, using the good olive oil from Azari in Nice, while she went to start her pastry.

She introduced Allie to Caterine, the teenage girl slowly washing lettuce at the sink, and to her temporary assistant, another Brit, who was preparing chicken for a fricassee, amongst a dozen other things.

All at once Allie found herself doing at least two jobs at once. There was no time to think. Customers were arriving and before she knew it she was discarding her work apron and wrapping herself in a starched white waiter's one and was out there taking orders.

To her astonishment the first customer was Red Shoup, glamorous in a flowing Pucci shirt.

"Ohh, hi," Allie said, smiling a welcome, pencil poised over her pad, ready to write their drinks order.

"Well hello! Petra's put you to work already, has she?" Red laughed. "I should have warned you, she does that to all her B & B guests. Tells them it's 'life experience.' "

"And they fall for it every time," the handsome, silver-haired mustached man with Red said, holding out his hand. "I'm Jerry Shoup. And you must be Mary Raycheck. Red told me all about you."

"I didn't know there was that much to tell."

"Don't worry," Red said. "I told him about the dog. And about the haircut."

"Oh God!" Allie put a worried hand up to her cropped locks. "It was one of those mad moments. My husband had gone off with another woman. It was a kind of retaliatory thing."

"Well, it kind of 'retaliated' on you." Red laughed. "But no need to worry, our local hairdresser will sort you out."

"That's what Petra said. And I can't thank

you enough for sending me to her. The Manoir is wonderful, so full of surprises."

"Yeah, well, Petra's known for taking in the waifs and strays," Jerry said. "Not that you are in that category, Mary." He too gave her a long considering glance. "In fact, far from it, I suspect."

Made nervous by his penetrating look, Allie asked about what she could get them to drink, writing their order of a bottle of local red quickly onto her pad.

"I'll be right back," she promised, whisking quickly away, starched apron crackling.

"Excuse me?"

She glanced to her left and saw a man sitting with a tall slender blonde at a table on the terrace just outside the French windows.

"Could we get a bottle of Badoit, please, and two vodka tonics?"

"Of course." Allie made a quick note of their order and the table number, though they were as yet only the second to arrive. "With lemon?"

"Lime, please," the blonde said. In her little white fitted linen jacket, black tank top and black linen pants, her blond hair pulled back, she looked the epitome of simple elegance.

"Thank you. Yes, of course." Managing a quick smile, Allie hurried back to the bar, placed her order then dashed into the kitchen to alert Petra that the Shoups were here plus an unidentified dark-haired man with a blonde.

"Oh, that'll be Robert Montfort, the local squire. Remember I told you he was dishy. And that's his latest girlfriend, from Paris." She heaved a sigh. "Somehow they are always from Paris."

"She's beautiful," Allie said, heading back to the bar to pick up her orders.

"So is he." Petra's voice floated after her.

Jean-Philippe, the bartender and would-be sommelier, had already set up the bottle of red for the Shoups and when Allie took it out to them the "dishy" squire was standing next to their table, chatting to them.

She hurried past with their vodkas. She placed one carefully in front of the blonde, who was alone at the table. "I've brought extra ice and extra lime, just in case," she said.

The woman nodded. *"Et bien, mademoiselle,"* she said, staring at her. Then switching to English. "Haven't you and I met before?"

Allie thought nervously the woman did seem vaguely familiar. "Oh, no. No, I don't think so. I only arrived today. I'm a guest at Petra Devonshire's B & B."

"Of course you are." The squire was back, looking at her with dark blue eyes under lowering brows. His black hair was brushed back from a widow's peak and his lean face had the hard tanned look of an outdoorsman. "Petra always takes in the attractive ones," he said, making Allie blush. "Despite the hair," he added, and he and the blonde burst into laughter.

Allie hurried back to the kitchen and told Petra what they had said.

"Robert's like that." Petra wrapped a fillet of beef, swathed in pâté, inside a lump of pastry. "There, let's hope the bloody thing doesn't come out looking like an underdone sausage or an overcooked rattlesnake," she said, cutting a series of little V-shaped slits in the pastry then brushing it with a beaten egg mixture. "To add a nice brown glaze," she informed Allie, who was watching interestedly.

"Anyhow, you be careful with her. She's something to do with TV. A journalist. One of those nosey parkers who likes to dig up scandal and gossip about celebrities for the delectation of we plebeian folk with nothing better to do than watch her. Not that she would find anyone to talk about here. Unless it's Robert, of course. Red told me she'd 'set her cap' at him, as the saying goes."

She wiped her hands on a kitchen towel and surveyed her kitchen. "Okay, love, better get out there and start taking orders. The tables are already filling up. Tell them the night's special is the beef Wellington and that it's very good. And remember to keep your fingers crossed behind your back as you say it."

Back outside Allie cast a wary eye at the blond TV journalist. She was leaning across the table, deep in conversation with Petra's "dishy" squire. They looked as though they didn't need to be disturbed by a waitress asking for their order, and still nervous of the woman's probing question, she decided to ask Jean-Philippe if he would take it instead, since she now had to run around catching up with six more tables that had suddenly filled up.

"That's the way it is, everybody always comes at once," Jean-Philippe told her. "Better get used to it."

Allie rushed from one table to the next, distributing menus, mentioning the night's special, taking drinks orders, then hurrying to the bar to get them filled.

Too soon, Petra was poking her head out of the kitchen and yelling, "Chef here. Or have you all forgotten? The first orders are bloody well ready so get a move on, Mary."

Allie had not worked so hard since the last week of her filming, when she'd had to be dragged by a runaway horse. She'd insisted on doing her own stunts and her body had ached for weeks after, but it had given her the same kind of personal satisfaction she felt now.

By ten-thirty most diners were finished and already drifting away, while others were on to dessert. The squire and his girlfriend were still ensconced beneath the wisteria bower and looked ready to spend the night there, when Petra emerged from the kitchen, wiping her hands on her apron, her perky chef's hat pushed to the back of her fluffy blond head.

"Is that you over there in that dark corner, Robert Montfort?" she demanded loudly, stalking toward him and pulling up a chair.

"It is," Allie heard him say in a resigned tone. "You know Félice de Courcy?"

"I know *of* you, of course." Petra shook Félice's hand. "I've watched your program. I don't think you'll find much scandal and mystery around here, though. But then, you're probably not here for that," she added, with a meaningful smirk at Robert.

Allie hurried back into the kitchen where she began rinsing off plates and stacking them in the dishwashers. She left the roasting tins to soak, wiped down the countertops and put ingredients back in the pantry. The assistant had left long ago and she was alone.

She stepped outside to check on Dearie. He was sitting on the grass. As soon as he saw her he came running and she gave him the big bone from the lamb roast.

Walking back into the kitchen, she poured herself a glass of wine, picking at the leftover Wellington and mulling over her first night as a waitress. All in all, with the exception of the too-perceptive TV woman, it had been a success. She hadn't once thought of Ron or of her world as a movie star. Or of Mac Reilly. She couldn't wait to do it again tomorrow.

Life was pretty good, here in Paradise.

CHAPTER 39

Mac was back in Malibu, having just flown in from Paris. He'd told Sunny he'd gotten exactly nowhere in France.

"I don't believe Allie has come to any harm, though," he said. "My guess is she's temporarily had enough of it all. I think she's gone looking for something better."

"Good," Sunny said, but not as though she meant it.

"Anyhow, the letters have stopped, and the stalker has disappeared into thin air. And so, by coincidence, have Jessie Whitworth and the blond Queen of England."

"No kidding?" Sunny's brows rose in surprise but still, she didn't ask what he meant by that. She was too preoccupied with her own guilt.

They were sitting on the sofa at Mac's place, eating take-out sushi and drinking a Gewürztraminer. It probably wasn't the appropriate wine to accompany spicy tuna hand rolls and yellowtail sashimi, but Mac thought it sure tasted good. Anyhow, he was glad to be home, with his boy, Pirate, lying at his feet gazing adoringly at him. Which was more than Sunny was doing. In fact, she wasn't looking at him at all. He might even have said she looked distinctly nervous.

"So what's up?" he asked, slipping an arm around her shoulders.

She looked sideways at him from under her eyelashes. "I have a confession to make."

He grinned. "Don't tell me you've bought another Chihuahua!"

"I wish . . ."

She seemed distinctly put out. Serious now, he said, "Okay, so let's have it."

"I found Ron Perrin."

He stared at her. "Go on."

Sunny told him the whole story, from finding the tax receipt in the pocket of her shorts to her talk with Perrin in the Bar Marinera, and the fact that she had let him get away.

Mac heaved a regretful sigh. There was no point in telling her she should have waited, what was done was done. And besides she was obviously upset.

"Hey, maybe you're just not cut out for this detecting business." He squeezed her shoulder sympathetically. "So the man got away. At least now we know he's alive and kicking. Anyhow, we'll be on his trail, don't you worry."

"What do we do next?" she asked, looking hopefully at him, wishing he would say let's forget all this and go to Vegas and get married.

"We'll go pay Demarco a visit in his fancy new desert mansion," Mac said. "Find out what he's all about."

"Be glad to see you," Demarco boomed when Mac gave him a call and told him they would be in the desert that weekend. "I'm having a party Saturday night. Why don't you join us?"

Perfect, Mac thought. It would give him the opportunity to observe the lion in his natural habitat.

"By the way, it's a costume party," Demarco added. "Come as the person you wish you were, is the theme."

With her confession off her conscience, Sunny was looking forward to getting away, just the two of them, no dogs allowed this time. Driving through the Coachella Valley, she was thrilled by the unexpected beauty of the desert, ringed by mountains that turned pink at sunset and fringed with groves of fluffy palm trees, like a storybook oasis. Flowers bloomed everywhere and the many golf courses were dotted with glamorous Mediterranean-style houses.

They checked in to the old La Quinta Resort, which had started as a small adobe getaway for Hollywood's famous in the late twenties. Now it was a wonderful sprawl of coral-roofed buildings surrounded by turquoise blue pools, grassy lawns and sparkling fountains. They had just enough time to get into their costumes for the party.

Demarco's house was in an expensive gated community. The guard took a long look at them in their vampire costumes, then, with a smirk on his face, waved them through.

What's wrong with him? Sunny tugged nervously at her blouse. "Do I look okay?"

"You look fantastic," Mac said. "I'll bet you win first prize."

"I didn't know there was to be a prize," she said, pleased.

"I was speaking figuratively."

"Oh!" She gave him a glare, but they were already pulling up in front of a massive single-story mansion, complete with marble steps and valet parkers standing to attention.

To Sunny, a party implied some kind of jolly revelry, but here there were just rich older folk in a large overdecorated room, knocking back hefty martinis and making polite conversation. Music droned in the background and a few of the women were in discreet costumes of the twenties flapper era but most looked as though they were wearing Bill Blass or St. John, and all the men, including Demarco, wore black tie.

All eyes turned to look when Sunny made an entrance in her Vampira outfit, flounced peasant blouse hanging off one shoulder, skirt a short sliver of artistically ripped and tattered leather. She wore her tall red Versace boots with the seriously high heels and pointy toes, and a ferocious set of fangs that glittered in the candlelight. Mac was looking pretty outrageous too, in his black leather Lestat outfit with fangs that were a match for Sunny's.

Sunny took in the sedate party group and grabbed Mac's hand for moral support, sucking in her stomach and looking as haughty as a girl could, wearing fangs.

A tall, silver-haired man detached himself from the crowd. "Good to see you, Reilly." Demarco held out his hand. "Though I almost didn't recognize you in that getup."

Sunny noticed Mac held his own hand out somewhat reluctantly, and she wondered why. That is until their host took her own hand and crunched it in his. She batted her eyelashes in an effort to keep the tears of pain from falling.

"Sam Demarco," he said, apparently unaware of his bone crusher. His smile was jovial, but his eyes were absorbing too much of her. Sunny hitched up her slipping peasant blouse. She felt maybe their roles should be reversed, with him as the vampire and her holding the cross while Mac hammered a silver stake through his heart. But this was just gut reaction and he was her host, so she smiled back and said, "Good to meet you."

"I thought it was supposed to be a costume party," she said.

"Well, of course it is," Demarco said. "It's just that my guests are a little too old to be dressing up. To tell you the truth, it's all some of them can do to get into their normal clothes." He surveyed his guests dispiritedly. "Let me get you a drink and introduce you to some people," he said.

They followed him, shaking various rather limp beringed hands. Sunny sipped a very strong martini and choked down a bite of very good caviar on very hard pumpernickel with a blob of crème fraîche, and unsuccessfully tried to merge with the St. John crowd.

"They're eyeing us like we're the hired entertainment," she whispered to Mac. "I think they're waiting for us to spin into a Fred and Ginger routine or a wild Apache tango."

They slipped out onto the terrace, pretending to be moon gazing, then decided in a quick consultation that Sunny would suddenly feel unwell. They went back inside and Mac told Demarco, who said he hoped it wasn't anything she had eaten there.

"Oh no," Sunny said, attempting to look pale and weak. "The caviar was divine. Thank you so much for a lovely evening and I'm sorry but I really must go."

She avoided Demarco's bone crusher, but to her astonishment, he bent his patrician lion head and kissed her on the cheek. Not an air kiss either, but a real buss that hovered somewhere near her mouth. She smelled his cologne, citrusy, sharp. It suited him.

Leaning pathetically on Mac, she tottered from the house, leaving behind a discreet murmur of conversation, the sound of Andy Williams singing "Moon River" and the scent of Joy hanging like a pall over the room.

"I feel stupid in this outfit," she fumed, waiting impatiently on the white marble steps for the parking valet to bring the Prius. "It wasn't worth holding my stomach in for. Why did we do this anyway?"

"Demarco invited us. I wanted to see how he lived."

"And?"

"RP's assistant lives in maybe even greater splendor than RP himself. This man is earning some serious money simply for being 'the right-hand man.' "

"Maybe RP gives him insider tips." She grinned at Mac. "Y'know like buy Yahoo! for ten dollars a share."

Tires screamed on the blacktop and the Prius stopped on a dime right in front of them. "Thanks a lot," Mac said scathingly to the maybe eighteen-year-old who had gotten off on driving the car with his foot to the metal, just to see how it felt. He tipped the kid, who ran to hold the door for Sunny, grinning down into her peasant blouse. Sunny wanted to trap his fingers in the door but she contented herself with a haughty glare.

"Anyhow, where are we going now?" she asked.

They were heading down a darkened road with nothing but desert rubble on either side. The headlights picked out an animal. Its eyes gleamed like gold mirrors at them for a second before it disappeared.

"Coyote," Mac told her, and in the distance she heard the baying of the pack. "We're going to Ron Perrin's place," he added.

"Great," Sunny muttered, annoyed with her silly costume and the fact that she had made a fool of herself in Desert Society. "Another fun location. Can't I just go and change first? Anyhow, you didn't tell me you had found Perrin. When did he get back?"

"I didn't and he hasn't." Mac swung the car toward a pair of lofty iron gates topped with spikes. The big stucco wall all around the property was studded with shards of glass and a foreboding shiver tickled Sunny's spine.

"What a way to live," she said. "Behind big walls and broken glass and iron spikes."

Mac took the electronic opener he'd filched from RP's Hummer and pressed the code into it. The gates swung open and they drove through.

"We're trespassing," Sunny said nervously. "There must be dogs, killer Dobermans or something."

"No dogs," Mac said calmly as the gates clanged shut behind them. He parked in front of the pink stucco house that had been built in another era, when glamorous movie stars of the thirties and forties fled L.A. to avoid the media and find solace and sex behind the hidden gates of Palm Springs.

And now it was Ron Perrin's home. Or at least, one of them.

CHAPTER 40

The moon gazed serenely down. Beyond the walls they could see the distant shimmer of lights in the houses built into the foothills, and above them stars twinkled in an un-smoggy desert sky. Mac had a key in his hand, and in a second they were inside and he had disabled the alarm.

Sunny stood nervously just inside the doorway. She had stepped into a wall of blackness. It pressed against her eyes. The shiver up her spine was no longer a tickle; it was a definite tremor.

"I don't think I like this," she whispered, groping in the darkness for Mac's hand. She couldn't find it. "Mac," she hissed, frantic.

"Keep quiet, Sunny," he said in her ear, making her jump.

She turned blindly to him. "Don't do that to me," she said, shaken. "I don't think I'm the big strong girl I thought I was."

"Oh yes you are, and for God's sakes,

honey, shut up."

Sunny seethed inwardly, wondering why she had agreed to come on this fool's expedition, and looking like a fool in her Vampira outfit. Besides, the pointy-toe boots were killing her.

After a while the darkness seemed to melt a little. Now they could make out the shapes of furniture, heavy-looking carved Spanish pieces. An antler chandelier with crystal drops swung in the tiny breeze they had made closing the front door. Other massive doors led off the hallway and there was a lot of art on the walls.

Sunny squinted interestedly at them. This was not modern stuff like in Malibu, nor the reputedly priceless Impressionist collection of Bel Air, but simple amateur-looking desert landscapes. She wondered whether RP had painted them himself, whenever he took time off from making money, that is. And then there was the wonderful miniature railway track, more elaborate than the one in Malibu, complete with stations and pretty little trains.

Mac walked through a door to the right and she scurried to keep up, heels clattering on the tiled floor. She heard Mac groan again.

"Jesus, Sunny." She caught the glitter of his

eyes. "We might as well just turn on all the lights and say, Well, folks, here we are, burglarizing your desert compound."

"Why don't we just do that?" she said wistfully. A little light seemed like a great idea.

She shadowed him, glancing nervously over her shoulder, as he went quickly from room to room. It was not a small house. Sunny counted seven bedrooms, each with bath; plus several main rooms. And through them all ran the metal tracks of the fantastic mini-railway. Kneeling to inspect it, she was so delighted she could have played with it all night.

In the office, Mac switched on the bank of computers, blinking in the green glow from the screens. He looked up puzzled by a sudden roaring noise. It sounded like an approaching express train.

Sunny stared, surprised, at the mini–rolling stock, half-expecting to see a train shoot by. There was a sudden hard jolt. The floor rolled beneath her feet and the whole world was shaking. Things flew off shelves and plaster crashed from the ceiling and she fell backward beneath a heap of tumbled masonry.

"Sunny," she heard Mac yell. Lifting her head, she stared foggily around. The ground began to shake again. She clung despairingly to a heavy table leg.

"Mac, it's an earthquake," she screamed, choking on the dust. *"Mac."*

The scream of rock plate against rock plate as the earth slipped and heaved suddenly stopped. In the eerie silence, the only sounds Sunny could hear were those of her own breathing and of dust trickling in little streams onto the rubble. Then with a sudden rumble and a toot a miniature train scooted tipsily past on its twisted metal track.

"Jesus, Sunny." Mac's arms were around her. "Are you all right, baby? Oh, God, tell me you're all right."

She leaned tearfully against his chest. It was almost worth getting half-killed to hear the tremor in his voice that meant he loved her.

An aftershock rolled through and arms around each other, they staggered through the broken glass and fallen artworks out into the night.

Moonlight dazzled down onto the fabulous cactus garden. Whoever had built the house in the thirties had been a collector and there were many wonderful old species, each with a metal tag, recording where it was native to and its age.

They stood in front of a tall cruciform saguaro cactus planted on top of a small sandy hillock, clinging together, waiting for the earth to stop shaking. The cactus speared upward into the night like a gigantic thorny branch. It looked green and healthy and very well nourished.

The ground shivered again. It was like standing on Jell-O. The sand piled around the base of the cactus began to slide in a miniature torrent. Faster, faster.

As they watched, a second cactus seemed to be growing. It emerged, slowly at first. Then suddenly, it snapped stiffly upright.

It was an arm. No flesh. Just bones. The radius, the ulna, and the hand. A diamond watch clasped around the wrist bone glittered in the moonlight.

"My God, oh my God," Sunny screamed. "Oh my God, there's a body under there. Oh God, Mac, tell me I'm dreaming, tell me this is not really the house of horrors."

Mac was already checking the watch. It was still ticking.

CHAPTER 41

Palm Springs is a cute little town, a leftover from the twenties and thirties with modern-day accents, and its own brand of charm. Sitting in an interview room at the Palm Springs Police Department, Sunny was sure nothing ever happened there, and even though she had removed her fangs she knew that she and Mac still looked highly suspect — she as Vampira and Mac like Johnny Depp playing a psychopath, and both of them battered and bruised and filthy. To say nothing of terrified. At least she was.

A short while later, Mac went back to Perrin's house with the detectives while Sunny sat drinking coffee under the skeptical eye of a young policewoman. They traded stories nervously about their jobs. Eventually Sunny ran out of conversation and coffee, then Mac came back to "rescue" her and tell her they were going back to Demarco's. The cops wanted to ask him some questions too.

Demarco's new house seemed to have withstood the earthquake well, mostly just broken martini glasses and a couple of heart-quakes amongst his more senior citizen guests that had needed defibrillation.

Demarco himself was remarkably cool, un-fazed by the questions, polite and enigmatic. He had no idea whose the body was, or where his friend Perrin was. He told the de-tectives that Mac could testify to that be-cause he'd already hired him to try to find Perrin.

"Feel free to search my house," Demarco said. "Of course you'll not find him."

"But," as Mac said later to Sunny, "this is the Mojave Desert. Perrin could be almost anywhere. Once you are out of the man-made oasis that constitutes the Palm cities, you can drive for miles and all there is, is rocks and rubble, sand and more sand, foothills and gigantic mountains, with occa-sional small homes and cabins tucked away. It would be like looking for a rat in the desert."

But he knew there was no holding back the police now. A body was buried in Perrin's backyard and Ron Perrin was "a person of interest" to the Palm Springs Police. And everybody knew what that meant.

"He has to be somewhere close by," Mac

said. "Remember OJ's friend sheltering him after the murder of his wife? Demarco has a lot to gain from remaining loyal to Perrin. He could be hiding him somewhere."

CHAPTER 42

Allie was deep into life at the Manoir and the Bistro. Every morning early, with Dearie ambling at her heels, she walked through green-lit birch groves and alongside fields bright with sunflowers. Pale buff-colored bunnies skipped in the hedgerows and the dog chased them, though he'd never caught one. She had avoided reading the newspapers, and the first hue and cry on TV about her disappearance had stopped.

With a tug at her heartstrings, she wondered why Ron had not tried to find her. She guessed he must still be with Marisa. Their affair must have been "love" after all.

Soon she found herself walking along a sandy lane where rows of leafy vines dangled enormous bunches of hard green grapes. A rosebush in full bloom stood at the end of each immaculately groomed row. The roses were so lush and overblown they reminded Allie of Petra and she took out her little

Swiss Army knife, bought for use on spur-of-the-moment picnics, thinking she would cut a few to take back to her landlady.

"Hey, what do you think you are doing? This isn't a flower garden open to the public." Robert Montfort's angry blue eyes stared at her. "Don't you know, madame, that the roses are there to protect the vines?"

"Ohh, hmmmmm . . . actually, no I did not." He'd spoken in French and automatically so did Allie, somehow better at it when she was agitated and not thinking too hard about conjugating the verbs. "I'm so sorry," she added, "I didn't think."

"Then next time perhaps you will. Those roses serve a purpose. The bugs are attracted to them, and it keeps them off the vines. It also allows us to know what pests might be infesting them."

She nodded, doing her best to look apologetic. He was giving her a long look and she glanced away, but not before noticing that he wore old jeans and a blue open-necked shirt and he looked, to quote Petra, "very dishy."

"The hair looks better," he said, walking toward her.

She put up a nervous hand to touch it. She said, "Jacqui in the village cut it properly for me."

A smile lit his lean face. "Sorry I was so

rude about it at the Bistro," he said. "I didn't mean it quite the way it came out."

She nodded again. "That's okay. I probably deserved it."

"Nobody deserves to be laughed at." He stood looking at her as she shuffled the roses from one hand to the other.

"My grandparents were farmers," she said, surprising herself since she had never met them and only vaguely remembered the story passed down in a more sober moment by her mother. "Sharecroppers really. Tobacco. They were from the South."

"Well, I'm a sort of farmer," he said. "Tell me, Mary Raycheck, have you ever seen a winery?" She shook her head, no. "Then would you like to see one now? I'll personally give you the 'Grand Tour.' Come on," he said, walking back down the lane to where a battered old Jeep was parked.

The dog followed them. "His name is Dearie," Allie said, hurrying to keep up with Montfort's long-legged stride. "I found him abandoned at an autoroute café."

"You're a dog lover then?"

Allie thought about Fussy. "Not all dogs," she said. And then she remembered Pirate. "Though there is one other special one I'm in love with."

They took off at a fast clip down the narrow lane. "I've never heard of anyone being 'in love' with a dog," Robert said, swinging onto a smaller road. In the distance a group of pale gold stone buildings rose to meet the sky.

Nor had Allie, and now she wondered if what she had really been thinking of was Mac Reilly. Was she in love with him? Did she still love Ron? And who exactly was this attractive stranger she was drawn to right this moment? She glanced at him out of the corner of her eye. Or was it that she was simply a lonely woman on the lookout for love?

Behind them, balanced on the narrow backseat, Dearie gave a complaining growl as Robert swung the Jeep under a sign that said in elaborate black script, *Château de Montfort.* Underneath were the words *Appellation Contrôlée.*

"Not many wines in this area are designated *appellation contrôlée,*" Robert told her. "I'm lucky enough to have one of the best *terroirs,* the best pieces of land in the area. This hillside is perfect for the grapes."

They walked into the winery where he showed her the enormous new stainless steel vats that, he said, compared very well with his older wooden ones. Then he took her into the *cave* where he punctured a cask and drew off a sample of wine for her to taste.

314

"This is still too young," he said when she tried it, though Allie thought it good. But then he said, "Now, taste this. It will be bottled next month and out on the market."

She tasted again, nodding her enthusiastic approval. Since she had been in France she was beginning to get the hang of this wine thing.

"I'm proud of it. It's one of my best," Robert said, taking her elbow as they walked outside. They stood under a striped canvas awning, looking at each other. Then he said, "Did anyone ever tell you you're a beautiful woman?"

Allie shoved the big square-framed glasses further up her nose. Remembering her other life, when she had been famous for her looks, she smiled. "Not for a long time," she admitted.

"Hmmmm, then whenever I see you, I must remember to tell you." His eyes linked with hers. "And I'll also tell you that behind those glasses your eyes are the bluest I have ever seen. The exact color of the Mediterranean at the very beginning of summer."

Allie felt the blush rise through the back of her neck to the tip of her head. "Thank you, monsieur," she said politely.

His laughter echoed around the paved courtyard, causing Dearie to cock his head wonderingly to one side. "After that," he said, "I must ask if you would do me the honor of having lunch with me. There's a café just down the road. It's simple: omelets, salads, that sort of thing."

Soon, they were sitting on the terrace, sipping glasses of the local wine — not Robert's, too expensive for this little café, he told her — and he was asking her to tell him all about herself. Who was she? Where was she from? What did she do before she became a waitress at Petra's bistro?

She looked at him over the rim of her wineglass. "I hate talking about myself," she said quickly. "I'll tell you only that I'm married, that my husband is in love with another woman, and that I'm getting a divorce."

"Foolish man," he said calmly. "To let a woman of your caliber go."

"And how do you know what my 'caliber' is?" She put down the glass. He was, she thought, almost too handsome. He probably had a dozen gorgeous Paris blondes running after him.

He lifted a shoulder in an easy shrug. "I'm not sure. It's just something about you. It's reflected in your eyes . . . a kind of simplicity, I think. And of course, there's your beauty." He studied her blushing face a little longer. "An honest kind of beauty I would

call it," he said finally. And to her astonishment he leaned across the table and took her
hand. Then he bent his head and kissed it.

"Now," he said, releasing her hand, all
practicality again. "What shall we order? I
recommend the *omelette aux cèpes* — the
wild mushroom omelet. They're good at this
time of year. And a little salad?"

"Sounds great to me," she agreed. "And
then it's your turn to tell me all about yourself."

Over their leisurely lunch, sitting on the
terrace of the Café Jeannette, Robert told
her that he had inherited the Château Montfort from his grandfather when he was only
twenty years old. "It's been my home ever
since," he said.

"Lucky you," Allie said, thinking of her
own soulless house in Bel Air. "I wish I could
find a place here I could call home."

"A cottage with wisteria climbing the walls
and roses of your own, in a garden with a little brook running through it," he said, and
she laughingly agreed.

And then he surprised her and said, "I
know just the place. Come on, let's go look
at it."

The cottage was hidden down a rutted white
lane where brambles and blackberries glis-

317

tened in the hedgerows. It listed drunkenly to one side and part of the roof had caved in. The gardens had grown wild in a riot of roses and weeds and a rain bucket stood under a crumbling drainpipe leading from the blue-tiled roof. But it fronted onto a small pond fed by the bubbling little stream Allie had dreamed of, and over which glittering, many-colored dragonflies danced.

The cottage was known as Les Glycines, named, Robert Montfort told her, for the wonderful old purple-flowering wisteria that twisted over the stone walls and an arbor in the back, and Allie suddenly wanted it more than anything she'd ever wanted in her life before, except for wanting to be an actress.

She peered through the windows at the dusty old-fashioned rooms, in dire need of restoration, thinking of what she could do to it.

Of course she could not even think about buying it. She was living a lie, and one day soon she would have to move on. No, this cottage was meant for some happy young couple who could breathe new life into it.

She thanked Robert Montfort for showing it to her, but said it was not for her.

Still she found herself returning there on her morning walks, peering again into the dusty windows and sitting by the dragonfly pond watching Dearie chase the fast-jumping frogs that he never caught, though he somehow always managed to end up chest deep in the mud. Later, Allie would have to turn the hose on him, laughing as he shook himself, sending her running and almost as wet as he was.

Finally, she called Sheila and told her about it.

"I've been so worried," Sheila said. "Where on earth are you?"

"In a French village, staying in a B & B manor house, working as a waitress and kitchen help in a simple country bistro, with my new dog, and looking longingly at a broken-down cottage by a pond . . ."

"Oh my God, you've gone native," Sheila said, but there was a relieved laugh in her voice.

"I think I have," Allie agreed, with that new lift to her voice that made her sound the way she had when Sheila had first met her, when she was young and optimistic and unscarred by life.

"I feel so good here," Allie said. "It's so peaceful and uncomplicated, and nobody cares who I am."

"So where are you exactly?"

"Oh, in the French countryside. . . . I'm not going to tell you exactly where, Sheila, because then if someone like Ron or Mac Reilly or the tabloids asks, you won't have to lie."

Sheila sighed. "Okay, I understand. So, are you buying the cottage?"

"I wish . . ." Allie's voice trailed off. Then, "Have you heard anything about Ron?"

Sheila thought quickly; she didn't want to be the one to tell Allie that her missing husband was wanted for questioning in a murder.

"No one knows where he is," she said, evading the issue. "I know Reilly was in France though, looking for you."

"Really?" Allie said, pleased. Then in a burst of confidence she said, "There's a man here, the local squire they call him. He owns a vineyard . . ."

"Do I interpret that to mean you're interested?" Sheila asked shrewdly.

"Well, not exactly. . . . At least, I don't think so." Allie was not sure herself about her attraction to Robert Montfort. "I have to go now," she said. "Time to put on my apron, I'm expected at the Bistro."

"Oh my God," Sheila said. "I don't believe it, Allie Ray, back where she started, as a waitress."

Allie laughed. "Hey, maybe I'll be promoted to hostess soon," she said. "I love you,

Sheila, just keep on being my friend."

"Of course I will, you know that," Sheila said as they rang off.

CHAPTER 43

Petra was reclining in her massive brass bed, the one with the huge flying sphinx finials that she'd told Allie came from Egypt and had once been owned by King Farouk. Since Petra had found it cheap in Bergerac's Sunday flea market, this was unlikely, but still Allie was inclined to believe her. After all, who else but Petra would own such a fantastical thing?

Pale green satin pillows were heaped behind Petra's head, while the matching coverlet kept sliding annoyingly onto the floor. Petra told Allie satin was always a slippery problem but she just loved the way it looked. A red bandanna was wrapped around her swollen jaw and tied in a perky bow on top of her head. That morning she'd had an aching wisdom tooth removed. It had hurt like hell and now her face looked like a football.

"Just look at my eyes," she wailed to Allie,

as best she could through tight-shut lips.

"They've disappeared," Allie said, offering her a glass of iced orange juice with a straw, which was all Petra had said she could manage.

Petra pushed aside the covers and wobbled to her feet. She sat down again quickly on the edge of the bed. Had she been able to frown, she would have. "I have to get to the Bistro, start prepping the food for tonight."

"Oh, no, you're not." Allie swung Petra's legs back onto the bed. She took off the annoying slippery satin coverlet, straightened the sheet and pulled a blanket over her. Walking to the window, she closed the shutters, turned a lamp on low, then went back and inspected Petra again.

"We'll have to close the Bistro then," Petra mumbled, and despite her slitty eyes Allie could tell she was looking hard at her. "Unless *you* could take over for me, of course."

"What? You mean me? Run the restaurant?" Allie was stunned. All she had done so far was follow orders, chopping, slicing, sautéing and serving.

"Why not? You've been there long enough to know how it all works. If you keep the menu simple: chicken, chops, grilled fish, that sort of thing, you'll be fine. And you know how to make sauces now. You can get

Caterine to make the salads and the veg. Gazpacho's a breeze, you do it in the blender. That or a goat cheese salad, or quails' eggs, for starters. A fruit pie for dessert — there's a couple in the freezer. With ice cream. Or simply berries and cream. The suppliers will have delivered the fresh produce."

Allie said nothing and despite the swelling Petra managed a frown. "Don't tell me you're afraid," she said. "A woman like you."

"A woman like *me?*" Allie repeated, sounding definitely scared.

"A woman like *you,* who can leave her cheating husband and come to France alone in search of a new life. I would have thought a woman like that could do almost anything. Including running a kitchen in a small local restaurant, where she's worked for the past few weeks."

Sitting up, Petra inserted the straw into her swollen pout of a mouth and took a sip of the cold juice.

Allie said nothing. If she took over the Bistro, everything would depend on her. But Petra needed to make money to keep going. Closing for a few nights would make a severe dent in her budget. And besides, Petra was her friend. Now it was Allie's turn to help her.

"Do it, love," Petra said, sinking back into

her pale green pillows and attempting to look fragile, even though she knew she resembled a small hippo. Facially that is.

"So okay," Allie said in a small voice. She hoped she was doing the right thing.

Of course the assistant chef, the young Brit, chose that night not to show up. He'd left a phone message saying he'd gone back to England. "Urgent family business," he'd said.

"Isn't it always?" Petra said resignedly, when Allie phoned from the Bistro to tell her. "Never mind, love, you've got jolly little Caterine to help you."

"Jolly little Caterine" was slow and methodical. Watching her languidly washing lettuces and snipping the tips off *haricots verts*, Allie despaired. There was nothing for it but to plunge in the deep end and prep the food herself.

She took the fruit pies out of the freezer and set them to defrost. She hulled a pile of tiny *fraises des bois*, absently popping a few into her mouth as she went. It was like eating fruity perfume, and she knew they would be wonderful with a touch of Cointreau and maybe some thick rich cream. She found tubs of Carte Noir ice cream in the freezer, the *chocolat gâteau* flavor she already knew was to die for, and *café* as well as vanilla. So much for dessert.

Now for the gazpacho. Checking Petra's recipe book, she peeled and seeded ripe tomatoes and red peppers, previously washed by Caterine, who was now employed slowly setting up the tables. Garlic, tons of it, a bunch of fresh herbs, anything she could get her hands on because she wasn't exactly sure of which ones to use, so she put them all in. Chervil, parsley, basil, chives, tarragon. She contemplated using the blender but then decided she wanted the nuggety texture and chopped all the ingredients by hand instead, sniffing up the herby aroma, and managing not to cut herself. She stirred in lemon juice and the good olive oil from Azari in Nice that Petra preferred. She added thinly sliced sweet onion and cucumber. Salt, a little paprika. And it was done.

Standing proudly back from her handiwork, she called Caterine over to taste. The girl's round brown eyes grew even rounder.

"Mais c'est superbe," she murmured, eyes reverently closed. *"Formidable."*

Allie heaved a satisfied sigh. Telling Caterine to get the salad ingredients together, she took the goat cheese from the fridge and sliced it into neat rounds. She ground hazelnuts in the blender, added fresh bread crumbs, a touch of nutmeg and a hit of olive oil, then rolled the cheese slabs in them, ready to be toasted under the grill and served on a bed of tiny fresh lettuces, scattered with the edible nasturtium flowers Petra grew in the garden behind the Bistro.

Pleased, she stood back and took a look at the big old railroad clock over the door. With a yelp of distress, she realized it was five o'clock and she had yet to create at least two main courses.

There could be no fresh fish tonight since Petra had not been able to get to the market, but the thin pork chops would grill up crisp and easy. The hanger steak needed a bit of help and she quickly chopped spinach, fried up some bacon, chopped that and mixed in some hearty blue cheese. She beat the steak with the meat tenderizer, spread the stuffing on it and rolled it up. Then she cut it into roulades. It looked terrific. Steak and blue cheese pinwheels. God, she thought, pleased, I'm getting good at this.

Fresh baguettes were delivered to the kitchen door and Caterine, back from setting tables, carried them over to the bread guillotine, where they could be quickly sliced and placed in the waiting red-napkin-lined baskets.

Caterine had already taken out the butter

and was making neat little pats out of it, laboriously stamping each one with Petra's favorite cow image. Eyeing the couple of mangoes in the fruit basket, Allie had an idea. She quickly peeled the soft ripe mangoes, mashed them with a fork, then began to blend them in with the butter. She placed the result in little round yellow pots and put them in the fridge to chill. *Mango butter.* She'd had it once in the Caribbean and loved it.

Jean-Philippe arrived, surprised to see her in charge of the kitchen. When she told him what was what, he said not to worry, he'd help with the serving as well as taking care of the bar, and he was sure everything would be okay.

Chopping chickens ready for the fricassee and arranging the piles of already prepped vegetables on the counter in front of her — a *mise en place,* as Allie knew a chef like Wolfgang Puck (a friend of hers) would have called it — Allie was exhausted. And scared. Playing around in the kitchen was one thing. Cooking meals — *many different meals* — and all at the same time scared the hell out of her.

So, okay, you took on the responsibility, she told herself, tying on a clean apron. Petra trusted you with her bistro. It's up to you to take care of things for her.

"*Eh bien,* what shall I do now, Mary?" Caterine's myopic brown eyes stared anxiously into hers.

Allie took a deep breath and pulled herself together. "Okay . . . I mean, *eh bien,*" she said firmly. "Caterine, you will be in charge of cooking the vegetables. You know exactly how many minutes for the *haricots verts* and the baby squash, correct?"

"Correct." Caterine stood up straight, spine stiffened with new responsibility.

"Petra is trusting us to take care of everything for her," Allie reminded the girl. "So we'll both do our best. Correct?"

"Correct," Caterine agreed again.

"First customers," Jean-Philippe yelled from the bar, and despite her resolve, Allie's heart sank. She glanced panicked around the small country kitchen. It was a long way from the Hollywood movie sets, with the catering tents and the professionals in charge of feeding several hundred people every day. And from her own enormous, immaculate Bel Air home, where Ampara ruled over the kitchen. She was on her own.

Jean-Philippe hurried in. "It's Robert Montfort," he said. "And he's with his mother."

"Oh my God," Allie said. Robert of all

329

people. *And* with his mother.

"He'd like the gazpacho, and the lady will have the goat cheese salad to start. One steak pinwheel, and one grilled pork with pepper sauce."

She had forgotten to make the pepper sauce . . .

Caterine organized the bread and the mango butter and took it out to their table, while Allie got the goat cheese under the grill and toasted the rounds of baguette on which it was to be served, atop a watercress and baby lettuce salad dressed with lemon vinaigrette with a hint of honey. She flung red peppers into the blender with garlic and herbs for the sauce to accompany the pork and added capers and cream for good measure. Then she garnished the gazpacho with a slice of lemon and floated chopped cucumber on top. She checked the goat cheese. It was ready. Caterine held the plate while she arranged the toasts artfully. First courses done. On to the next.

"More people arriving," Jean-Philippe called out.

Peeking through the bead curtain that divided the kitchen from the restaurant, Allie saw that there were. Lots of them. She glimpsed Robert on the terrace sitting with a handsome older woman. They were tasting their first courses and seemed pleased.

Back in the kitchen, she prepared for a long night.

By ten o'clock Allie was sweaty and tired. Her hair hung in short limp strands around her face and her glasses were pushed on top of her head. At least there had been no complaints.

Robert thrust his way through the bead curtain. "Jean-Philippe just told me it was you, here all alone," he said. "I had no idea you could cook."

"I can't." Allie slid the glasses back onto her nose and ran damp hands through her miserable hair. "I wasn't asked for my culinary ability. I was the only one available."

"The food was excellent," he said. "Ask Maman. She knows good food when she eats it. Come." He held out his hand. "You've finished for the night. There are no more customers. Join us over a glass of wine, and perhaps some more of those delicious *fraises des bois.*"

"Oh . . . but . . ." Looking desperate, Allie ran her hands through her ragged hair again, making him laugh.

"You look just fine," he said softly "Don't you know by now that you always do?"

"Ohhh . . ." she said again, but Robert was

already untying her apron. "Come on," he said, taking her by the hand and leading her out onto the terrace.

"This is my mother, Céline Montfort," he introduced them. "Mary Raycheck."

Madame Montfort was a still pretty woman, tall, as her son was, with her black hair turning silver at the temples in the most elegant way Allie could have imagined. She wore a blue linen skirt and a white silk shirt with a little scarf at her neck, large pearls, and those chic cream Chanel sling backs with the black toes. She looked, as Allie knew Ron would have said, like a million bucks. While Allie, in her black T-shirt and jeans, felt closer to a mere ten.

"I'm happy to meet a friend of my son's. And such a talented chef." Madame Montfort gave her a long searching look that Allie felt sure missed nothing.

"But I'm only an amateur, madame, standing in for my friend, who had her wisdom tooth removed this morning."

"A painful business," Madame Montfort agreed. And then, over glasses of champagne, she asked Allie a few questions, about how she was enjoying France, and did she find her life here very different from California.

Dearie came wandering over. Madame

Montfort caressed his thick shaggy neck, and the dog gazed adoringly up at her. The moon shone down, the little white lights twinkled in the trees and the wine tasted delicious. Allie turned to look at Robert and caught him looking at her. She smiled, feeling good. It was the perfect end to a surprisingly perfect day.

Chapter 44

Half an hour later, after they had said good-bye, Allie found herself alone again, humming softly as she washed off the counter-tops.

She stood for a moment gazing around the quiet kitchen with its white half-tiled walls, its black-beamed ceiling, its enormous royal blue chef's stove — Petra's favorite color — and the steel grills and cooktops. Battered pots and pans were stacked on wooden shelves, alongside the simple cream plates and dishes stamped in blue with the name BISTRO DU MANOIR.

Outside, Dearie sprawled on the step and a cool breeze mixed with the aroma of chicken fricasseed in good sweet butter and the perfume of wild strawberries. Leaning against the counter, arms folded across her chest, Allie felt a deep sense of contentment. Tonight she had made a success out of what had promised to be total chaos. On her own, she had fed thirty-five customers and there had been no complaints. Only compliments. Flushed with a new kind of success she thought this kitchen felt like home, in a way nowhere else ever had. She couldn't wait to get back and tell Petra all about it.

But first she and Jean-Philippe had to cash up the night's takings. She put the money and credit card receipts in an envelope and stashed them in her big canvas tote, then walked round the restaurant, turning out the lights, making sure all the doors were locked. She said a fond good night to Jean-Philippe, hauled Dearie into the car and drove home to the Manoir. She liked the sound of that word *home.*

Petra was waiting up for her. Her room smelled of the blue irises in a white pottery jug on her bedside table, and a lamp, shaded with a red silk scarf to prevent the glare bothering her poor swollen eyes, cast a rosy intimate glow. Allie had stopped to brew tea in the big brown pot, making sure to place a packet of the chocolate digestives on the tray before she carried it upstairs.

"Here I am," she said, putting the tray down on the bed.

"Oh, goodie, tea!" Petra tried to beam but her mouth got stuck.

"Do you think you'll need a straw?" Allie said.

"What? Tea through a straw? Never. What are you thinking, girl!" Petra unraveled the bandanna that strapped up her jaw, groaning as the swelling hurt even more.

Allie handed her the Tylenol. "Take some with the tea and biscuits. You need food of some sort," she said as Petra's tummy rumbled loudly.

"So?" Petra said. "Tell *all!* How did it go?"

And, over tea and biscuits in the cozy pinkly lit bedroom, with the faraway song of a blackbird caroling in the woods, Allie described the night's events. She told Petra how panicked she had been by the sous-chef's unexpected defection back to England, and how helpful Caterine and Jean-Philippe were. About the food and her new invention of the blue cheese pinwheel steaks that had, as Petra would have said, "gone down a treat" with the diners. She told her about the fabulous *fraises des bois* steeped in a little Cointreau and served with a huge dollop of fresh cream.

And then she said casually, "Oh, and by the way, Robert Montfort was there tonight."

Through the blue slits that were her eyes, Petra gave her a sharp glance. "Oh? With the

Paris girlfriend, I suppose?"

"Actually, he was with his mother."

"Ahhhh! The *maman!*"

"After I'd finished in the kitchen he asked me to join them for a glass of wine."

"And did you?"

Allie grinned. "Of course I did. And Maman was very nice, kind of quietly elegant in that French way, you know with the scarf and the pearls and the Chanel shoes, hair turning silver at the temples."

"Like her son," Petra said, yawning. "I wonder if he'll show up again tomorrow night when he hears you'll be cooking again."

"I will?"

"Well, love, I can't go out with my face like this, can I? The customers would think I'd been poisoned."

Surprised by how thrilled she felt at the news that she was to be chef again, Allie tidied away the tea tray, plumped Petra's pillows, made sure she had a glass of water and kissed her good night. Then she and Dearie drifted down the broad wood-floored landing to her room.

She'd been on such a high she hadn't noticed that her back ached and her feet hurt. Standing in the tiny plastic shower cubicle she let the hot water wash away the smells of the kitchen and her aches and pains.

For once she was not thinking of Ron when she climbed into the soft bed and closed her eyes. She wasn't thinking of Mac, either, nor of Robert Montfort. She was thinking about grilled fresh asparagus with a crust of Parmesan. Perfect for tomorrow's starter.

CHAPTER 45

Mac and Sunny had managed to dodge the TV camera crews and the paparazzi lurking at the Colony's gates, hot on the news about the body at Perrin's Palm Springs house. They were sitting on Mac's deck, the sun was shining in a clear sky and a soft breeze blew. It was one of those heavenly late afternoons in Malibu.

Sunny was stretched out on the old metal lounger, hands behind her head, gazing skyward from behind large rose-colored aviators. Mac thought the color appropriate. Rose-colored specs matched Sunny's optimistic view of life and people.

She was wearing an orange bikini and her toenails were painted to match. He'd thought she was sensational as Vampira in the red boots, but now she was something else again.

There was no need to keep an eye out for paparazzi because Pirate had stationed himself at the top of the beach steps and, as though he knew what was expected of him, was on the lookout for strangers. Mac gave him a pat then stretched out on the chair next to Sunny. He had no doubt that the skeleton at Perrin's Palm Springs house was that of Ruby Pearl, though it was yet to be confirmed.

"So, what's *your* take on this?" he asked Sunny, catching her hand in his.

Sunny thought about the man she had talked to that night in the Bar Marinera in Mazatlán. She remembered his sad brown eyes and him saying, *"It's love gone wrong . . ."* Of course with that he'd had her on his side in a minute, and even though he'd disappeared on her, she didn't like to think of him being charged with murder.

"I don't believe it was Ron," she said.

"That's because you don't want to believe it."

"True," she admitted. "But even if Ruby Pearl was blackmailing him, I still don't see him as a psychopath, killing a woman and burying her under his prize cactus. It takes a different kind of man to do something as evil as that. And anyway *why* would he kill her? He's a powerful man with a battery of lawyers. He could simply have given her a chunk of money, had her sign a legal contract, and then he could have said goodbye and thank you very much."

The house phone beeped and Mac got up and went to answer it. He came back a minute later and said, "Demarco's here. He wants to talk."

Sunny sat up quickly. "Why?"

"If I knew I'd tell you."

She slipped on a yellow cotton cover-up. "Do I offer him coffee?"

"If he's coming to talk, my guess is he'll need something stronger than coffee."

He was right. Demarco was looking cool as ever, although this time he was wearing a casual white golf shirt and beige pants. He said he would prefer vodka. On the rocks. He took a long look around the place, then, one disdainful eyebrow lifted, followed Mac outside.

"Good to see you again," he said, though he was obviously surprised to see Sunny. He glanced questioningly at Mac.

"Sunny is my partner," Mac said, thrilling her. "What I hear, she hears."

Demarco nodded. "Then I'd better tell you it's true that the FBI is investigating Perrin's . . . *our* . . . business."

Demarco wasn't telling Mac anything he didn't already know, but he caught the odd look in Demarco's eyes. He puzzled over it

341

while Perrin's right-hand man talked. What was it exactly? Insecurity? Fear? Surely not, in a man like that. But there was also the matter of the body in the cactus garden.

"I wonder," Mac said. "Do you think Perrin was money laundering?"

Demarco coughed into his drink. "I surely hope not," he said. "Because as his partner I'm afraid that might implicate me."

"Yes, it might," Mac said with a smile.

"Of course, that would be entirely untrue." Demarco looked Mac in the eye. "If Ron is guilty of something, besides murder, it would be up to you to prove I had nothing to do with it."

"And how will I do that?"

Demarco took a long swig of his vodka. He put down the glass and looked from Mac to Sunny and back again. "I'd pay you well, Mr. Reilly. *Very well.* More than you've ever been paid in your life."

There was a long silence. Mac was aware that Demarco was trying to gauge their reaction. His own face was expressionless, while Sunny looked stunned by such an overt bribe.

Mac's cell rang. He took the call, listening while Demarco sat in silence, staring out to sea, though Mac knew he wasn't watching the brown pelicans doing their high diving act.

He clicked off the phone, then said, "I think you'll be interested to know that the remains found at Perrin's Palm Springs compound have been identified by dental records as those of Ruby Pearl. The diamond watch was traced to a Beverly Hills jeweler. It was purchased by Perrin."

Demarco shrugged his shoulders. "What can I say? Now you see why you have no choice but to help me."

When Demarco had left, with a promise from Mac to try to find Ron and sort out the mess, Mac called Lipski to confirm the news about Ruby Pearl. In a choked-up voice Lipski thanked him. He said now Ruby would be able to rest in peace and he would try to get on with his life. "Just get that killer Perrin for me, will ya?" he begged Mac.

"He's not named as the killer yet," Mac told him. "He's still only 'a person of interest' to the Palm Springs PD. And remember, just because her body was buried in his garden, doesn't mean he killed her."

"Then who else would do it?" Lipski asked.

Mac thought he had a point.

CHAPTER 46

Ampara was worried. She and the dog had been staying with friends since the scary events at the house, but she was conscientious, and twice a week she went back to the Bel Air house to open up the windows and air out the place, dust the furniture and make sure everything was okay.

Mac Reilly had kept her up-to-date but whoever the perpetrator was, he had worn gloves and there were no fingerprints, other than the normal ones. In Ampara's view there should have been no fingerprints anyway. She kept a clean house and was proud of her work. Anyhow, so far the cops had no suspects and nor did Reilly, though he'd told her he felt sure it was safe for her to go to the house to do her work. "After all, sir, I'm being paid for it," Ampara had said when she called him about it. And besides, she didn't like just doing nothing except walk the dog.

She was there one morning when the usually silent phone rang. Surprised, she stopped her vacuuming and stared at it. It had not rung, at least when she was there to hear it, since that awful night, and now she was scared to answer it. It occurred to her, though, that it might be Miss Allie, and dropping the vac she ran to answer it.

"Perrin residence," she said, cautiously.

"Ampara?"

She recognized the voice. "Miss Whitworth," she said relieved. "I'm glad it's you."

"Were you expecting Allie to call?"

"Well, sort of hoping, y'know how it is . . ."

"I sure do, and that's why I'm calling. Listen, I know — we all know — that Allie's gone missing and I'm really worried about her. You know how I feel about Allie. How we *both* feel, Ampara. I'm hoping I can help her. She needs someone who knows her, y'know what I mean? And since I just happen to be here in Cannes on vacation, with a friend, I thought if you knew where she was, if she had been in contact . . . maybe I could go and help her. Allie needs someone who knows her *well*, Ampara, another woman who really cares . . ."

"Oh, I agree, Miss Whitworth. Miss Allie's such a lonely woman, and now she's run off and disappeared, and with what happened here at the house . . ."

"What do you mean? What happened at the house?" Jessie's voice was sharp with anxiety, but just in time Ampara remembered that Mr. Reilly told her not to talk about it to anyone.

"Oh, I just got nervous being on my own, so me and Fussy are sleeping over at a friend's. And anyway, Miss Whitworth, I don't know where she is. Nobody does. And nobody knows where Mr. Ron is either."

"Hmmm, well I'll call back soon, and if you hear anything, Ampara, promise you'll let me know."

"Okay, Miss Whitworth. I sure will."

Ampara went back to her vacuuming. Outside the window she saw the pool guy wielding his long net across the water, sending it rippling like blue silk over the infinity edge, and heard the familiar whine of the gardener's leaf blower and the hum of the John Deere tractor mower. Life went on as normal at the Perrin house. Except its owners had disappeared.

The phone rang again. She stopped and stared at it. Could it be Miss Allie this time? Or maybe Mr. Ron?

She picked it up. "Perrin residence?"

"Oh, Ampara, this is Sheila Scott."

A relieved smile crossed Ampara's face. She knew Sheila Scott and liked her.

"How are you, Ampara?"

"Well, Miss Scott, as good as can be expected, I guess, after what happed here . . . Ohh . . ." Again, she stopped herself just in time.

"What do you mean? What happened?"

Ampara heard the concern in Sheila's voice. She was Miss Allie's best friend, surely she could trust her. Deciding she could, she told her about the break-in, about the beautiful dresses, all slashed with a knife and thrown onto the floor . . .

"God knows what might have happened if Miss Allie had still been here," she ended with a sigh that was almost a sob.

Sheila was silent for a moment, taking in the full horror of the situation. Then she said, "But surely the police . . . ?"

"They don't know nuthin', Miss Scott, and nor does that Mac Reilly, who was looking after her. And I sure miss Lev, that bodyguard, being around."

"I'll bet you do. Then you haven't heard from her, I suppose?"

"Not a word, miss. And nor has no one else. Not even Mr. Reilly, and I know she trusted him. You too, miss," she added.

Sheila thanked her, and Ampara said she would call her if she happened to hear from Allie.

"You're the second one today calling asking about her," Ampara said, just as Sheila was about to ring off. "Jessie Whitworth called just a few minutes ago. You remember, she used to be Miss Allie's assistant."

"I remember."

"Well, she's concerned too. She thinks Miss Allie must be all alone somewhere and needs another woman, a friend to help her. She said she also happens to be in Cannes on vacation and did I know where Allie was."

"And what did you say?"

"Of course I told her I had no clue. Only that she was in the South of France when she disappeared."

"Hmm. Quite a coincidence," Sheila said thoughtfully. "Well, thanks, Ampara. You know if you hear from her, to give me a call?"

"I will, Miss Scott. And thank you for calling."

Feeling better now that Allie's friends were rallying round, Ampara went back to her vacuuming. Life at the Bel Air mansion went on almost as usual.

CHAPTER 47

Sheila could not get the vision of Allie's gowns slashed by a knife-wielding madman out of her mind. Even in Bristol Farms supermarket buying fruit and flowers and a good French cheese to go with the freshly baked bread, it was still on her mind. The stalker had finally gotten into her house, and it could have been Allie stabbed instead of just those dresses. From Allie's cheerful phone call from somewhere in France, Sheila realized she did not know about the incident and she thanked God for that. Still, she was worried. What if the stalker found her? He was clever enough to get into the house, wasn't he? Scared now, she decided she had to tell Mac Reilly about Allie's phone call.

Mac was surprised to get a call from Sheila Scott, who described herself as one of Allie's closest friends. As far as he'd known, Allie had no close friends. But when she explained to him that she'd heard about the break-in, that she was worried and needed to talk, he arranged to meet her at home in Venice Beach.

Miss Scott's house was in one of those charming little walk-streets lining the canals that gave L.A.'s Venice its name. It was a cottage, larger than his own and set in a pretty rose-filled front garden, with diamond-paned windows and a sturdy wooden front door. With Pirate at his heels, he opened the wrought-iron gate in the shape of a peacock, stopping en route to sniff the lavender-colored Barbra Streisand rose, which had a spicy, sweet aroma. He wished he could grow roses out at the beach but the salt spray would be too much for them. He climbed the two wide steps and rang the bell, hearing it peal a little song that he recognized but couldn't identify.

When Sheila Scott opened the door to him, she said, "I'll bet you're wondering what the song was."

He grinned. "How did you know?"

"Everyone always does. It's 'Nessun dorma,' from Puccini's opera *Turandot*. You're probably more familiar with Pavarotti singing it at World Cup finals, or the three tenors, Pavarotti, Domingo and Carreras, in the televised concert from Rome. It's harder to identify it without the words." She waved him inside. "Please, come on in. And the dog too."

She stooped to give Pirate a pat, then, taking in his injuries, glanced back up at Mac. "I can tell you are a very kind man."

He shrugged. "I only did what anybody else would. Anyhow, I wouldn't be without him."

"You're very fortunate."

She was the second woman to tell him that. Allie had been the first. He liked Sheila Scott already.

She took him into her kitchen, a lovely room, low-ceilinged with French doors thrown open to a vista of more roses and shady pepper trees and eucalyptus and with a narrow flagged terrace overlooking the peaceful green canal. The houses were crammed next to each other in a mishmash of styles, jockeying for space, as they did at the beach. But here, there was also a view of the gardens and houses opposite, with little pleasure boats tied up outside.

"Lovely home you have," Mac said appreciatively.

"I bought it thirty years ago. It was a wreck — as was almost everything else in Venice. Nobody wanted to live here then, too close to the hood, no infrastructure — y'know, no supermarkets, restaurants, boutiques, cafés. Now you can't walk two blocks without falling over all of that."

"And now it's worth a couple of mil, I'll bet," Mac said. "You made a shrewd investment."

Sheila laughed. "There was nothing 'shrewd' — or 'investment' — about it. It was all I could afford, and besides I loved being on the canal. It has its own magic, as you can see. Almost," she added with a twinkle, "as good as the beach.

"Coffee?" She held up the pot. Two mugs already awaited on a pewter tray.

Mac thanked her and they went outside and sat on the terrace in a pair of white basket-weave chairs. Mac took her in as she sat quietly, looking around her garden. She was a handsome woman, very much in charge of herself, but her casual confidence was appealing.

"Are you a musician then?" he said, thinking of the doorbell.

She laughed. "Absolutely not. No. I'm a voice coach. That's how I met Allie, years ago when she first came to L.A. I coached the Texas twang out of her. We've been friends a long time," she added, looking directly at Mac. "And that's why I'm so worried."

Mac sat back, in his usual wait-and-see-what-they-say-to-incriminate-themselves fashion.

"I heard from Allie a while ago," Sheila said, surprising him.

He put the coffee mug carefully down on the small tiled table between them. "Yes?"

"She was still in France but wouldn't tell me exactly where. She said it would save me having to tell a lie if anyone asked me if I knew. She sounded happy, said she had made some new friends, she had changed her appearance and was certain nobody knew who she really was. And that she was working as a waitress. Maybe she'd even met a man . . ."

"Sounds promising." Mac picked up his mug and took another sip of coffee.

"Let me get you some more." Sheila hurried into the kitchen and came back out with the pot. "It's still hot," she said, pouring it. She looked at him. "I haven't heard from her since, though she promised to call. I was worried, so this morning I telephoned Ampara, the housekeeper. She told me something even more worrying. About the stalker getting into the house, the slashed dresses . . ."

Mac nodded. "We have to be glad Allie wasn't there."

Sheila shuddered. She didn't want to think about it. "There was something else Ampara told me, though, that I thought was unusual. It's about Jessie Whitworth."

"Allie's personal assistant? The one she fired a few months back?"

Sheila's dark brown eyes met Mac's. "Exactly. And what Ampara told me was that just this morning, she'd had a phone call from her. Jessie told her she was concerned about Allie, that she felt she needed a friend, another woman, someone who knew her, who was close to her. Jessie said she felt strongly that Allie needed help, and since she happened already to be in the South of France, if Ampara would tell her where Allie was, she would go to her aid."

Mac put down the coffee mug and sat up straight. "About what time did she call?"

"Ampara said it was just before I did. So that would make it around eleven this morning."

Mac got to his feet. There was no time to waste. He said, "You have no idea what a big help you've been."

Sheila was also on her feet, looking anxiously at him. "Do you think Jessie has anything to do with it? I know Allie fired her, and that she used to have a key to the house."

"We're certainly going to find out," Mac

said. And then, because Sheila's kind face looked about to crumple into tears, he gave her a hug.

"You're a good friend, Sheila," he said, holding her away from him, smiling. "And don't you worry, I have just the man to find out where Jessie is, and exactly what she's up to."

CHAPTER 48

Mac was in the car, driving back down Main Street, Venice. As Sheila had said, it was crammed with boutiques and clothing stores, and cafés, with young people toting surfboards and beautiful girls pushing strollers with cute babies. Stalled at the light, Mac called Roddy and told him the news.

"Miss Whitworth, the perfect secretary?" Roddy said. "I always thought she was a dark horse. I wonder if she's still with the Queen of England."

"That's a thought," Mac said, already busy thinking in fact. "I'm on my way back to the beach," he said. "Why don't you meet me there."

"Okay. Maybe you should send me to Cannes to find her. After all, I'm the only one who's actually met her, the only one who knows what she looks like."

Mac laughed. "Maybe," he said, just as he had to Allie when she had asked him to go to

Cannes with her. "I'm calling Lev right now, getting him over too. We'll talk then."

He speed-dialed Lev, told him what was going down and asked him to meet him at the Malibu house.

"I'm in Hollywood, I'll be there in forty," Lev said. "God and traffic willing," he added.

In the event, he was there on time. Roddy and Mac were already out on the deck where the velvety sunshine dappled the ocean with sequins of silver. Cold beers were handed round and the three men sat for a moment, enjoying the view of the ocean and the blond girls — somehow they always seemed to be blond — cavorting at the edge of the waves, as well as the surfers "boldly going where few men dare to go," Roddy said, though in fact he happened to be a great surfer himself.

"Okay, so Allie is still in France." Mac laid out the scenario as told him by Sheila and the two men listened attentively.

"The Whitworth woman left her apartment early on the morning after the break-in at the Bel Air house. She told the apartment manager she was going on a vacation to Cancún."

"She and her friend, the blonde?"

"Yes. And who, by the way, she said,

looked a bit like Allie."

"The security guard at Mentor Studios told me the same thing," Roddy said.

"So maybe Whitworth didn't go to Cancún. Maybe she went to France instead," Mac said.

"Looking for Allie," Lev said.

Mac turned to Lev. "You know your way around that part of the world?"

"I used to work in Paris. I have contacts there. Whitworth will have needed to rent a car, and for that you need a driver's license and credit card. We can check that."

"Okay, I want you to get on to it right away. We don't know where Allie is, but then, neither does Jessie."

Lev glanced at his watch. "I'll be in Paris tomorrow. Trust me, I'll find her."

Roddy looked at Mac. "Can't I go with him?"

Mac laughed. "I think Lev is a man who likes to work alone. Besides, you don't have any contacts in Paris."

Roddy looked aggrieved. "I could make some."

"Okay, so when we find her, you'll go to France."

"I always like celebrations," Roddy said gloomily.

CHAPTER 49

Sunny was fixing a salad with mini lettuces, fresh herbs, avocado and sliced strawberries, sprinkled with toasted walnuts, while Mac barbecued a couple of steaks and opened a bottle of Caymus, a deep rich Cabernet that was one of his favorites. Pirate snored loudly, competing with Bryan Ferry's album playing in the background. Sunny had bought it thinking of Ron. They had just settled down to eat when the phone rang.

"Hi," Marisa said. "It's me."

"I know." Mac put her on speakerphone so Sunny could hear. "I recognize your voice."

"I read about the murder in the *Herald Tribune,*" Marisa said in a trembly voice. "About Ronnie being suspected of killing that woman, about him still being missing, and about . . . oh, everything. Then someone broke into my apartment. Nothing was stolen just everything turned over. Now I'm scared. I think it might be Ronnie."

"Maybe it was the paparazzi," Sunny said to Mac. "They could have found out she was Ron's girlfriend," she added.

He shook his head. "Paparazzi don't break in. And anyhow what could they have been looking for?" He asked Marisa the same question.

"I didn't tell you everything," she confessed. "I took some documents from Ronnie's place that night. They're to do with offshore bank accounts, serial numbers, that sort of things. I know Ronnie wants them. And I think now he's killed one woman and maybe I'm next on his list. I'm scared, Mac. I'm really *really* scared. Please, oh *please,* you have to come out here and help me. I'll give you the papers. And I'll tell you everything I know about Ronnie."

Mac's eyes met Sunny's.

"I'm already packing," she said.

"We'll be there tomorrow," Mac promised Marisa.

CHAPTER 50

The following day Sunny and Mac were installed in their usual Rome hotel. There was a message to meet Marisa at her apartment. Mac called to tell her they were on their way. There was no reply but he assumed she had gone out shopping and decided to go anyway.

They took a cab to Trastevere, the old section where artisans and manual workers once lived crammed together in the narrow cobbled alleys and ancient piazzas. Now, though, parts of it had been gentrified and there were smart little bars and cafés with tiny flowered sidewalk terraces.

Marisa's apartment was on the first floor of a tall narrow stucco building whose green-shuttered windows looked out onto the cobbled street. TV aerials sprouted from the roof and a black cat gazed down at them from an ugly iron balcony that was definitely not meant for Juliet.

"Hmm." Sunny sniffed disparagingly. "I would have thought the famous movie producer could have done better for her."

Marisa's apartment was to the left of the hall. A yellow Post-it note was stuck on the door. "Mac, meet me at Bar Gino, in the piazza at the end of the street," was all it said.

They walked to the bar, a long dim narrow cavern of a place with glass shelves lined with bottles of cheap red wine, and sat on the tiny terrace, sipping a glass each of the red, waiting for Marisa. After a while, Mac got worried and called her again. Still no reply. He called their hotel to see if she had left a message, but there was nothing. They decided to wait a little longer and Mac went back into the bar to order more wine and a pizza. Without anchovies.

Left alone Sunny was drinking the last of her wine, listening to the hungry rumble of her tummy — she hadn't eaten on the plane — when she spotted Marisa at the top of the street.

How could she miss her, even if she was draped in a black pashmina shawl, attempting to look like an old Italian peasant woman and definitely not succeeding? And it wasn't only her Prada platform shoes that were a dead giveaway. A tall man was walking with her, a priest in a black soutane and one of those wide-brimmed flat hats. He was holding on to Marisa's arm and seemed to be guiding her firmly away from the piazza. *Too* firmly, Sunny thought, alarmed. In fact he was pushing her.

Leaping to her feet, she yelled for Mac, then took off after them, trotting in her high heels over the tricky cobbles. In front of her she saw the priest drag Marisa down the dark narrow alley and heard her cry out as they disappeared round a corner.

Panting for breath, Sunny whizzed after them. Her heel caught in between the cobbles, and she went sprawling. A pain shot through her ankle and tears stung her eyes. When she looked up, Marisa and the priest had disappeared.

She saw Mac at the end of the street, running toward her. "What happened?" he asked anxiously.

Sunny told him the story. "Looks like Marisa knew she was being followed and she tried to lose him, and you almost caught up to them," he said.

Her ankle was swelling rapidly and he helped her up. "At great self-sacrifice," she muttered, clenching her teeth against the pain.

Mac said there was no point looking for Marisa now, she might be anywhere in that maze of alleys, and with his arm around her, they hobbled back to the bar, where a shot of grappa temporarily eased the agony. Then they took a cab to a *farmacia* where they inspected her ankle and prescribed ice packs and an elastic support stocking.

" Terrific," Sunny muttered, surveying the piggy-pink elastic support miserably. "It'll look just great with a little black dress." She had already decided that Marisa Mayne was not worth it.

Mac had called the police and told them what had happened. They said there was not much they could do since Marisa was not yet officially a missing person.

"She'll turn up," the cops told him. "These women always do."

Mac left Sunny at the hotel nursing her swollen ankle, while he returned to Marisa's apartment. The yellow Post-it note was still stuck on the door. He put it in his pocket, then tried the lock. He was surprised to find it open.

Marisa's tiny apartment was immaculate. Her clothes were neatly arranged in the closet, shoes lined up, sweaters stacked. There was a coffee mug on the sink and a couple of magazines on a side table next to the sofa, along with a remote for the TV. And that was about it. No Marisa. No photos, no letters, and no personal documents, not even her passport. And certainly no incriminating papers with RP's offshore account numbers. There was just the yellow Post-it pad near the phone with an address written on it.

"Villa Appia, Gali, Tuscany."

Mac called Hertz, arranged for a rental car, then went back to the hotel to tell Sunny he was off to Tuscany. Of course, despite the injured ankle, she said she was definitely going with him.

CHAPTER 51

It was Saturday night, Robert Montfort was at the Bistro again, sitting at the same table under the arbor opposite the same Paris blonde. Allie, her crackling starched white apron wrapped twice around her narrow waist, bustled outside, saw them and stepped hurriedly back in again. She asked Jean-Philippe to take their order and ran to the kitchen to tell Petra.

"It's Robert Montfort," she said. "He's out there with the Paris blonde again."

"The TV gal?" Petra raised an eyebrow at her. "Should've thought he'd know better, after having met someone like you. Don't you worry about it," she added, briskly sautéing up chicken pieces in butter with garlic and shallots and pouring in a dollop of good heavy cream. Petra never spared a thought for calories or cholesterol. Good food was good food. "I've no doubt he's saying goodbye to her." She threw Allie a penetrating glance. "That's what you want, isn't it? Robert Montfort, I mean."

Allie stared back at her. An image of Ron flew into her mind: his beetling brows, his powerful frame, his arrogance that concealed the vulnerable man inside. There was nothing vulnerable about handsome Robert Montfort. "I don't know," she said, honestly.

"Well, then." Petra sloshed the chicken around in the sauce, seasoned now with cumin and thyme. "You can't grumble, can you, if he has a blonde here on a date."

"True," Allie agreed. And she strode outside, order pad at the ready.

"*Bonsoir,*" she said, wafting away the new wispy bangs that were sticking to her glasses, and giving them a smile. "Lovely to see you both again. Vodka tonics, isn't it?"

Robert smiled at her. "It is," he agreed.

The blonde took out her cell phone. "Has anyone ever told you you should get rid of those glasses?" she said. "Try some contacts. It would make a difference." And suddenly she reached over and snatched off Allie's glasses, holding her cell phone up to her surprised face.

"There, that's better, isn't it, Robert?" she said. She put down her cell phone, handed Allie back the glasses and said, "I'm dying for that drink."

"Do you have any idea how rude you are, Félice?" Robert's voice was icy. "How dare you do that to Mary."

She shrugged. "It was just a helpful little gesture, that's all. Woman to woman. And you have to admit she looks better."

Allie rammed the glasses on her nose and, furious, hurried back indoors. She knew Félice de Courcy had photographed her and there was nothing she could do about it.

She sent Jean-Philippe out with the drinks. When she complained to Petra, Petra went out and took their order and served them herself.

"There's something about that Félice woman I don't like," she told Allie later, when they were tidying up the kitchen together. "She's too nosy by far, always asking questions about you. I don't know what she's up to, but it's something, you can bet on that."

Allie's heart sank. All was not well in Paradise after all.

CHAPTER 52

The drive to the Villa Appia in Tuscany was a long one, and it was dark when Mac finally found the large rose-colored house behind iron gates. A perfectly kept gravel driveway led to the massive front door. No lights were on and it appeared to be deserted.

"It looks like a fortress," Sunny whispered.

Mac tried the gate. "Maybe not," he said, opening it.

He rang the bell. There was no answer. He rang again, waiting. When no one answered, he tried the door. It was locked. They walked around the side of the house until they came to the kitchen door. Surprisingly, this one was open.

Sunny hobbled indoors after Mac. She didn't want to be left behind in that silent garden. There was not even a bird singing, for God's sakes.

She hated breaking into places, it gave her the creeps. Like in Palm Springs, she had spooky tremors up and down her spine as they stood waiting for their eyes to adjust to the dark. When they did, she saw they were in an enormous kitchen. Very modern, with everything built in behind dark wooden doors with no handles. It might have been a library there was so much wood. The only way you could tell it was meant for food preparation was the double sink beneath the window and the six-burner stove top. Not a coffeepot, not a canister, not a refrigerator or dishwasher in sight.

"Wait a minute," she whispered. "Aren't Tuscan villas supposed to have nice old-fashioned kitchens, all white tiles and ancient cast-iron cooking utensils and a big old fireplace for roasting wild boar or something?"

Mac was standing at the polished granite slab that served as a table, looking at an envelope by the thin beam of a pencil flashlight.

"Not this one," he said. "This is a movie mogul's country house. Renato Manzini."

"Manzini," Sunny breathed, stunned. "Could *he* have kidnapped Marisa?"

But Mac had already disappeared through a door. It swung silently closed behind him, creating an evil little draft, and Sunny shivered again.

She shifted uncomfortably on her painful foot, hating to be left alone yet afraid to go through that door and try to find Mac in the dark. Why did she always let herself in for these kinds of things anyway? She should have learned by now.

She had a raging thirst and she peered round, wondering which of the doors might be hiding a refrigerator, hoping for some cold bottled water. She tried one, then another. No luck. She stopped and listened, heard the low hum that said fridge, stepped toward it and pulled open the door.

The interior light switched on.

She was staring into the blank green eyes staring back at her. All the shelves had been removed and Marisa Mayne's slender body was folded into the space like a ventriloquist's dummy. A trail of blood streaked from her mouth, and her tumbling red hair sparkled with a coat of frost.

For a numb second, Sunny was spellbound. Then fear surged into her throat, hot as battery acid. A scream gurgled somewhere inside her but she couldn't get it out. She turned to run . . .

A hand slammed over her mouth. Her arms were pinned at her sides. She felt his breath on her neck . . .

She bit down on his hand, gagging as her

teeth sank into his flesh. She bit harder, teeth seeking bone, red-hot with panic.

The killer ripped his hands away, and now he wrapped them around her throat . . .

She had not known fighting for your life could feel like this, fighting for every breath.

Her tongue was forced out of her mouth . . . She choked and gurgled.

"Sunny?" Mac's voice came from the door. "Sunny, where are you?"

So this is what it feels like to die, Sunny thought, as the killer let go and she dropped like a stone to the floor.

"Oh my God, Sunny." Mac knelt over her, his mouth on hers. "Tell me you're okay. Please, *please,* Sunny . . ."

She could still feel the painful imprint of the killer's fingers on her neck, and her tongue was twice as big as normal. When she finally managed to speak her voice came out in a hoarse whisper.

She pointed a trembling finger at the wall of dark wooden cabinets. "Look," she managed to say.

"Who was it?" Mac demanded.

"Just open it," she whispered. . . . *"Open the door."*

Mac opened the refrigerator door, the interior light clicked on. And Marisa's icy body slid out and landed at his feet.

Oddly, the one thing Sunny noticed then was that Marisa's ring was missing.

CHAPTER 53

If Petra was surprised when Robert Montfort casually dropped by the Manoir, she
kept her silence. He pushed through the
door calling *"Bonjour"* and strode into her
kitchen, making himself at home. He took a
seat at the vast table, doing as Petra did and
clearing a space with his arm while she
brewed up tea in the big brown pot and produced jam tarts fresh from the oven, that
along with the chocolate digestives were her
favorite snack.

Allie's cheeks were pink from embarrass-
ment as Petra gave her a long inquiring
smirk, when she came and sat at the table.

"You two going steady or what?" Petra
asked, impatient with all this dodging
around the issue. "If not, Robert, then it's
time you took her out on a proper date. Tell
you what, I'll give her the night off. She's
earned it. Take her to Monpazier, let her see
how beautiful a thirteenth-century fortified

village is. Show her the square and the stone arcades, have dinner at the Hôtel de France. It's always nice there."

Robert laughed. "You should be in the matchmaking business."

"I thought I was," Petra replied, passing him a hot-from-the-oven jam tart.

Robert looked at Allie. "Would you please accept my invitation to dinner tonight?" he asked with mock solemnity.

"Only if you tell me you weren't bullied into it," she said, and they both laughed.

"And Mary love," Petra said to Allie, "wear something pretty tonight, instead of those everlasting jeans."

So Allie put on the black pants and the white taffeta shirt she had worn for the movie premiere in Cannes. Her hair had grown from the Jean Seberg *Breathless* crop into a spiky bob, though she still colored it a drab brown. The heavy framed glasses were not exactly an asset, but she was afraid to go without them. As it was, in her pretty outfit, wearing a little makeup and with gold hoops in her ears, she hoped she would not be recognized.

"But you are gorgeous tonight, Mary," Robert said, his eyes gleaming with appreciation. Actually, Allie thought Robert didn't look so bad himself, in a blue shirt that set off his dark good looks.

As they drove he told her about the history of the countryside: about the wars when Edward of England, known as the Black Prince because of his dark suit of armor and his black stallion, had ruled over this part of France. Monpazier remained as one of the most beautiful of the fortress villages in the area.

The tiny ivy-covered Hôtel de France was tucked away behind one of the stone arcades. Tables were set outside and the bar was doing good business. Robert had requested a corner table indoors and they were greeted by the owner, who of course knew him, and settled them in with a bottle of Clos d'Yvigne white. A rival vineyard, Robert explained, though he liked their wines very much.

Relaxed, Allie found herself telling him about her Malibu home, and how much she liked the temperamental weather and the sound of the ocean always roaring in the background.

"So tell me the truth, why are you really here?" Robert asked, surprising her.

"So, okay," Allie said, thinking. "I'd lost myself," she said eventually. "I didn't know who I was anymore. Perhaps I never had known. I needed to get away, try something new, in some place where no one knew me, and where I would have to function completely on my own." She smiled with the relief of saying it. "And I did. I found myself a dog. I found Petra and the Manoir. I got myself a job at the Bistro . . ."

"And tell me, what did you do before you became a waitress and now a chef?"

She could see he wanted the whole truth and nothing but, and he knew he wasn't getting it.

"I guess I was just a Hollywood wife," she said finally. "I loved my husband very much," she added, surprising herself. "He was the only man I ever met who saw the real me."

"Until now, perhaps," Robert said with a meaningful look.

Worried that this might be progressing faster than she wanted, Allie avoided his eyes.

"All my life I've had to make plans, to show up and please people," she said. "When I came here I made a new vow. I would only please myself. I'm happy the way I am. For now," she added.

He lifted his glass in a toast. "Then let us drink to that. I admire a woman who knows her own mind."

Robert didn't try to kiss her on the way

home, though he did hold her hands in both his as they stood on the steps of the Manoir, saying good night.

"Tell me we can do this again, Mary," was what he said, and Allie replied that she would love to.

It proved to be the first of several dinners together, each more enjoyable than the last, where Robert taught her the history of the area and they enjoyed the local gourmet produce, until Allie protested that she was getting fat.

"Never," he said, kissing her this time. "You'll always be beautiful."

Allie had kissed a lot of men in her movie career, but she really enjoyed that kiss.

CHAPTER 54

In the South of France, Jessie Whitworth was driving a silver Peugeot convertible with the top down. She wore large dark glasses and an expensive designer dress that belonged to Allie. She had taken it from Allie's over-stuffed closet a few months ago, because she liked it and saw no reason why she should not have it. After all, Allie had so much, she wouldn't miss it.

She had been taking things for a long time, knowing that Allie was caught up in her work and also in her emotional turmoil. And besides, Jessie knew she was unlikely even to notice. Allie was not into clothes, she was into "people" and finding what was "missing" from her life. Jessie knew exactly what was missing from *hers*. And it was exactly what Allie possessed.

Driving along the Promenade des Anglais in Nice, Jessie thought appreciatively that the Peugeot was a better car than the Sebring she had used in L.A. to stalk Allie, though of course it was not in the same league with the Porsche she'd rented every now and then, when she would get dressed up in one of Allie's outfits, with one of her designer handbags, and shoes. (Jessie preferred Vuitton, because it was so recognizably expensive and she liked people to look at her and admire her and imagine how rich she must be.) Sometimes she would go shopping in Beverly Hills, exploring Neiman's and Barneys and Saks, though she could not afford to buy anything. Still, she'd try things on and the saleswomen were more than nice to her, and for a few hours she'd feel important. She had felt like Allie Ray.

Her own car was a down-to-reality Ford Explorer, five years old and bought used. Somehow, making money had never been one of Jessie's talents.

It was true, what she had told that nosey parker friend of Mac Reilly's. Roddy Kruger, that was his name. She had indeed had ambitions to be an actress. More than mere ambition, she'd wanted it desperately. She was different, with her anonymous half-plain, half-attractive look, and with makeup and clothes she could become almost anyone. But somehow nobody had taken her seriously. Worse, they had rejected her. Time

380

and again they had rejected her. Every audition she went to she faced those same blank dismissive looks, the "Thanks for coming, miss. . . ." They never even remembered her name. The scars of those years were visible now in her tightened mouth and hard eyes.

Jessie so wanted to be Allie, sometimes she believed she really was her. Sometimes she would have her blond hair, her look, her clothes. Then her whole persona became Allie.

But Allie had rejected her too. She had fired her. At first Jessie had thought her boss had found out she had been stealing. But no. It was the same old same. Allie did not want her around anymore. Three months' salary and a reference and that was it.

Jessie had wanted to kill her right there and then, but the timing was bad. And then there had never been the opportunity. And Allie had already left for France the night she went to the house in search of her, the big kitchen knife in her hand. So she'd slashed the gowns instead, taking with her the ones she fancied, and which were now in her luggage ready for her next public appearance.

She was so busy thinking of Allie, she almost missed the turn onto the autoroute. She swerved into it, saw the car in front of her and braked hard. Too late. She plowed into the back of it, saw it lift up into the air and come crashing down on its roof. The Peugeot shuddered to a halt, the air bag burst in her face and she fell back, unconscious.

CHAPTER 55

Lev was having lunch at a café in the Marais district of Paris with his old friend from their Israeli Army days. Zac Sorensen was now a French citizen and had for years been a top detective. Like Lev, he was lean and mean and tough, and much too dedicated to his job ever to marry. He rarely even took a vacation.

"Too many crimes," he told Lev over a fast ham and cheese baguette and a beer, keeping an eye on the traffic whizzing past just in case he might spot something of interest.

"Still the same, I see," Lev said with a grin.

Zac glanced at him. "You too. So why are you in Paris anyway? I can tell it's not a vacation."

Lev told him the story and saw Zac's eyes light up.

"You really think Allie Ray is here in France?"

"She might even be here in Paris for all I

know," Lev said. And then he went on to tell him the details of the anonymous letters, the stalker, the break-in and slashing of Allie's gowns.

"We're looking for a potential killer," he said, "and I have reason to believe it's a woman named Jessie Whitworth. Seems she's over here now, on vacation with a friend of hers, by the name of Elizabeth Windsor."

Zac glanced at him, one skeptical eyebrow cocked, but he made no comment. "If she's over here the odds are she's rented a car."

"Correct."

"It's a tough job to check the driver's license of every tourist who comes to France," he reminded Lev.

"True." Lev bit into his sandwich, waiting for Zac to come up with an answer.

"Of course, there's always immigration," Zac said thoughtfully. "You said she was in Cannes?"

"We believe so."

"Then that's probably where she rented the car. I know someone there, an old friend . . ."

Lev grinned at him. Now he was getting somewhere.

"Perhaps I could make a call," Zac said, and this time Lev laughed.

"You old bastard," he said, punching him on the shoulder. "I knew I could rely on you."

Lev didn't have to wait long for his answer. He got the call from Zac two hours later. A woman named Elizabeth Windsor had been involved in a crash just outside of Cannes. She had hit a car. It had rolled over and the driver was killed instantly. "She's practically untouched, except the face where the air bag hit her. Red and bruised. You know how it is. Her license was false."

"What does she look like?" Lev asked.

"About forty, wearing a blond wig and expensive designer clothes, though she was staying in a cheap hotel near the railroad station."

"You might want to check out the name Jessie Whitworth," Lev said. "And by the way, what happened in the crash?"

"She was hospitalized for one night. Currently she's in police custody in Nice on vehicular manslaughter charges."

"Keep her there," Lev said. "I think you'll find she's wanted for more than that."

CHAPTER 56

The local Tuscan carabinieri seemed even less thrilled than the Palm Springs PD to have a couple of housebreakers find a body in the refrigerator at an important movie director's villa. They were even less happy with the story Mac told them, but when the paramedics arrived and saw Sunny's throat, they were more inclined to believe them. A search of the area was instigated and Sunny was carted off to the local Cruz Rosa hospital for further inspection.

Following in the rental car, Mac groaned out loud. His Sunny might have been killed. First Ruby Pearl was dead. Had she been blackmailing Perrin? Now poor Marisa. She'd said she had incriminating documents. Had she been trying to blackmail Perrin too? Who next? he wondered.

The name flashed into his head. Allie Ray. Who else would have better access to Perrin's private files than his wife?

The next day, when Sunny was released from the hospital and Mac had completed his questioning by the police, they drove back to their hotel in Rome. After a bowl of soup and half a bottle of wine, with her bruised throat wrapped in a silk scarf (Hermès of course), Sunny was sprawled in front of the TV, watching the news to see if there was any mention of the body in the villa, when to her surprise Allie's face appeared on the screen. She quickly called Mac over.

The Italian newscaster was saying it was a "scoop." A French TV correspondent had discovered Allie Ray, the "lost" movie star, working as a waitress at the Bistro du Manoir near the small town of Bergerac. The Italian station had preempted the French channel and was the first to broadcast the news.

"Jesus," Sunny yelled, then wished she hadn't because it hurt her throat.

But Mac was already dialing the airport in search of a private plane for hire.

Half an hour later, Sunny was ready, dressed in jeans and Mac's green cashmere sweater with cute little Manolo boots into which she'd managed to stuff her swollen ankle. Actually, it felt quite good, she thought, testing it by walking back and forth across the room. The boots supported the ankle, and the scarf hid the bruises on her neck.

"I'm a wreck," she complained. And Mac said yes she was and it was all his fault and he was sorry, and why didn't she just stay here at the hotel and take it easy.

"Room service, TV, anything you want," he added.

"Are you kidding?" Sunny looked scornful.

"Think you're up to it, babe? After last night?"

"Too right I am," she said. She was darned if she was gonna let Mac Reilly go to the rescue of the beautiful movie star all on his own.

They were ready to leave when Mac's cell rang. With a what-the-hell-can-it-be-this-time look on his face, he answered it. His brows rose.

"Where are you?" he asked.

"Here in Rome. Reilly, we need to talk."

"You bet we do. And — it had better be soon. Like right now."

"The Café del Popolo, on the piazza. I'll be there in fifteen."

Mac clicked off the phone. "You'll have to go on ahead," he said to Sunny. "Make sure she's safe. I'm meeting Ron Perrin in fifteen minutes."

388

CHAPTER 57

Mac put Sunny in a taxi to the airport then walked to the café in the Piazza del Popolo.

He almost didn't recognized Perrin. He'd shaved his head and was wearing large very dark sunglasses. In a striped T-shirt and a gold earring he looked like a cartoon of an old-time burglar. All he needed was a sack labeled SWAG slung over his shoulder.

Mac took a seat and ordered an espresso. Perrin was already drinking grappa.

"So, how are you?" Mac sat back, taking him in.

"Not good." Perrin lifted his dark glasses and his sorrowful brown eyes met Mac's. "How could I be? First my wife disappears. Then my girlfriend." His heartfelt sigh shook his strong frame.

"Marisa is dead," Mac said.

For a long silent minute Perrin stared at him. Then he lifted his hand and summoned the waiter. "Another grappa," he snarled.

"Rapido."

When it came he tossed it down. "How do you know that?"

"I was the one who found her. In Renato Manzini's Tuscan villa. She had been strangled."

Perrin stared into his empty glass. "I want you to know I didn't do it," he said quietly.

He had put back his dark glasses and it was impossible for Mac to read his eyes. Perrin downed the second grappa. "Marisa was like a reincarnation of Rita Hayworth," he said quietly. "All tumbling red curls, flashing green eyes, and that sexy mouth." He signaled for another grappa. "Her real name was Debbie Settle. She's from Minnesota though how that cold state could have bred such a fiery young woman beats me. I'm telling you so you can contact her family out there in Minnesota. God, I feel so sorry for them."

"So what are you doing here in Rome anyhow?"

"What d'ya think I'm doing? Takin' a fuckin' vacation? I'm avoiding the FBI of course. And trying to find out what's happened to my money. A lot of which seems to have gone missing."

"You know we found Ruby Pearl buried under the saguaro cactus at your Palm

Springs place."

"Jesus Christ," Perrin, said, white to the lips.

Mac hoped that was a prayer, he was gonna need it.

"You're on the hook for two murders now," he told him. "It's only a matter of time, Perrin, so you'd better tell me the truth."

Perrin eyed him. "You on my side?"

"Tell me your version of the story, and then I'll give you my answer."

Despite the grappa, Perrin was sober.

"So I met a couple of women on the Internet," he says. "Look, I don't do drugs, I don't drink to excess — except under special circumstances." He signaled for another grappa. "And so what if I shifted a little money around here and there? So does everybody else in my position, don't they?"

"The FBI claims you committed fraud."

"I'd like to see them prove that," Perrin said angrily.

"Maybe they can," Mac cut through Perrin's bluster. He could see he was a frightened man. He had taken off the sunglasses now and those worried puppy eyes told him so.

"The man following you was not the FBI. It was Ruby Pearl's ex-boyfriend. He told me you gave her the diamond watch. The jewelers confirmed you bought it. He believes you killed her."

Perrin shook his head. "That's not true. I didn't really even know the woman. Demarco got me to employ her as my secretary. Ruby stole some documents from my private files in Malibu, with the coded account numbers for offshore banks. I discovered it by chance when I went to look for one and it wasn't there. When I looked again the next day, it had been replaced. I figured she was working for the FBI and told her goodbye, get out of my life. She'd only been with me a couple of weeks. It was about then I met Marisa." Tears moistened his eyes as he looked at Mac. "I don't have to love her to mourn her, now do I? I'm human, y'know."

"The 'engagement' ring was missing," Mac said.

"You mean somebody *killed* Marisa for the ring I gave her?" Perrin put his head in his hands. "Then maybe in a way I did kill her," he muttered, half to himself.

Mac watched him. He was sure now Perrin had not killed Marisa. Nor Ruby Pearl.

"Why didn't you come to Rome right away, to be with Marisa?"

"I thought if the FBI was after me, they'd be able to trace us. I needed to be 'a lone

392

wolf' at that point, get my affairs sorted out. I made sure, via Demarco, she was okay financially. I didn't want to involve her with the FBI."

"What if it wasn't the FBI?"

"Who the hell else would want to steal my offshore account numbers? And who else would kill Ruby Pearl?"

"How about the man who employed her?" Mac said.

Perrin jerked back in his chair as though he'd been shot. *"Demarco?"* he said. *"Demarco?* Sumavabitch, you're telling me he's been stealing my money, then laundering it himself?"

Perrin thought for a minute, then he said, "Here's the truth. The missing papers did not have the important account numbers. Only Allie has access to them. I hid them in code on her laptop. She was the only one I trusted. I told Demarco that."

Realization came to them at the same time. They stared at each other across the table. "Then we'd better find Allie," Mac said. "Before Demarco does."

CHAPTER 58

Demarco was in a standard room at a cheap chain motel outside of Florence when he heard about Allie, in a broadcast earlier than the one caught by Sunny and Mac.

The TV news blared in rapid Italian, interspersed with variety revues consisting of almost-naked showgirls and comics he didn't understand. He was polishing off a fifth of vodka and wondering what to do next.

He had not thought he would have to kill two women to achieve his goal. It had almost been three, but he hadn't realized that the woman in the kitchen at the Villa Appia was Reilly's assistant, and that Reilly was with her.

For years Demarco had been committing fraud and laundering the money stolen from the business into his own offshore accounts, leaving a trail that could incriminate only Perrin. But what he'd really wanted was access to Perrin's private offshore accounts. He'd known Perrin kept them in his files in his bedroom safe at the Malibu house where he could never access them. So he'd wooed Ruby Pearl with expensive gifts, culminating in the diamond watch that he'd charged to Perrin. The bills were always sent to the office and Demarco took care of them. Too bad that this time things had gone wrong and the bill for the watch had been sent to Allie by mistake. That had taken a little fancy explaining but he had managed.

Of course he'd used Marisa in the same way. One thing those two women had in common though, they both wanted money. And once they'd got it they wanted more. Marisa had been more than willing to work for him.

Marisa was often alone in the Malibu house and he'd ordered her to access the safe and find the current offshore account numbers. She'd told him later that somebody had come into the house and interrupted her, and that when she finally looked, there were no such papers.

And then, just when he thought he had her neatly tucked out of the way, playing at being a movie actress in Rome, she had pulled the same blackmail trick as Ruby Pearl. The priest disguise had come in handy when he'd followed her in that street in Trastevere, then stunned her and got her into his car. He'd had to kill her of course, she couldn't be left around to tell her story to the world.

Stuffing Marisa into that refrigerator when he'd seen the car's lights approaching the villa had been tougher than actually strangling her. He had strong hands: that was no problem. He had meant to bury her under the chestnut tree on the hill in back of Manzini's villa, but Sunny Alvarez had put paid to that. If she and Reilly had not shown up it would have worked out fine. But he still hadn't won. The papers retrieved from Marisa were the same ones he already had.

Demarco had finished the vodka. The TV was still blaring, only now it was some newscaster talking fast.

Breaking news, he said. *"A woman's body has been found in famous movie producer Renato Manzini's villa."* Demarco couldn't understand the details but he got the picture, and also that Ronald Perrin's name was mentioned several times.

How convenient, he thought, settling back with a grin. They obviously suspected Perrin not only of killing Ruby but now also Marisa. He took the yellow diamond ring from his pocket, twisting it in his fingers, watching it sparkle. He couldn't just leave it on her finger after he'd killed her. It must be worth over a hundred thou. Besides, nobody really knew she had it. He'd be able to sell it secretly, later, when Perrin was doing life for her and Ruby's murders and the fuss had died down.

There was more breaking news, though. *"A scoop, as yet undisclosed on any other show. Not even the as yet unbroadcast show by the French TV journalist who had discovered her.*

"Allie Ray, the movie star wife of Ronald Perrin, missing since the Cannes Film Festival, has been found working as a waitress at the Bistro du Manoir, near Bergerac in France."

A map flashed on the screen showing the region and pointing out the exact village.

Demarco stared blankly at the TV. With Perrin missing and now a prime murder suspect, only one person stood between himself and all the money in Perrin's private offshore accounts. And only that person could know the correct numbers.

He packed his bag, checked out of the motel and drove to the airport. He took a flight to Bordeaux. It was less than a couple of hours' drive from there to Bergerac and the village where Allie Ray was living.

He didn't know it, but he had a good start on the other people on Allie's trail.

CHAPTER 59

Sunny got Mac's text message on the Cessna flying to Bergerac:

Mac to Sunny: DEMARCO IS THE KILLER. DO NOT KNOW HIS WHEREABOUTS. ALLIE HAS ACCOUNT NUMBERS DEMARCO WANTS. HE MAY BE ON HER TRAIL. TAKE GREATEST CARE. IN OTHER WORDS SUNNY PLEASE DO NOT DO ANYTHING SCARY AND RISKY. WHEN U FIND ALLIE STAY WITH HER. HIRING FASTEST AVAILABLE JET WILL MEET U THERE. CALL ME AND WAIT TILL I GET THERE. LOVE U . . . MAC

Sunny to Mac: SINCE WHEN DID I EVER DO ANYTHING FOOLISH OTHER THAN BREAKING AND ENTERING (THREE TIMES) AT YOUR REQUEST; SPRAINING AN ANKLE WHILE ON TRAIL OF A KILLER; ALMOST GETTING MYSELF KILLED IN AN EARTHQUAKE (WITH U) AND OPENING A REFRIGERATOR WITH A BODY IN IT (ALSO

WITH U) AND ALMOST GETTING STRAN-
GLED. WHAT DO U EXPECT FROM GLAM-
OROUS PI (IN TRAINING)? WILL CALL
WHEN MISSION COMPLETED. ANYWAY
WHAT HAPPENED WITH PERRIN? AND YES
SINCE U DIDN'T ASK I DO LOVE U
THOUGH THERE ARE MOMENTS WHEN I
WONDER WHY. ANYHOW HOW DO U KNOW
DEMARCO IS KILLER AND NOT PERRIN?

Mac to Sunny: TRUST ME I KNOW. MUST
GET TO ALLIE BEFORE DEMARCO FINDS
HER. URGENT AND DANGEROUS. I WISH
I'D NEVER SENT U ALONE.

Sunny to Mac: HAH! I'LL TAKE CARE OF
IT DON'T WORRY. AM CALLING BISTRO
DU MANOIR SOON AS I GET OFF THIS —
RATHER NICE — PLANE. COULD GET
USED TO THIS. BY THE WAY WHO IS PAY-
ING FOR IT? DIDN'T I TELL U YOU'D
NEVER GET RICH?

Mac to Sunny: BE SERIOUS. BE CARE-
FUL. BE MINE.

Sunny to Mac: OH MY GOD . . . I
THOUGHT YOU'D NEVER ASK . . .

Mac to Sunny: I THOUGHT U ALREADY
WERE MINE . . . SERIOUSLY SUNNY THIS IS
DANGEROUS.

Sunny to Mac: FORGET DANGER.
THERE'S BEING YOURS AND "BEING
YOURS." WE MUST TALK ABOUT THIS. I'LL

BE CAREFUL.

Mac to Sunny: THANK GOD. CALL ME WHEN U ARRIVE.

CHAPTER 60

Allie had been invited to dine at the Château Montfort. It was Saturday and Petra had given her the night off, "for good behavior," she'd said. Then she'd winked and added, "Not that I expect you to keep to that."

However, this was not the rendezvous for two Petra and Allie had expected. It turned out that Robert had invited eight other guests.

Allie was in her room, getting ready. Dearie was sprawled on the chintz-cushioned window seat, his favorite place to sleep, keeping a reproachful eye on her, as though he already knew that this time he was not included on the guest list.

Allie twisted and turned in front of the spotty triplicate mirror over the dresser. She had traveled light when she ran away and did not have much choice. Now she was wearing a cream skirt that hit just above the knee and a thin black cashmere sweater that left her shoulders bare. Her only jewelry was the gold hoops and her wedding band. Somehow she still could not bear to take that off. It was too final.

She brushed her hair, which now looked the way Audrey Hepburn used to wear hers, falling in short bangs over her forehead. Then she put on the disguise glasses and sprayed on a little Chanel. She peered into the mirror again, wondering if she still looked like Mary Raycheck. Feeling suddenly lost, neither one thing nor the other, she went and sat on the edge of the bed, wishing she had said no to tonight and didn't have to go. She was remembering the woman she used to be when she was Ron's lover and then his wife. It seemed a lifetime ago.

Petra had already left for the Bistro and the house seemed too empty. Allie hated to leave Dearie, but she gave him a farewell kiss and made for the door. Of course the dog followed her. She stopped in the kitchen and gave him a chew bone. "I'll be right back," she told him.

It was the first time she had been to the Château Montfort and she found it hidden at the end of a tree-lined drive, emerging like a surprise in all its symmetrical pale limestone glory. There was a columned entry atop a shallow flight of steps and tall French windows that reflected the lights within. Other cars were parked in the gravel circle.

Allie pulled down her skirt, smoothed her hair, took a deep breath and marched up the steps to meet her fate. If Robert Montfort was to be her fate, that is. She still didn't know.

"Welcome, *chérie,*" he said, kissing her three times, first on one cheek then the other, then back again, in that welcoming way the French have for intimate friends. He looked appreciatively at her, then said, "Come with me, beautiful woman. I want to introduce you to my friends. In fact you already know one of them. Félice de Courcy. She surprised me by showing up unexpectedly."

Oh God, the Paris blonde was here. Remembering Félice's behavior last time they had met, Allie's heart sank. So, okay . . . "I'm happy to meet all your friends," she said.

Taking her hand in his, Robert walked with her into the grand *salon* where the rest of the guests were drinking champagne. He made the introductions and mentioned that Mary might be interested in buying a cottage with a small vineyard attached.

People smiled welcomingly and the talk was general and mostly, for her sake, in English. Allie relaxed. Nobody was looking at her like she was anything special, just another pretty woman of which there were at least two others that night. Not counting the Paris blonde, who had taken up a stance near the fireplace, vodka tonic in hand, and who was watching her through narrowed eyes.

Fortunately, Robert had placed Félice at the far end of the table, with Allie on his right, and all seemed well until the end of the evening, when they assembled once more in the beautiful pale-paneled *salon* for coffee.

"So, Madame Raycheck." Félice was quickly at Allie's side. "Tell me, how are you enjoying the quiet of the Dordogne? After Hollywood it must come as quite a culture shock."

"Actually, it's the other way round," Allie said quietly. "Hollywood is always a culture shock."

"And do you miss it? That other life?"

"Not especially."

There was a cunning look in Félice's narrowed eyes and a tight smile on her lips that made Allie nervous. Félice glanced at her watch, then turned to face the room, clapping her hands for silence.

"Messieurs et Mesdames, I have a surprise

for you. Normally, I would not disrupt a party but this is special. It is time for my TV show."

She pressed the button and a panel slid back revealing a TV set. Switching it on, she went to stand by Robert. He glanced at her, puzzled, then at Allie who was watching the lead-in to the program.

THE FÉLICE DE COURCY SHOW was imprinted over the journalist's face and then she came into view.

Allie thought Félice looked pretty good on TV, blond hair swinging free, eyes narrowed in that habitual knowing stare, elegant in a low-cut black silk jacket showing plenty of cleavage.

"Tonight, my friends, I have something very special for you. A 'scoop' you might call it. In fact two 'scoops.' Like ice cream only better."

The other guests glanced at each other, smiling, but Allie didn't understand completely what she was saying, only the word *scoop.*

"Missing billionaire Ronald Perrin is now wanted for questioning in the murders of two women, both reputed to be his ex-girlfriends."

Allie's face drained of color. She stood rooted to the spot, not even hearing as Félice's voice droned on.

"The first woman is Ruby Pearl." A picture

of a pretty dark-haired woman came to the screen. *"The second is Marisa Mayne."* It was the glamorous redhead's turn.

It wasn't Ron, Allie was thinking . . . *It couldn't be him. . . . Ron could never kill a spider . . . he wouldn't do that . . .* she'd swear to it . . .

Then suddenly her own face flashed onto the screen. *"Ron Perrin's wife, the movie star Allie Ray, glamorous at the Cannes Festival on the arm of her director,"* Félice said.

"This is the last picture taken of the famous movie star before she too disappeared. Some said forever. And after the critical reviews of her last movie" — Félice shrugged dismissively — *"who could blame her? Or could Ronald Perrin have killed her too? But then . . ."* She was smiling into the camera now as another picture flashed onto the screen. *"Take a look at this. This is Mary Raycheck, a waitress at the Bistro du Manoir in the Dordogne."*

Allie gasped. It was the photograph Félice had taken with her cell phone that night at the Bistro. "Oh God," she said, turning to run.

"Wait." Robert grabbed her hand, glaring at Félice.

"So, my friends, who is *Mary Raycheck? Well you and I and most of the world know her*

as Allie Ray. The 'missing' movie star. So Ronald Perrin did not kill her after all."

The stunned faces of the other guests swung Allie's way, but she didn't wait for more.

The last thing she heard was Félice's laugh as she fled, and Robert calling out to her to wait and that it was okay.

It was not okay. She was Allie Ray again and now everybody knew it. And Ron was wanted for murder.

Paradise was lost.

CHAPTER 61

Allie drove back through the narrow country lanes as thunder rumbled all around. Then quite suddenly the rain came down, sloshing across the windshield in a mini-waterfall the wipers had trouble keeping up with. Allie's tears matched the rain and she was forced to a crawl.

Lightning lit up the beautiful valley bright as a carnival fairground, then darkness settled over her again and she peered through the windshield, looking for the safe lights of "home."

Dearie heard the car and was waiting for her, tail wagging enthusiastically as she rushed in through the kitchen door, brushing rain and tears from her face. She stopped, surprised.

Petra and another woman were sitting at the kitchen table, drinking wine, deep in conversation. Their backs were toward her and their heads swiveled as they heard her come in.

"Oh, hi, Allie," Sunny Alvarez said, sounding relieved. "Am I glad to see you." But then Sunny's face fell. "Oh my God," she said in a kind of breathless way that sounded scared.

Allie stared at them puzzled. Both women were looking past her. She turned to look and saw Demarco standing there.

She said, astonished, "What are you doing here?"

CHAPTER 62

Demarco took a step toward her. "Ron asked me to find you, Allie," he said. "He's my friend and now he's in trouble. I'm helping him. He wants me to get his private account numbers, the ones you have on your laptop. He needs them now more than ever. If you give them to me, I'll be on my way back to him."

Allie looked at Sunny and Petra. They were sitting perfectly still, eyes glued on Demarco. She looked back at him and this time noticed the hunting rifle slung over his shoulder. She took a quick wary step backward, still not quite understanding.

"That's my ex-husband's hunting rifle," Petra said suddenly. "Where did you get it?"

Demarco shrugged. "You should not leave your doors open, madame. And rifles should be kept in locked cupboards." He turned to Allie.

"Come on, Allie, let's just go get them,"

he said.

"Where's Ron?"

He sighed. "I'm afraid Ron's wanted on two counts of murder. That's why he'll need his money. It's better if you and I cooperate, my dear. We'll sort out all the problems together. All I want from you are those account numbers."

Allie knew what he was talking about now. Ron had put the coded numbers on her computer a long time ago. So long ago, in fact, that she had erased them.

"I erased them," she said truthfully.

Demarco's cold eyes told her he didn't believe her. He said, "Well isn't that too bad. Then I suggest we go look again." In a quick movement he grabbed her arm.

Sunny screamed and Dearie gave a warning growl.

"He's the killer, Allie," Sunny yelled. "He almost killed me a couple of days ago, and now he'll kill us all if necessary. He's *crazy,* Allie —"

Demarco took a step toward Sunny and backhanded her so hard her head snapped.

"Shut up, you interfering Mexican slut," he snarled.

Dearie gave a warning growl and jumped to his feet. Sunny bit her lip hard in an effort not to cry, and Petra clutched her hand

411

under the table.

"You can't do that," Allie yelled and Demarco turned and smacked her head back too.

Dearie leaped for his throat. Demarco had the rifle at the ready. He didn't even have to aim, the dog was so close.

The shot cracked through the room, and Dearie slid to the floor.

"Oh, God, oh God . . ." Allie screamed. She hurled herself at Demarco. *"Bastard,"* she screamed. "You *bastard.*"

She'd learned what to do at self-defense classes and now she jabbed her fingers at his eyes. He dodged her, caught her arms, held them in a tight grip behind her back.

The rifle clattered to the floor and in a second Sunny had grabbed it. Now Sunny held the rifle and Petra hefted a handy meat cleaver.

"One more move, you evil louse, and I'll slit you in two," Petra said in a tone that spelled business.

Demarco let go of Allie. He began to back toward the door.

"Stay right there," Petra warned.

Demarco had his hand on the doorknob when the wail of police sirens cut through the silence. The women's heads shot up expectantly.

Demarco flung open the door. The rain was still slicing down and lightning lit the sky. And in that second, Demarco was gone.

Outside, gravel spurted as the police cars squealed to a halt.

"Here comes the cavalry," Petra said calmly.

A few seconds later, Ron Perrin burst through the door, followed by Mac.

"Allie, Allie, are you all right?"

Ron's strong arms were around her and she was pressed to his fast-beating heart. "Oh, Allie, what did I do to you? How did we ever let things get to this?"

Mac looked at Sunny. "I'm okay," she said, but she could see he was shocked. "Demarco just ran out the door," she added urgently, as the police swarmed into Petra's kitchen. "He has a rifle."

Mac and the cops ran back outside, leaving two detectives with the women and Perrin. Mac heard a car start up near the bottom of the drive. Demarco was getting away. He leaped into the cop car and they took off after him. It was just like when he was that city reporter in Miami, all those years ago.

Allie was on the floor, kneeling next to Dearie. The dog's eyes were rolled up in his head but when she put her face to his mouth she felt a faint breath.

"Get me a big towel," Ron said to Petra. "And one of you who speaks French tell this detective that he has to get us to the nearest vet."

Petra gave Ron a towel and he carefully wrapped the dog in it. Then she rang the vet, woke him up and told him it was an emergency.

Ron and Allie took off with the cop driving and the dog on Ron's lap, and Sunny and Petra were left alone again under the watchful eye of the second detective.

Sunny spotted something sparkling on the floor. She picked it up. It was Marisa's ten-carat yellow diamond ring. She put it in her pocket to give to Mac later. Demarco must have dropped it.

Petra gave an aggrieved sigh, and put the kettle on to boil. "All I thought was Mary was going to a nice dinner party at Robert Montfort's," she said. "Will someone please tell me what's going on?"

"I'm sorry," Sunny said. "I was in the middle of telling you when Allie got back and then Demarco showed up. He's the killer you see. In fact he almost killed me the other night and . . . Ohh." She shuddered. "It's just too much to tell."

"Of course I knew she was Allie Ray," Petra said, pouring the boiling water onto

414

the tea leaves (Darjeeling, her favorite in moments of stress). "Everybody did. Well, those who counted anyway. Nobody else really cared, they just knew her as Mary. You know, the new waitress."

"And you never told Allie?"

"Of course not. She'd come here seeking sanctuary, and until Félice got the story and then Mr. Demarco showed up, she had found it. Now she'll probably lose her dog and regain her identity."

"And maybe even her husband," Sunny said thoughtfully. But she was worried. Mac was out there with the killer and anything could happen.

"How about a nice cup of tea?" Petra said, pouring out.

CHAPTER 63

Demarco saw the police car's lights behind him. He swung off the main road and onto a narrow lane that ran alongside the river. The signpost read TRÉMOLAT. The road was difficult at night, muddy and completely dark, but that was in his favor. He doused his own lights and sped on. He could no longer see the police car's and he sighed with relief. He'd given them the slip.

He stared ahead into the blackness and the rain. Over the roar of thunder he could hear the Dordogne River rushing past and the trees cracking in the gale-force gusts of wind.

He had been a fool. He should never have risked it. He would have been home free with Ron guaranteed to be accused of the murders. Now he had blown it.

He slowed down. No matter how hard he ran there was no way out. And anyway there were the police sirens again, *blah blah-blah*

blah, that up-and-down wail that meant they were coming closer. He could see the blue lights flashing now.

The muddy road widened as it merged into a corner. Next to him the Dordogne roared into a weir, white foamed and swirling. A power-generating plant loomed a hundred yards ahead.

The sirens were coming closer . . . the lights were turning the corner behind him . . .

Demarco swung the car suddenly to the left. He put his foot down and headed directly into the surging river.

Mac heard the thud as the car hit the water. He was out of the police vehicle before it had even stopped. He ran to the edge of the river, staring down at the vortex caused by the car's descent into the depths.

He nodded. It seemed an appropriate ending for an evil man.

CHAPTER 64

Allie walked wearily back into Petra's kitchen, followed by Ron. The two women looked hopefully at her.

"So, okay . . . Dearie didn't make it," she said in a choked voice. And then she sank into a chair and put her head in her hands and wept.

Petra quickly took the seat next to her. She put her arm around Allie's shaking shoulders. "Poor Dearie," she said softly. "And poor Allie. There's only one way to look at it though, love. Dearie came to you at a time when you needed him. It's as though he were meant for you, just for that moment. Now he's gone, but he did his work. He gave you the love and companionship you needed. And then he gave you his life. He was a noble fellow and we'll miss him. So cry all you want, love, if it makes you feel better."

She got up and went to pour the tea. She put the cup in front of Allie and said, "A nice cup of tea. They say it's good for all that ails you but this time I'm not so sure." And throwing Perrin a grim look, as though daring him to say anything, she sat next to Allie again, waiting for her to stop her sobbing.

The door opened and Mac strode in, followed by the two cops and two plainclothes detectives who had pulled up next to them.

Mac was soaked. He ran his hands through his wet hair, taking in the scene. He walked over to Sunny and she got up and went into his arms. Tears were also running down her face. "It's okay, Sunny honey," he said, not minding the awful rhyme. "It's all over now. Demarco's dead."

The two detectives strode over to Ron Perrin, who was standing near the big Welsh dresser that held about a hundred blue and white plates and a collection of teapots and carved wooden Welsh marriage spoons, as well as Petra's usual junk.

"Ronald Perrin, we have a warrant for your arrest," the first detective said.

Just like in a play, Sunny thought. She stopped crying and watched, amazed as the two collies came bustling in from the rain. They ran to the detectives and shook themselves, spattering the two men with about half a gallon of muddy rainwater.

Fastidious, they exclaimed with horror and

took a step back.

"It's just the dogs," Petra explained.

Allie's head was up. Her eyes were lost behind two red swollen circles and her nose was running.

The kitchen door slammed back and Robert Montfort strode in. He stared, stunned, at the collection of people. He looked at Allie's blotched face; at the big bald man in the tight striped T-shirt, the earring and the handcuffs; at Petra pouring tea and the handsome couple staring back at him amid an assortment of cops and detectives.

Puzzled, he spread his hands wide. "What's going on?" he said.

"You're too late to join the cavalry, Montfort," Petra said. "It's all over."

CHAPTER 65

A week later, they were all back in L.A. Ron Perrin was in a downtown jail awaiting trial on tax evasion, but the FBI had not pursued the allegations of fraud. Demarco was the one guilty of that. Allie was cloistered in her Bel Air mansion dodging the press, with Lev back in charge of security, and her friend Sheila for company. And Mac was busy giving depositions and being interviewed by the police, and walking Pirate along the beach at midnight. Sunny was incommunicado in an expensive spa, having her frazzled nerve ends soothed by hot stone massages and mud baths, the bill for which she was sending to Mac since she said it was all his fault she was in this state anyway.

Right now though, Mac was on his way to the pet rescue center in Santa Monica. Leaving Pirate in the car, he explained what he needed then walked along the rows of sad-eyed caged animals. He wanted to take them all home, but he could take only one.

He spotted the dog halfway down the line, but walked on, still checking. Its soft taupe-colored eyes followed him. He thought they were not so different in expression from Ron Perrin's when he'd last visited him in jail. "Get me outta here," the dog's eyes seemed to say. Which is exactly what Perrin had said.

The assistant took the dog out of the cage for Mac to see. It was young and battle-scarred. A line where stitches had sewn its head together lay unfurred and pink from ear to ear, one of which stuck up, the other down. It was a sort of taupe brown, short haired, with maybe a bit of Weimaraner. That explained those soft taupe eyes that almost matched its coat.

"I'll take him," Mac said. And apologizing to all the other rescue animals for not taking them as well, he made his way quickly out of there.

He put the new dog on its lead in the back-seat and Pirate turned for a quick suspicious sniff. He needn't have worried. The dog immediately curled up, closed its eyes and went to sleep.

Mac drove back to Malibu, smiling at the blue and gold beach scene as he descended the incline onto PCH. It felt good to be home.

He punched a number into his cell. "Hey," he said when Allie answered. "Can you escape? Come out to Malibu for lunch?"

"I'll do it," she said, sounding relieved. "I'm alone, Sheila's gone to the hairdresser, but Lev will get me out from under the paparazzi's noses."

An hour later she was there. Almost the old Allie Ray this time, only with short blond hair and no glasses — well, only the de rigueur shades.

"It's you," Mac said with a grin.

"The original," she said. "Back to me again."

"That's funny," he said. "Somehow I thought you were always just you."

She laughed as they walked out onto the deck. The two dogs were sitting next to each other inspecting the beach. They turned to look as Allie stepped through the door.

"Ohh, hi, Pirate," she said.

"Meet Frankie," Mac said, and the dog came running over.

"He's sweet," Allie said. Then with a catch in her throat, "He has that same look in his eyes . . ."

"As Dearie and Pirate you mean."

"Yes." She sighed, still stroking the dog's smooth short fur.

"That's why I got him for you." Mac was

holding his breath. If she didn't want the dog then it was his and though Pirate was okay with Frankie, his home was a bit crowded for two people and three dogs. He was counting the Chihuahua in that group.

"Fussy's not your kind of dog," he said referring to Allie's snippy Maltese. "You told me yourself she's happiest living with the housekeeper. I thought you needed Frankie to take care of you."

She looked up at him, surprised, and then she was laughing. "You sure know the way to a girl's heart," she said.

He'd brought in some sandwiches and they sat on the deck enjoying them with cold beer drunk from the bottle. "What's happening with Jessie Whitworth?" she asked.

"She's still in jail in France, facing vehicular manslaughter charges, as well as driving with a false license. Seems she'd also run up a few quite hefty credit card debts."

"She wanted to be me more than I wanted it," Allie said. "The blond wig, wearing my clothes."

Mac felt Allie's eyes on him. He looked at her. "That night when she cut up my gowns, did she . . . I mean . . . ?"

"Was it you she meant to slash? Yes, I believe it was. And once the French police are through with her, it's our turn. You'll have to press charges against her, you know. She's guilty as hell, and I've no doubt she'll do time in France and then here."

She nodded. "I'm just glad it's over."

"You can go back to living again."

She smiled. "I guess I can," she said.

"What'll you do next?"

"I've decided to go back to France. Petra's keeping my job open, and there's a sweet little cottage I could buy."

"You won't be lonely?"

She shrugged. "Perhaps. I don't know. There's someone I'm friendly with . . ."

"Montfort," Mac guessed with a grin.

She shrugged again, grinning back. "Maybe," she said.

He didn't ask her about Ron. He figured it was her business.

He took the yellow diamond ring from his pocket and gave it to her. "I guess you should have this," he said. "It was Marisa's."

She turned it over and over, letting the sunlight catch it in a thousand scintillating prisms. Then she heaved a sigh. "Poor Marisa," she said. "I'll give it back to Ron. He can sell it. He's going to need the money with all those lawyers."

Mac nodded okay. It seemed appropriate.

They took a walk to the beach with the dogs, then Lev came to pick Allie up and drive her back to Bel Air.

She turned at the door, leaning against it, her hands behind her back. "I don't know what I would have done without you," she said softly.

Her eyes were that deep turquoise blue that searched the depths of a man's soul. Mac felt their pull, saw her beauty. Her gentleness.

"You'll manage," he said, kissing her on the lips.

He stepped back and they looked into each other's eyes one last time. Then, "Goodbye," she said, taking the dog's leash and walking outside to the car.

Neither she nor the dog looked back.

CHAPTER 66

Allie went to see Ron at the downtown jail before she left for France. They met on either side of a glass partition and Ron was wearing the regulation orange jumpsuit. She thought he looked tired and sad.

"Why did you do it, Ron?" she asked.

He shook his head. "Because I could, I guess. It was just a game to me. Demarco was the one who took it seriously. Enough to kill for." He shook his head again. "Those poor girls."

Their eyes met through the glass partition that separated them.

"You look wonderful," he said. "Like my old Allie Ray."

"I'm going back even further than that," she said. "Back to Mary Allison Raycheck. No more movies. No more Hollywood. I'm putting the houses up for sale."

"Malibu as well?"

She nodded.

Looking relieved, Ron said, "Just save my trains, though."

"I'm going back to France."

He looked startled. "For how long?"

"I don't know. Maybe forever."

Their eyes met through the glass wall searching for the truth that somewhere along the line had gotten lost.

"Will you write to me?" he asked, and Allie promised she would.

The next day Allie and the new dog, Frankie, were on a flight to France.

Back at the Manoir, Petra welcomed her with kisses and cups of tea, and Robert Montfort welcomed her with open arms.

Later, Petra accompanied her to a *notaire's* office in Bergerac and within a couple of weeks Allie found herself signing the documents that made her the new owner of five hectares of land with a two-bedroom, one-bath dwelling, a bunch of rickety farm buildings and a broken-down barn. The garden was still a riot of weeds, the rain bucket still stood under the drainpipe to catch the rainwater, and dragonflies still danced over the pond. Of all the splendid homes she had owned, Allie had never before felt this thrill of ownership.

Petra said she would be sad to see her leave

the Manoir but glad that she was to be a permanent neighbor.

"I've gotten used to having you around, love," she said, dabbing her teary eyes with a pink silk scarf that just happened to be handy. "And don't you worry about the garden," she told her after inspecting it with critical blue eyes. "We'll have that knocked into shape in no time. To tell you the truth I think it looks just gorgeous as it is. Old Madame Duplantis lived here for as long as anyone can remember, until she went to stay with her only daughter in Bergerac a couple of years ago. It's been empty ever since.

"It'll need a bit of tarting up inside, though," she added, noticing the peeling flowered wallpaper and the rustic kitchen, which consisted of a small two-burner stove, a stone sink with a scrubbed wooden draining board, a table covered with red-flowered oilcloth and a couple of cupboards. It did have a nice stone-flagged walk-in larder though. Good, Petra told Allie, for keeping cheeses and eggs cool, as well as fresh veggies.

Robert came too, with an architect friend, who planned on knocking out a couple of the interior walls to open up the space, and who also redesigned the kitchen and the only bathroom. He mentioned that eventually

Allie might want to think about renovating the barn. He had no doubt she was going to need more room. The work on the cottage would take a couple of months, they said, which made Petra laugh.

"Think more like six," she whispered to Allie. "You'll be stuck at my place for a while yet."

Meanwhile, there was a tired-looking old vineyard at the back of the property, a mere three hectares, but Robert said the land wasn't bad, and he could get it cleared and replanted for her.

Allie was thrilled. All of a sudden, she was a new homeowner and a future winemaker. And Robert Montfort's eyes were telling her things she half liked, half didn't.

A few weeks later Allie and Robert were having a picnic on the grass near her pond. The sound of hammering and the whine of tiles being cut had stopped for the day and they were finally alone. Robert uncorked the wine and poured her a glass.

"To us," he said, smiling.

"To us, Robert," she said, but she turned away from him as she said it.

"Mary," he said, making her smile this time because he still refused to call her Allie. "I never knew Allie Ray," he'd said.

Epilogue

The Colony homes are mostly empty in winter, when their owners flee all of ten miles inland to the more temperate climate of Beverly Hills. But Mac loved it then, even more than in summer. He loved it when it was gray and wild and lonely, when the waves surged at his little house demanding entrance and the wind rattled the shutters and blew smoke down the chimney.

It was kind of cozy here tonight, though, in the master bedroom, with a Bach CD playing and the fire lit and the curtains closed and the wind whipping the waves into storm force outside, and the candles flickering in the ever-present draft. Sunny was wearing his old green cashmere sweater and a pair of white sweat socks and nothing else. She was snuggled up beside him in bed, keeping a watchful eye on Tesoro, to whom Pirate and Mac had reluctantly granted visitation rights.

The Chihuahua had paced for the last hour, back and forth, back and forth. When she wasn't pacing, she was tugging at the curtains with her tiny paw and peering mournfully outside, tail drooping, ears back, the very picture of an unhappy exile. Meanwhile, Pirate sat on his haunch, his one eye following her every move.

Patient, Mac thought, that's what my dog is.

"Trust me," he said to Sunny, "I know Pirate. He's just biding his time. Like all men, he's waiting for that right moment to make a move."

Gazing at his Sunny, he wondered again why women always wanted to get married. Weren't the two of them perfectly good the way they were? Anyhow, he wasn't sure he was up to the job of being Tesoro's father. Plus Pirate wouldn't take too well to having a bossy female permanently around, infringing on his hard-come-by territory. Like him, Pirate had come up in life the hard way. Tesoro was like her mother, a lady — and you had better not forget it. Besides, two females in one small Malibu house? He sighed. It was simple math. Two into one just didn't go.

Tesoro finally settled at the very foot of the duvet, making it clear through the angle of her whiskers and her tipped-back ears that she was not about to move a single inch.

But then they spotted Pirate's black mutt nose sniffing over the end of the bed as he crawled sneakily closer. Pirate lay down next to Tesoro, nonchalantly making like he hadn't noticed her, though Mac could see a flinch waiting on his face in case he had to use it.

Sunny and he looked at each other, then back at their dogs lying side by side. Twenty . . . thirty long seconds ticked by. They held their breaths.

Pirate glanced at Tesoro, then slowly, hesitantly, he lay his mutt head across the Chihuahua's tiny aristocratic paws. His eye watched her hopefully. Another ten seconds passed. Tesoro slid Pirate a sideways glance then burrowed her head into the duvet, flicked her mini-tail over her nose, adjusted her ears back to normal and closed her eyes. Peace reigned supreme in Malibu.

"So that's it," Sunny said with a sigh of relief, cozying up under the duvet.

"That's what?" Mac asked, burying his face in her fragrant neck, ready to devour her.

"There's no excuse anymore. Tesoro and Pirate are best friends. They are sleeping together on our bed and nobody is killing anybody."

"So?"

"So there's no excuse for not marrying me." She turned her face to his.

"Maybe a wedding on the beach," Mac said, kissing her pouty mouth, clasping the soft cashmered length of her against him until he felt every heartbeat, every pulse, every throb of longing. "Yeah, a moonlit wedding with just a few good friends. And the dogs of course."

"What?" she said.

"Sunny," he whispered, in between kisses.

"Mmmmmmm?"

"Marry me," he said more firmly.

"I do," she said, and then they were tangled up in the sheets, laughing and crying together. Outside was a stormy moon and the slur of the ocean hitting the shore.

It was just one of those Malibu nights.

But this is where we came in.

ABOUT THE AUTHOR

Elizabeth Adler is the internationally acclaimed author of twenty-two novels, including the recent *New York Times* best-seller *Meet Me in Venice.* She lives in Palm Springs, California. Visit her at www.elizabethadler.com.